Spectrum Estimation
and
System Identification

S. Unnikrishna Pillai   Theodore I. Shim

Spectrum Estimation and
System Identification

With 74 Illustrations

KEN KOSNIK

Springer-Verlag
New York Berlin Heidelberg London Paris
Tokyo Hong Kong Barcelona Budapest

S. Unnikrishna Pillai
Department of Electrical Engineering
Polytechnic University
Five MetroTech Center
Brooklyn, NY 11201 USA

Theodore I. Shim
Department of Electrical Engineering
Polytechnic University
Five MetroTech Center
Brooklyn, NY 11201 USA

Mathematics Subject Classifications (1991): 62Mxx, 93Gxx

Library of Congress Cataloging-in-Publication Data
Pillai, S. Unnikrishna, 1955–
    Spectrum estimation and system identification / S. Unnikrishna
Pillai, Theodore I. Shim.
        p.   cm.
    Includes bibliographical references and index.
    ISBN 0-387-94023-5. — ISBN 3-540-94023-5
    1. Signal processing — Mathematics.   2. System identification.
    3. Power spectra.   I. Shim, Theodore I.   II. Title.
TK5102.9.P55   1993
003′.54 — dc20   93-12289

Printed on acid-free paper.

Production managed by Ellen Seham; manufacturing supervised by Jacqui Ashri.
Camera ready copy prepared from the authors' LaTeX files.
Printed and bound by Edwards Brothers, Inc., Ann Arbor, MI.
Printed in the United States of America.

9 8 7 6 5 4 3 2 1

ISBN 0-387-94023-5 Springer-Verlag New York Berlin Heidelberg
ISBN 3-540-94023-5 Springer-Verlag Berlin Heidelberg New York

To My Mother
— Unni

and My Parents
— Inshig

Preface

Spectrum estimation refers to analyzing the distribution of power or energy with frequency of the given signal, and system identification refers to ways of characterizing the mechanism or system behind the observed signal/data. Such an identification allows one to predict the system outputs, and as a result this has considerable impact in several areas such as speech processing, pattern recognition, target identification, seismology, and signal processing.

A new outlook to spectrum estimation and system identification is presented here by making use of the powerful concepts of positive functions and bounded functions. An indispensable tool in classical network analysis and synthesis problems, positive functions and bounded functions are well understood, and their intimate one-to-one connection with power spectra makes it possible to study many of the signal processing problems from a new viewpoint. Positive functions have been used to study interpolation problems in the past, and although the spectrum extension problem falls within this scope, surprisingly the system identification problem can also be analyzed in this context in an interesting manner. One useful result in this connection is regarding rational and stable approximation of nonrational transfer functions both in the single-channel case and the multichannel case. Such an approximation has important applications in distributed system theory, simulation of systems governed by partial differential equations, and analysis of differential equations with delays.

This book is intended as an introductory graduate level textbook and as a reference book for engineers and researchers. The material presented here can be readily understood by readers having a background in basic stochastic processes and linear system fundamentals. A preliminary course in matrices, though not essential, could make the reading easy. This book also can be used in a one- or preferably two-semester course. Concepts are explained in a leisurely manner and illustrated in detail along with necessary mathematical techniques. As usual, reasonably complete proofs of all major results are included, and the book is essentially self-contained. Problems at the end of each chapter have been chosen to supplement the text material.

On reflection, this book should have been written by Professor Dante C. Youla, my distinguished colleague here at Polytechnic. Working with Professor Youla on some of these problems has been a most enjoyable and rewarding experience. His many contributions in this context are gratefully acknowledged. The contributions to this work by my student, Theodore

Shim, have been substantial, and it is a pleasure to acknowledge him as a co-author and colleague. My student, Won Cheol Lee, also deserves great credit for checking out various proofs and computations in this book.

Special thanks to Michael Rosse for his expert editing of the manuscript and Brig Elliott, S. Radhakrishnan Pillai, Henry Bertoni, Farshad Khorrami and Shivendra Panwar for their kind encouragement. I would like to acknowledge the support received from the Office of Naval Research and the National Science Foundation, since the research under these contracts and grants has prompted me to write this book.

Finally, I dedicate this book to my mother, who, together with my father, made it all possible for me.

Brooklyn, 1993   S. Unnikrishna Pillai

Contents

Spectrum Estimation
and
System Identification

1

Introduction

1.1 Introduction

Spectrum estimation refers to analyzing the distribution of power or energy with frequency of the given signal, and it dates back to at least Newton and his prism experiments with sunlight. Newton used these experiments to demonstrate that sunlight consists of a band of colors and each color corresponds to a particular wavelength. The modern-day approach to spectral analysis has its roots in Fourier's series representation of periodic functions.

In a wide variety of situations, information collected through various means needs to be analyzed to determine the underlying mechanism or system behind that data. Such an identification allows one to predict the system outputs, and, as a result, this problem has considerable importance in speech processing, pattern recognition, target identification, seismology, communication theory, and signal processing.

The given signal or waveform can originate from a deterministic source or may represent a particular realization of a stochastic or random process. Further, the signal can also exhibit underlying periodicities in its description. Thus, for example, if a deterministic and periodic signal $f(t)$ is continuous together with its derivative and possesses only a finite number of jump discontinuities in any interval, then from Fourier's theorem, such a signal can be represented as

$$f(t) = \sum_{k=-\infty}^{\infty} a_k e^{-jk\omega_0 t}, \tag{1.1}$$

where $\omega_0 \overset{\triangle}{=} 2\pi/T$ denotes the fundamental frequency, T represents the period of the signal, and

$$a_k \overset{\triangle}{=} \frac{1}{T} \int_0^T f(t) e^{jk\omega_0 t} dt.$$

The infinite series in (1.1) converges for all t, and the sum of this series equals $f(t)$ at every point of continuity, and it equals $(f(t-0) + f(t+0))/2$ at every point of discontinuity. If $f(t)$ is continuous everywhere, then the series converges absolutely and uniformly. Further, if the signal is also of finite power

$$P = \frac{1}{T} \int_0^T |f(t)|^2 dt < \infty, \tag{1.2}$$

then

$$\sum_{k=-\infty}^{\infty} |a_k|^2 = P < \infty, \tag{1.3}$$

with $|a_k|^2$ representing the power contained at frequency $\omega = k\omega_0$. This implies that the total power is distributed only among the fundamental frequency ω_0 and its overharmonics $k\omega_0$, $k = -\infty \to \infty$. This also follows by noticing that the Fourier transform $F(\omega)$ of $f(t)$ is given by

$$F(\omega) = \mathcal{F}\{f(t)\} \triangleq \int_{-\infty}^{\infty} f(t)e^{j\omega t}dt = \sum_{k=-\infty}^{\infty} a_k\delta(\omega - k\omega_0), \tag{1.4}$$

and the discrete nature in the frequency domain is evident from (1.4). Schuster used this observation to suggest that the quantity $|a_k|^2$ in (1.3) be displayed against $\omega = k\omega_0$ as the spectrum. Schuster used the term Periodogram to represent this method. For finite power periodic signals, from (1.3) since $a_k \to 0$, eventually the overharmonics become of decreasing importance and the periodogram can be always displayed almost entirely.

More generally, a continuous-time signal $f(t)$ with finite energy

$$E = \int_{-\infty}^{\infty} |f(t)|^2 dt < \infty$$

also can be given a similar frequency domain interpretation through its Fourier transform

$$F(\omega) = \int_{-\infty}^{\infty} f(t)e^{j\omega t}dt. \tag{1.5}$$

From Parseval's theorem, since

$$\frac{1}{2\pi} \int_{-\infty}^{\infty} |F(\omega)|^2 d\omega = \int_{-\infty}^{\infty} |f(t)|^2 dt = E, \tag{1.6}$$

$|F(\omega)|^2$ represents the distribution of energy in the frequency domain.

The meaningful present-day analysis of power distribution with frequency for stochastic signals started with Norbert Wiener in his classical work on generalized harmonic analysis [1], where a firm statistical foundation to random processes is given using the notion of ensemble average and other key concepts.

Stochastic processes are everywhere; the received signal at the output of a sonar transducer, seismic recordings, stock market fluctuations, sample speech waveforms, temperature readings at a weather laboratory, all these constitute examples of such a phenomenon. Although any portion of such a phenomenon appears to be nothing but chaos, apparently there could exist order behind these observations. For example, in the case of the sonar signal, the received signal could represent noisy returns from a target such as a submarine. If such is the case, identification of the object or mechanism behind the observations is an important problem, and this essentially forms the system identification task. In this context, a new approach is developed here by making use of the class of all extensions for power spectra, together with several key ideas from classical network theory. The problem of obtaining the class of all extensions for power spectra from partial information regarding itself is also known as the trigonometric moment problem, and it has been the subject of extensive study over a long period partly because of its considerable significance in several areas such as interpolation theory, power gain approximation theory and spectrum estimation. This basic problem also forms part of our study here.

Finally, systems with several inputs and several outputs — multichannel systems — where every input participates in generating every output in a deterministic manner, are studied here and the power spectral extension problem as well as the system identification problem are addressed in this context also.

1.2   Organization of the Book

Chapter 2 introduces stationary stochastic processes, their power spectra, and associated Wiener factors beginning with continuous-time processes and specializing in discrete-time processes. Linear prediction of stationary processes is next examined in detail, and various useful results are reviewed in this context.

The concept of positive functions is introduced, and their one-to-one correspondence with power spectral density functions is established. Since positive functions play a crucial role, at least conceptually, in the rest of the study, their properties are examined in some detail, and, in this context, another useful concept — that of bounded functions — is introduced. Two key results in the form of Schur algorithm and Richards' theorem are explored in some detail.

Chapter 3 deals with the power spectral extension problem: Given a finite set of autocorrelations from a stationary finite power stochastic process, obtain the class of all power spectra matching these requirements, i.e., every such spectrum must interpolate the given autocorrelations in addition to generating a finite power regular stochastic process. Starting with a geometric approach, explicit formulas for the class of all spectra and their associated Wiener factors are derived here by making use of the Schur algorithm. The parametrization of such spectra is in terms of the Levinson polynomial associated with the given autocorrelations and an arbitrary bounded function that satisfies certain mild restrictions. In this context, the role of Levinson polynomials and reflection coefficients are examined in detail and the autoregressive form of the maximum entropy solution follows as the most robust solution from a geometric viewpoint. Since maximization of entropy is equivalent to maximization of the one-step minimum mean-square prediction error, the immediate generalization of maximizing the multistep minimum mean-square prediction error is examined next, and the solution to this problem is shown to be of autoregressive-moving average form. Details for the two-step case are worked out in detail that shows the usefulness of the arbitrary bounded function in the power spectral extension formula.

Chapter 4 addresses two problems: identification of rational systems and rational and stable approximation of nonrational systems. An examination of the positive function associated with every rational transfer function shows that every rational transfer function has an irreducible representation in terms of certain bounded functions that are invariant with respect to their degrees. This is made possible with the help of Richards' theorem described in Chapter 2 and together with the class of all spectral extension formula derived in Chapter 3, this observation forms a powerful tool for ARMA-system identification. A test criterion for model order selection is derived in this context together with procedures for evaluating the system parameters. Unlike the statistical criteria, the key feature of such systems — their rational nature — is exploited here in a systematic manner and this is used for both model order identification and system parameter evaluation.

Rational approximation of nonrational transfer functions is examined in a Padé-like manner by requiring the Fourier coefficients (autocorrelations) associated with the square-magnitude functions (spectra) of the nonrational transfer function as well as the rational approximation match up exactly to a maximum extend. However, such a matching allows this problem to be examined using the class of all extension formula associated with the autocorrelations that match. Since the above class is parametrized in terms of bounded functions, it follows that the desired rational approximation, if it exists, must follow for a specific bounded function with certain degree restrictions. When such is the case, because of the finite power constraint on the spectrum, the corresponding Wiener factors are automatically stable and minimum-phase. This is unlike the Padé approximation, where the rational approximations so obtained need not be stable *irrespective* of the stability of the original system. The usefulness of the present method is finally illustrated through various worked out examples.

The generalization to the multichannel case is addressed in Chapter 5, where the spectrum extension problem, the system identification problem and the rational approximation problem are worked out in detail. After a leisurely discussion of rational matrix function, positive and bounded matrix functions, the notion of matrix order is shown to be a useful working tool in place of the ordinary degree in the scalar case. Using this, the multichannel version of the Schur algorithm and a weak form of Richards' theorem are introduced and the class of all spectral extension formula is derived. The left- and right-forms of Wiener factors and the noncommutativity of matrices bring in additional subtleties that must be taken into account in all derivations. Once again, rationality is exploited in a useful manner to obtain a new technique for model order identification, system parameter evaluation and multichannel rational approximation. Illustrative examples are once again provided to cover various aspects of this new approach.

1.3   Notations and Preliminaries

Throughout this book, scalar quantities are denoted by regular lower-case or upper-case letters. Lower-case and upper-case letters with bold type faces are generally used for vectors and matrices, respectively, although occasionally lower case with bold type faces is also used for matrices. Thus a (or A), $\mathbf{a}$ and $\mathbf{A}$ (or $\mathbf{a}$) stand for scalar, vector and matrix in that order. In addition, $\mathbf{d}(z)$ and $\boldsymbol{\rho}(z)$ have been used to represent matrix functions of z. Similarly $\mathbf{A}^T$, $\mathbf{A}^*$, $tr\,(\mathbf{A})$, $det\,\mathbf{A} = |\mathbf{A}|$ represent the transpose, complex conjugate transpose, trace and determinant of $\mathbf{A}$, respectively.

A principal minor of $\mathbf{A}$ is a determinant of a submatrix of $\mathbf{A}$ formed with the same numbered rows and columns. If the rows and columns involved in forming a principal minor are consecutive, then the determinant is said to be a leading principal minor. Thus an $n \times n$ matrix $\mathbf{A}$ has n leading principal minors given by

$$a_{11}\,, \quad \begin{vmatrix} a_{11} & a_{12} \\ a_{21} & a_{22} \end{vmatrix}\,, \quad \begin{vmatrix} a_{11} & a_{12} & a_{13} \\ a_{21} & a_{22} & a_{23} \\ a_{31} & a_{32} & a_{33} \end{vmatrix}\,, \quad \ldots \quad det\,\mathbf{A}\,,$$

where a_{ij} represents the $(i, j)^{th}$ element of $\mathbf{A}$. A square matrix $\mathbf{A}$ is said to be nonsingular (singular) if its determinant is nonzero (zero).

A square matrix $\mathbf{A}$ of size $n \times n$ is said to be Hermitian if $\mathbf{A} = \mathbf{A}^*$, i.e., $a_{ij} = a_{ji}^*$, i, $j = 1 \to n$. The matrix $\mathbf{A}$ is said to be nonnegative definite if for any $n \times 1$ vector $\mathbf{x}$, the quantity $\mathbf{x}^*\mathbf{A}\mathbf{x} \geq 0$. When strict inequality holds, i.e., $\mathbf{x}^*\mathbf{A}\mathbf{x} > 0$, for $\mathbf{x} \neq \mathbf{0}$, the matrix $\mathbf{A}$ is said to be positive difinite. A square matrix $\mathbf{U}$ is said to be unitary if $\mathbf{U}\mathbf{U}^* = \mathbf{U}^*\mathbf{U} = \mathbf{I}$. Two $n \times n$ matrices $\mathbf{A}$ and $\mathbf{B}$ are said to be unitarily similar if there exists a unitary matrix $\mathbf{U}$ such that $\mathbf{A} = \mathbf{U}\mathbf{B}\mathbf{U}^*$. A classical result due to Schur states that every $n \times n$ matrix $\mathbf{A}$ is unitarily similar to a lower triangular matrix [2, 3]. Thus

$$\mathbf{A} = \mathbf{U}\mathbf{L}\mathbf{U}^*\,,$$

where $\mathbf{L}$ is lower triangular. In particular, it follows that if $\mathbf{A}$ is Hermitian, then $\mathbf{L}$ is diagonal, and if $\mathbf{A}$ is positive definite, $\mathbf{L}$ is diagonal and positive definite and hence all its principal diagonal entries $L_{ii} > 0$. Thus $\mathbf{L} = \mathbf{D}^2$ where $\mathbf{D}$ is a real diagonal matrix with

$$D_{ii} = \pm\sqrt{L_{ii}} \tag{1.7}$$

and

$$\mathbf{A} = \mathbf{U}\mathbf{L}\mathbf{U}^* = \mathbf{U}\mathbf{D}^2\mathbf{U}^* = \mathbf{U}\mathbf{D}\mathbf{U}^*\mathbf{U}\mathbf{D}\mathbf{U}^* = \mathbf{C}^2\,, \tag{1.8}$$

where $\mathbf{C} \overset{\triangle}{=} \mathbf{UDU}^*$. Notice that $\mathbf{C}$ is Hermitian and it represents the square root of $\mathbf{A}$. Clearly, from (1.7), $\mathbf{C}$ is not unique. However, it can be made unique by choosing all $D_{ii} = \sqrt{L_{ii}} > 0$. In that case, $\mathbf{C}$ is positive definite. Thus for every Hermitian positive definite matrix, there exists a unique Hermitian square root that is also positive definite.

For Hermitian matrices, the necessary and sufficient condition for it to be nonnegative definite (positive definite) can be given in terms of the signs of its principal minors. To be specific, a Hermitian matrix is nonnegative definite (positive definite) if and only if all its principal minors are nonnegative (positive). More interestingly, a Hermitian matrix is positive definite if and only if all its leading principal minors are positive [2, 3].

A matrix is said to be Toeplitz if its entries along every diagonal are the same. Thus if $\mathbf{T}$ is Toeplitz, then $T_{ij} = t_{i-j}$. The inverse of an invertible Toeplitz matrix is not Toeplitz unless it is 2×2 or lower (upper) triangular. Alternatively, any two lower (upper) triangular Toeplitz matrices commute and their product is again (lower (upper) triangular and) Toeplitz! However, lower (or upper) triangular *block* Toeplitz matrices do not commute. If an $n \times n$ Toeplitz matrix is also Hermitian, then it has only at most n independent entries, namely, those along the first (last) row (column).

A matrix $\mathbf{H}$ (or scalar) whose entries are functions of a variable z is denoted by $\mathbf{H}(z)$. For any such matrix, following the standard convention, let

$$\mathbf{H}_*(z) = \mathbf{H}^*(1/z^*) \tag{1.9}$$

denote its para-conjugate form. Notice that $\mathbf{H}_*(z)$ is also a function of z (and not of z^*) and since

$$\mathbf{H}_*(e^{j\theta}) = \mathbf{H}^*(e^{j\theta}), \tag{1.10}$$

the para-conjugate form represents the ordinary complex conjugate transpose operation on the unit circle. Thus the para-conjugate form can be considered as an extension of the familiar complex conjugate operation on the unit circle. A matrix $\mathbf{H}(z)$ is said to be para-conjugate Hermitian if $\mathbf{H}(z) = \mathbf{H}_*(z)$. Of course, para-conjugate Hermitian matrices are Hermitian when evaluated on the unit circle. Similarly a matrix $\mathbf{U}(z)$ is said to be para-conjugate unitary if $\mathbf{U}(z)\mathbf{U}_*(z) = \mathbf{I}$. If $\mathbf{H}(z)$ is real for real values of z, then $\mathbf{H}_*(z) = \mathbf{H}^T(1/z)$ and in that case if $\mathbf{H}(z) = \mathbf{H}_*(z)$, then $\mathbf{H}(z)$ is said to be para-Hermitian. Similarly, a real matrix $\mathbf{U}(z)$ that satisfies $\mathbf{U}(z)\mathbf{U}_*(z) = \mathbf{I}$ is said to be para-unitary. Using this notation, it is easy to write down the all z-extension formulas for functions evaluated on the unit circle. For example, the all z-extension of $|\mathbf{H}(e^{j\theta})|^2$ is given by $\mathbf{H}(z)\mathbf{H}_*(z)$. The usefulness of this notation, of course, will become clearer in subsequent chapters.

2

Power Spectra and Positive Functions

2.1 Stationary Processes and Power Spectra

Let $x(t)$ represent a continuous-time stochastic process. For any fixed time instant t, the process $x(t)$ defines a random variable and let $\mu(t)$ represent its mean value. Thus,

$$\mu(t) = E[x(t)] = \int_{-\infty}^{\infty} x f(x,t) dx, \qquad (2.1)$$

where $f(x,t)$ represents the probability density function (p.d.f.) of the random variable $X = x(t)$. The correlation between the random variables associated with any two time instants t_1 and t_2 is defined as the autocorrelation function $R(t_1, t_2)$ of the stochastic process $x(t)$. Hence

$$R(t_1, t_2) = E[x(t_1)x^*(t_2)] = \int_{-\infty}^{\infty} \int_{-\infty}^{\infty} x_1 x_2^* f(x_1, x_2; t_1, t_2) dx_1 dx_2$$

$$= R^*(t_2, t_1). \qquad (2.2)$$

Here, $f(x_1, x_2; t_1, t_2)$ represents the joint p.d.f. of the random variables $X_1 = x(t_1)$ and $X_2 = x(t_2)$. In general, the autocorrelation function depends on both the variables t_1 and t_2. For any arbitrary set of numbers a_k, $k = 1 \to n$, define

$$Y = \sum_{k=1}^{n} a_k x(t_k).$$

Clearly, Y is a random variable and $E[|Y|^2] \geq 0$ gives

$$E[|Y|^2] = \sum_{i=1}^{n} \sum_{k=1}^{n} a_i a_k^* E[x(t_i)x^*(t_k)] = \sum_{i=1}^{n} \sum_{k=1}^{n} a_i a_k^* R(t_i, t_k)$$

$$= \begin{bmatrix} a_1^* & a_2^* & \cdots & a_n^* \end{bmatrix} \begin{bmatrix} R(t_1, t_1) & R(t_1, t_2) & \cdots & R(t_1, t_n) \\ R(t_2, t_1) & R(t_2, t_2) & \cdots & R(t_2, t_n) \\ \vdots & \vdots & \ddots & \vdots \\ R(t_n, t_1) & R(t_n, t_2) & \cdots & R(t_n, t_n) \end{bmatrix} \begin{bmatrix} a_1 \\ a_2 \\ \vdots \\ a_n \end{bmatrix} \geq 0$$

or

$$\begin{bmatrix} R(t_1, t_2) & R(t_1, t_2) & \cdots & R(t_1, t_n) \\ R(t_2, t_1) & R(t_2, t_2) & \cdots & R(t_2, t_n) \\ \vdots & \vdots & \ddots & \vdots \\ R(t_n, t_1) & R(t_n, t_2) & \cdots & R(t_n, t_n) \end{bmatrix} \geq 0.$$

But, using (2.2), this simplifies into

$$
\left[
\begin{array}{cccc}
R(t_1,t_1) & R(t_1,t_2) & \cdots & R(t_1,t_n) \\
R^*(t_1,t_2) & R(t_2,t_2) & \cdots & R(t_2,t_n) \\
\vdots & \vdots & \cdots & \vdots \\
R^*(t_1,t_n) & R^*(t_2,t_n) & \cdots & R(t_n,t_n)
\end{array}
\right] \geq 0,
\tag{2.3}
$$

i.e., every $n \times n$ Hermitian matrix in (2.3) is nonnegative definite for $n = 1 \to \infty$ [4]. Thus, in particular,

$$
\left[
\begin{array}{cc}
R(t_1,t_1) & R(t_1,t_2) \\
R^*(t_1,t_2) & R(t_2,t_2)
\end{array}
\right] \geq 0
$$

or

$$
|R(t_1,t_2)|^2 \leq R(t_1,t_1)R(t_2,t_2)
\tag{2.4}
$$

which also follows from Schwarz' inequality.

A stochastic process can exhibit certain stationary patterns in its behavior. For example, if its first-order description, namely, the first-order p.d.f. $f(x,t)$ in (2.1) is independent of the time index t, then the process is said to be first-order stationary. In that case, every random variable obtained from $x(t)$ for any specific value of the time index has the same p.d.f. and from (2.1), it follows that the mean value of such a process is a constant, since $E[x(t)] = \int x f(x)dx = \mu$. Similarly, if the joint p.d.f. of the random variables $x(t_1)$ and $x(t_2)$ depends only on the difference of time indices, i.e.,

$$
f(x_1,x_2;t_1,t_2) = f(x_1,x_2;\tau = t_1 - t_2),
\tag{2.5}
$$

then $x(t)$ is said to be a second-order stationary process. Clearly, in that case, with (2.5) in (2.1), we have

$$
R(t_1,t_2) = \int_{-\infty}^{\infty} \int_{-\infty}^{\infty} x_1 x_2^* f(x_1,x_2;\tau)dx_1 dx_2 = R(\tau) = R^*(-\tau).
\tag{2.6}
$$

However, a process can have constant mean and an autocorrelation function that depends only on the difference of time indices as in (2.6) without being second-order stationary.[1] Such processes are said to be wide-sense stationary (W.S.S.) processes.

Thus, if $x(t)$ is a W.S.S. process, then

$$
E[x(t)] = \mu
\tag{2.7}
$$

and

$$
E[x(t_1)x^*(t_2)] = R(t_1 - t_2) = R^*(t_2 - t_1).
\tag{2.8}
$$

---

[1] Consider the process $x(t) = a\cos(\omega_0 t + \phi)$, where a and ϕ are independent random variables with ϕ uniformly distributed in $(-\pi,\pi)$. Then, clearly, $E[x(t)] = 0$, a constant, and $R(t_1,t_2) = E(a^2)\cos\omega_0(t_1 - t_2) = R(\tau)$.

Since, without loss of generality, the constant μ in (2.7) can be assumed to be zero, hereonwards, we will assume $x(t)$ to be a zero-mean wide-sense stationary stochastic process. For such processes, the distribution of power vs. frequency ω, i.e., the power spectrum $S(\omega)$ can be related to its auto-correlation function. To see this classical result, consider any finite segment of the process $x(t)$ for $-T < t < T$. The Fourier transform based on this segment is formally given by

$$X_T(\omega) = \int_{-T}^{T} x(t)e^{j\omega t} dt.$$

The power spectral distribution based on the above segment is

$$\frac{|X_T(\omega)|^2}{2T} \geq 0.$$

The ensemble average of the above power spectral function is

$$P_T(\omega) = E\left[\frac{|X_T(\omega)|^2}{2T}\right] = \frac{1}{2T}\int_{-T}^{T}\int_{-T}^{T} R(t_1 - t_2)e^{j\omega(t_1 - t_2)} dt_1 dt_2$$

$$= \int_{-2T}^{2T} R(\tau)\left(1 - \frac{|\tau|}{2T}\right)e^{j\omega\tau} d\tau. \tag{2.9}$$

Clearly, $P_T(\omega)$ represents the average power distribution vs. frequency based on the information available about the process in $(-T, T)$. The above description becomes complete as $T \to \infty$ and hence the power spectral density $S(\omega)$ of the process $x(t)$ is given by

$$S(\omega) = \lim_{T\to\infty} P_T(\omega) = \int_{-\infty}^{\infty} R(\tau)e^{j\omega\tau} d\tau \geq 0. \tag{2.10}$$

This relationship is known as Wiener-Khinchine theorem [5] and it states that the power spectral density $S(\omega)$ and the autocorrelation function $R(\tau)$ of any W.S.S. stochastic process form a Fourier transform pair. From (2.10), the inverse relationship gives

$$R(\tau) = \frac{1}{2\pi}\int_{-\infty}^{\infty} S(\omega)e^{-j\omega\tau} d\omega \tag{2.11}$$

and, in particular,

$$\int_{-\infty}^{\infty} S(\omega)d\omega = R(0) = E[|x(t)|^2] = P. \tag{2.12}$$

Since $E[|x(t)|^2]$ represents the average power P of the process, from (2.12), $S(\omega)$ so defined in (2.10) truly represents the power spectral density function. If $x(t)$ is a real process, then $R(\tau) = R^*(-\tau) = R(-\tau)$ and from (2.10),

$$S(\omega) = 2\int_0^{\infty} R(\tau)\cos\omega\tau d\tau = S(-\omega) \geq 0, \tag{2.13}$$

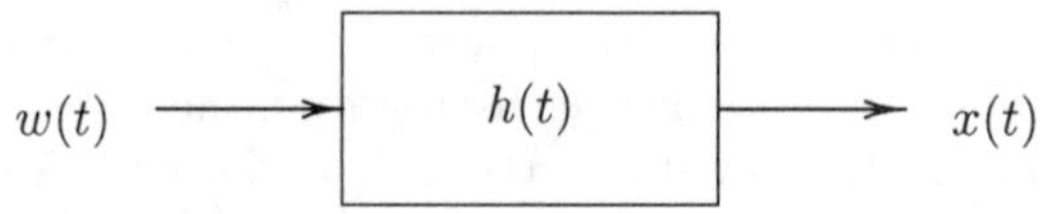

FIGURE 2.1. An LTI system.

i.e., $S(\omega)$ is also an even function.

Suppose $x(t)$ is the output of a linear time-invariant (LTI) system with impulse response $h(t)$ that has been excited by an input $w(t)$ (Fig. 2.1). In that case,

$$x(t) = \int_{-\infty}^{\infty} h(\tau)w(t - \tau)d\tau = w(t) \star h(t), \qquad (2.14)$$

where $\star$ represents the convolution operation. Then,

$$E[x(t)] = \int_{-\infty}^{\infty} h(\tau)E[w(t - \tau)]d\tau, \qquad (2.15)$$

and the cross-correlation function is given by

$$R_{xw}(t_1, t_2) = E[x(t_1) w^*(t_2)] = \int_{-\infty}^{\infty} h(\tau_1)E[w(t_1 - \tau_1) w^*(t_2)]d\tau_1$$

$$= \int_{-\infty}^{\infty} h(\tau_1)R_{ww}(t_1 - \tau_1, t_2)d\tau_1 = R_{ww}(t_1, t_2) \star h(t_1). \qquad (2.16)$$

Similarly the autocorrelation function of $x(t)$ becomes

$$R_{xx}(t_1, t_2) = E[x(t_1) x^*(t_2)] = \int_{-\infty}^{\infty} h^*(\tau_2)E[x(t_1) w^*(t_2 - \tau_2)]d\tau_2$$

$$= R_{xw}(t_1, t_2) \star h^*(t_2). \qquad (2.17)$$

Clearly, if $w(t)$ is a W.S.S. process, then (2.15) gives

$$E[x(t)] = \mu \qquad (2.18)$$

and with $t_1 - t_2 = \tau$, (2.16) and (2.17) translate into

$$R_{xw}(t_1, t_2) = R_w(\tau) \star h(\tau) \overset{\triangle}{=} R_{xw}(\tau) \qquad (2.19)$$

and

$$R_{xx}(t_1, t_2) = R_{xw}(\tau) \star h^*(-\tau)$$

$$= R_w(\tau) \star h(\tau) \star h^*(-\tau) \overset{\triangle}{=} R_x(\tau). \qquad (2.20)$$

Thus, wide-sense stationarity of the input $w(t)$ implies that $w(t)$ and $x(t)$ are jointly W.S.S. and, further from (2.18) and (2.20), $x(t)$ is also W.S.S. Using (2.10), this gives the power spectral density of the output process $x(t)$ to be

$$S_x(\omega) = |H(j\omega)|^2 S_w(\omega), \qquad (2.21)$$

where $H(j\omega) = \mathcal{F}\{h(t)\}$ represents the Fourier transform of the impulse response $h(t)$ and $S_w(\omega) = \mathcal{F}\{R_w(\tau)\}$ represents the power spectral density of the input process $w(t)$. If $w(t)$ is a W.S.S. white noise process with unit intensity, then

$$R_w(\tau) = \delta(\tau) \quad \text{or} \quad S_w(\omega) = 1$$

and

$$S_x(\omega) = |H(j\omega)|^2. \qquad (2.22)$$

If $h(t)$ is real, then $H^*(j\omega) = H(-j\omega)$ and (2.22) becomes

$$S_x(\omega) = H(j\omega)H(-j\omega) = H(s)H(-s)|_{s=j\omega} = S_x(-\omega). \qquad (2.23)$$

If $H(j\omega)$ is also rational, then so is $S_x(\omega)$ and from (2.23) in that case it is a function of ω^2. Thus, any real rational system excited by a white noise process generates a real process with rational power spectral density that is only a function of ω^2 as in (2.23).

Conversely, if $x(t)$ is a real process with a rational spectral density, then from (2.13),

$$S(\omega) = S(-\omega) = \frac{B(\omega^2)}{A(\omega^2)} = \left.\frac{B(-s^2)}{A(-s^2)}\right|_{s=j\omega} \geq 0, \qquad (2.24)$$

where $A(\cdot)$ and $B(\cdot)$ are real polynomials in s^2 satisfying (2.24). Thus, if s_k is a zero of either A or B off the $j\omega$-axis in (2.24), then so is $-s_k$, s_k^* and $-s_k^*$. Moreover from the nonnegativity constraint in (2.24), every zero of A and B on the $j\omega$-axis must occur with even multiplicity. Thus, every zero of A or B on the strict right-half plane is accompanied by its mirror image on the strict left-half plane, and moreover those on the $j\omega$-axis are of even multiplicity and can also be paired in a similar manner. This allows the representation

$$A(-s^2) = M(s)M(-s), \quad B(-s^2) = N(s)N(-s), \qquad (2.25)$$

where $M(s)$ and $N(s)$ are real Hurwitz polynomials.[2] Thus,

$$S(\omega) = \left.\frac{N(s)N(-s)}{M(s)M(-s)}\right|_{s=j\omega} = H(s)H(-s)|_{s=j\omega} = |H(j\omega)|^2, \qquad (2.26)$$

---

[2]A Hurwitz polynomial has no zeros on the open right-half plane (or interior of the unit circle, in the z-plane). A strict Hurwitz polynomial has no zeros in the strict right-half plane (or none on the closed unit circle in the z-plane).

where

$$H(s) = \frac{N(s)}{M(s)} \tag{2.27}$$

represents the minimum-phase factor[3] of the spectral density $S(\omega)$. Notice that if $S(\omega)$ in (2.24) is already in the irreducible form, then $M(s)$ and $N(s)$ are relatively prime polynomials. As remarked earlier, *a priori*, rational spectra can possess zeros and poles (of even multiplicity) on the $j\omega$-axis. However, if $x(T)$ has finite power, then from (2.12), $S(\omega)$ is integrable and necessarily $S(\omega)$ must be free of poles (i.e., $A(s)$ is a strict Hurwitz polynomial) and $S(\omega) \to 0$ as $\omega \to \infty$ (i.e., the degree of $B(s)$ is strictly less than the degree of $A(s)$).

To summarize, the power spectral density of any real rational continuous-time process with finite power is free of poles on the $j\omega$-axis and its minimum-phase factor $H(s)$ is proper.[3]

A stochastic process $x(t)$ is said to be band-limited if it its power spectral density $S(\omega)$ satisfies

$$S(\omega) = 0, \quad |\omega| > B_0, \tag{2.28}$$

for some $B_0 > 0$. As usual, letting $T_0 = \pi/B_0$ represent the Nyquist sampling period for a band-limited signal, in that case the samples $x(nT)$, with $T \leq T_0$, contain complete information about the process $x(t)$ in the mean-square sense. This follows easily, since

$$x(t) = \sum_{n=-\infty}^{\infty} x(nT) \, \frac{\sin \omega_0 (t - nT)}{\omega_0 (t - nT)}$$

in the mean-square sense, where

$$\omega_0 = \frac{\pi}{T}.$$

Thus, the above sampling theorem states that the entire process can be reconstructed in a mean-square sense from its samples $x(nT)$, provided $T \leq T_0$. The samples $x(nT)$, $n = -\infty \to \infty$, by themselves represent a discrete-time W.S.S. stochastic process.

More generally, let $x(nT)$ represent a zero-mean W.S.S. discrete-time stochastic process. Then

$$E[x(nT)] = 0$$

and the autocorrelation sequence is given by

$$E[x((n+k)T) \, x^*(nT)] = r_k = r_{-k}^*, \quad k = -\infty \to \infty. \tag{2.29}$$

---

[3]A rational function is said to be minimum-phase if it is free of poles and zeros in the open right-half plane (interior of the unit circle, in the z-plane). A rational function with numerator degree strictly less than the denominator degree is said to be proper. More generally, a function that is analytic together with its inverse in $|z| < 1$ is said to be minimum-phase.

In this case, the nonnegative definiteness of the Hermitian correlation matrices in (2.3) translates into

$$
\mathbf{T}_k =
\begin{bmatrix}
r_0 & r_1 & r_2 & \cdots & r_k \\
r_1^* & r_0 & r_1 & \cdots & r_{k-1} \\
r_2^* & r_1^* & r_0 & \cdots & r_{k-2} \\
\vdots & \vdots & \vdots & \cdots & \vdots \\
r_k^* & r_{k-1}^* & r_{k-2}^* & \cdots & r_0
\end{bmatrix}
\geq 0 ,
\tag{2.30}
$$

for $k = 0 \to \infty$, i.e., every such Hermitian Toeplitz matrix $\mathbf{T}_k$, $k = 0 \to \infty$ is nonnegative definite. Thus, the autocorrelation function of such a process is given by

$$
R(\tau) = \sum_{k=-\infty}^{\infty} r_k \delta(\tau - kT)
$$

and from (2.10), the power spectral density of a discrete-time W.S.S. process has the form

$$
S(\omega) = \sum_{k=-\infty}^{\infty} r_k e^{jk\omega T} \geq 0 .
\tag{2.31}
$$

As before, let $\omega_0 = \pi/T$. Then from (2.31),

$$
S(\omega) = S(\omega + 2m\omega_0), \quad m = 0, \pm 1, \pm 2, \ldots,
\tag{2.32}
$$

i.e., $S(\omega)$ is periodic[4] with period $2\omega_0$. Define

$$
\theta = \frac{\pi\omega}{\omega_0} = \omega T ,
\tag{2.33}
$$

then (2.31) becomes

$$
S(\theta) = \sum_{k=-\infty}^{\infty} r_k e^{jk\theta} = S(\theta + 2m\pi), \quad -\pi < \theta \leq \pi .
\tag{2.34}
$$

Thus,

$$
r_k = \frac{1}{2\pi} \int_{-\pi}^{\pi} S(\theta) e^{-jk\theta} d\theta ,
\tag{2.35}
$$

and in particular for finite power processes,

$$
E[|x(nT)|^2] = r_0 = \frac{1}{2\pi} \int_{-\pi}^{\pi} S(\theta) d\theta < \infty ,
\tag{2.36}
$$

---

[4]Note that the power spectral density of a discrete-time process is always periodic. If the discrete-time process has been obtained by sampling a continuous-time band-limited process with two-sided bandwidth $2B$, then from (2.32), there is no spectral overlap or aliasing, provided the sampling frequency is greater than $2B$.

i.e., the power spectral density is an integrable function ($S(\theta) \in L_1$). If $x(nT)$ is also real, then $r_k = r^*_{-k} = r_{-k}$, so that (2.34) reduces to

$$S(\theta) = r_0 + 2 \sum_{k=1}^{\infty} r_k \cos(k\theta) = S(-\theta) \qquad (2.37)$$

and

$$r_k = \frac{1}{\pi} \int_0^{\pi} S(\theta) \cos(k\theta) \, d\theta . \qquad (2.38)$$

The nonnegative definite property of every Hermitian Toeplitz matrix $\mathbf{T}_k$ in (2.30) and the nonnegativity of the power spectral density $S(\theta)$ in (2.31) are intimately coupled. In fact, they imply each other. To see this [4, 6, 7], first assume $S(\theta) \geq 0$ in (2.31); then from (2.30), letting

$$\mathbf{a} = [\ a_0, \ a_1, \ a_2, \ \dots, \ a_k\ ]^T , \qquad (2.39)$$

we have, using (2.35),

$$\begin{aligned}
\mathbf{a}^* \mathbf{T}_k \mathbf{a} &= \sum_{i=0}^{k} \sum_{m=0}^{k} a_i^* a_m r_{i-m} \\
&= \sum_{i=0}^{k} \sum_{m=0}^{k} a_i^* a_m \frac{1}{2\pi} \int_{-\pi}^{\pi} S(\theta) e^{-j(i-m)\theta} d\theta \\
&= \frac{1}{2\pi} \int_{-\pi}^{\pi} S(\theta) \left| \sum_{m=0}^{k} a_m e^{jm\theta} \right|^2 d\theta \geq 0 .
\end{aligned} \qquad (2.40)$$

Since $\mathbf{a}$ is arbitrary, this gives

$$S(\theta) \geq 0 \Longrightarrow \mathbf{T}_k \geq 0, \quad k = 0 \to \infty . \qquad (2.41)$$

Conversely, assume that every $\mathbf{T}_k$, $k = 0 \to \infty$, are nonnegative definite and for any ρ, $0 < \rho < 1$, and θ_0, $0 < \theta_0 < 2\pi$, define [7, 8] the vector $\mathbf{a}$ in (2.39) such that

$$a_k = \sqrt{(1 - \rho^2)} \rho^m e^{-jm\theta_0} .$$

Then, $\mathbf{T}_k$ nonnegative definite implies that (see (2.40))

$$0 \leq \mathbf{a}^* \mathbf{T}_k \mathbf{a} = \frac{1}{2\pi} \int_{-\pi}^{\pi} (1 - \rho^2) \left| \sum_{m=0}^{k} \rho^m e^{jm(\theta - \theta_0)} \right|^2 S(\theta) d\theta .$$

Letting $k \to \infty$ and using the dominated convergence theorem [9], we have

$$\lim_{k \to \infty} \frac{1}{2\pi} \int_{-\pi}^{\pi} (1 - \rho^2) \left| \sum_{m=0}^{k} \rho^m e^{jm(\theta - \theta_0)} \right|^2 S(\theta) d\theta$$

$$= \frac{1}{2\pi} \int_{-\pi}^{\pi} \left(\frac{(1-\rho^2)}{1 - 2\rho \cos(\theta - \theta_0) + \rho^2} \right) S(\theta) d\theta$$

$$= \lim_{k \to \infty} \mathbf{a}^* \mathbf{T}_k \mathbf{a} \geq 0. \tag{2.42}$$

The left-hand side is the Poisson's integral and its interior radial (ray) limit as $\rho \to 1 - 0$ equals $S(\theta_0)$ for almost all θ_0 [10]. Hence

$$S(\theta) \geq 0 \quad a.e., \tag{2.43}$$

provided $\mathbf{T}_k \geq 0$, for every $k = 0 \to \infty$. Thus, (2.41) together with (2.43) gives

$$S(\theta) \geq 0, \; a.e. \iff \mathbf{T}_k \geq 0, \; k = 0 \to \infty. \tag{2.44}$$

From (2.44), $\mathbf{T}_k$'s can be singular and for some k_0 if $\mathbf{T}_{k_0}$ is singular, then from its structure in (2.30), there must exist an $n \leq k_0$ such that $\mathbf{T}_{n-1}$ is nonsingular and $\mathbf{T}_n$ is singular. Then it can be shown that (see equations (3.90)–(3.92) in section 3.3.1)

$$r_k = \sum_{i=1}^{n} P_i e^{-jk\theta_i} \tag{2.45}$$

where $P_i > 0, 0 \leq \theta_i < 2\pi, \theta_i \neq \theta_j, i \neq j$. In that case, every $\mathbf{T}_k$ is singular for $k \geq n$, and

$$S(\theta) = \sum_{i=1}^{n} P_i \delta(\theta - \theta_i) \tag{2.46}$$

represents a degenerate discrete spectrum. Clearly, (2.46) corresponds to a sum of uncorrelated complex sinusoidal signals and the underlying process is essentially shape deterministic.

More interestingly, subject to the additional constraint, known as the Paley-Wiener criterion [6]

$$\frac{1}{2\pi} \int_{-\pi}^{\pi} \ln S(\theta) d\theta > -\infty, \tag{2.47}$$

every $\mathbf{T}_k, k = 0 \to \infty$, must be positive definite. This follows easily from (2.40). In fact, if some $\mathbf{T}_k$ is singular, then there exists a nontrivial vector $\mathbf{a}$ such that $\mathbf{T}_k \mathbf{a} = 0$ and, from (2.40),

$$\mathbf{a}^* \mathbf{T}_k \mathbf{a} = \frac{1}{2\pi} \int_{-\pi}^{\pi} S(\theta) \left| \sum_{m=0}^{k} a_m e^{jm\theta} \right|^2 d\theta = 0.$$

Since $S(\theta) \geq 0, a.e.$, the above expression gives

$$S(\theta) \left| \sum_{m=0}^{k} a_m e^{jm\theta} \right|^2 = 0, \quad a.e.$$

and $\sum_{m=0}^{k} a_m e^{-jm\theta} \neq 0$, a.e., implies

$$S(\theta) = 0, \quad a.e.$$

and

$$\int_{-\pi}^{\pi} \ln S(\theta) d\theta = -\infty,$$

contradicting (2.47). Hence subject to[5] (2.47), every

$$\mathbf{T}_k > 0, \quad k = 0 \to \infty. \tag{2.48}$$

The integrability condition (2.36) together with (2.47) permits the factorization of the power spectral density in terms of a specific function with certain interesting properties [4]. More precisely, there exists a unique function

$$B(z) = \sum_{k=0}^{\infty} b_k z^k, \quad b_0 > 0, \ |z| < 1, \tag{2.49}$$

that is analytic together with its inverse in $|z| < 1$ (minimum-phase) satisfying

$$\sum_{k=0}^{\infty} |b_k|^2 < \infty \tag{2.50}$$

and

$$S(\theta) = |B(e^{j\theta})|^2, \quad a.e., \tag{2.51}$$

iff $S(\theta)$ as well as $\ln S(\theta)$ are integrable functions.[6] Here $B(e^{j\theta})$ is defined as the interior radial limit of $B(z)$ on the unit circle, i.e.,

$$B(e^{j\theta}) = \lim_{r \to 1-0} B(re^{j\theta}) \tag{2.52}$$

and it may not be analytic or even continuous on $|z| = 1$. Even though $B(z)$ is free of poles on $|z| = 1$, it can still possess essential singularities[7]

---

[5]Unlike the integrability condition, $S(\theta)$ that satisfies (2.47) alone, can possess isolated poles (and zeros) on the unit circle. This follows since $\int \ln (1/x) dx = -\int (\ln x) dx = -(x \ln x - x)$ and as $x \to 0$, the above integral tends to zero. Clearly, for (2.47) to hold, $S(\theta)$ cannot be equal to zero over an interval of nonzero measure.

[6]Notice that integrability of $\ln S(\theta)$ implies $\int_{-\pi}^{\pi} |\ln S(\theta)| d\theta < \infty$ and this is equivalent to $-\infty < \int_{-\pi}^{\pi} \ln S(\theta) d\theta < \infty$. Since the right-hand side of this inequality automatically follows from (2.36), the integrability of $\ln S(\theta)$ is equivalent to (2.47).

[7]At an essential singularity, the function has a pole of infinite order (i.e., the power series about that point has an infinite number of terms with negative power). For example, in (2.53), the all-pass function $e^{-1/(1-z)}$ is analytic in $|z| < 1$ and has an essential singularity at $z = 1$. The functions $(z - 1)e^{-1/(1-z)}$ as well as $B(z)$ in (2.53) are analytic in $|z| < 1$ and continuous in $|z| \le 1$.

on $|z| = 1$. For example, the function

$$B(z) = 1 + \frac{(z-1)}{2} e^{-1/(1-z)} \qquad (2.53)$$

is analytic in $|z| < 1$ and free of zeros in $|z| < 1$. Thus, $B(z)$ is minimum-phase. Clearly, $B(z)$ is also continuous everywhere in $|z| \le 1$, except possibly at $z = 1$. However, since its interior radial limit at $z = 1$ is

$$\lim_{r \to 1-0} B(r) = \lim_{r \to 1-0} \left[1 + \frac{(r-1)}{2} e^{-1/(1-r)} \right] \to 1 + 0 \cdot e^{-\infty} = 1$$

and its tangential limit at $z = 1$ $(\theta = 0)$ is

$$\lim_{\theta \to 0} B(e^{j\theta}) = \lim_{\theta \to 0} \left[1 + j\, e^{-1/2}\, \sin\left(\frac{\theta}{2}\right) exp\left(j\left[\frac{\theta}{2} - \frac{1}{2}\cot\left(\frac{\theta}{2}\right)\right]\right) \right] \to 1,$$

it is also continuous at $z = 1$. Moreover,

$$S(\theta) = |B(e^{j\theta})|^2$$

$$= 1 + e^{-1/2} \sin\left(\frac{\theta}{2}\right) \left[e^{-1/2} \sin\left(\frac{\theta}{2}\right) - 2\sin\left(\frac{\theta}{2} - \frac{1}{2}\cot\left(\frac{\theta}{2}\right)\right) \right] \le 8$$

$$(2.54)$$

satisfies the integrability condition (2.36) as well as the Paley-Wiener criterion (2.47). Hence from (2.49), the unique $B(z)$ so obtained from the above $S(\theta)$ matches with (2.53) and it has an essential singularity at $z = 1$. Since this function is analytic in $|z| < 1$ and $B(e^{j\theta})$ is continuous for every θ in $(0, 2\pi)$, this essential singularity is not visible in $|z| \le 1$. However, in the vicinity of $z_0 = 1$, outside the unit circle, as z approaches the point $z_0 = 1$ through different paths, the limit $B(z_0)$ takes on every possible value. In that case, the radius of convergence of $B(z)$ is strictly less than unity.

However, if $S(\theta)$ and hence $B(z)$ are rational, i.e.,

$$B(z) = \frac{Q(z)}{P(z)} = \frac{\beta_0 + \beta_1 z + \cdots + \beta_q z^q}{1 + \alpha_1 z + \alpha_2 z^2 + \cdots + \alpha_p z^p}, \qquad (2.55)$$

then the absence of poles for $B(z)$ in $|z| \le 1$ implies that $P(z)$ is strict Hurwitz and $B(z)$ possesses a power series expansion valid in $|z| \le 1$ and in particular from (2.49), $\sum_{k=0}^{\infty} |b_k| < \infty$, i.e., such a $B(z)$ represents a bounded input-bounded output (BIBO) stable filter in the rational case.

In general, the minimum-phase factor $B(z)$ is free of zeros in $|z| < 1$ (by construction) and free of poles in $|z| \le 1$ (analyticity of $B(z)$ in $|z| < 1$ guarantees the absence of poles there and the integrability condition guarantees it on $|z| = 1$) and it is known as the Wiener factor of the given process and represents a causal (one-sided) filter with square summable impulse response. Its physical meaning is not difficult to grasp. When driven by

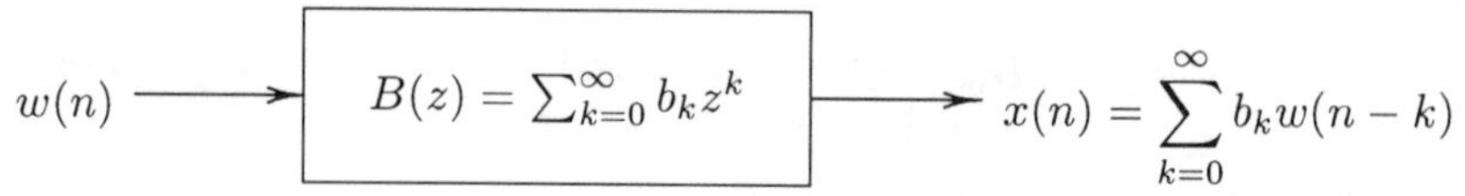

$$w(n) \longrightarrow \boxed{B(z) = \sum_{k=0}^{\infty} b_k z^k} \longrightarrow x(n) = \sum_{k=0}^{\infty} b_k w(n - k)$$

FIGURE 2.2. Wiener filter for a stationary process.

a stationary white noise process of unit spectral density, this filter generates a regular stochastic process $x(nT)$, entirely from the past samples of the white noise process (Fig. 2.2). Thus, any regular process has the representation

$$x(n) = \sum_{k=0}^{\infty} b_k w(n - k) \qquad (2.56)$$

and its power spectral density is given by (2.51).

If $B(z)$ represents a rational system as in (2.55), then, from Fig. 2.2 for any input $w(n)$, the output is given by

$$X(z) = B(z)W(z) = \frac{Q(z)W(z)}{P(z)}$$

or

$$x(n) = - \sum_{k=1}^{p} \alpha_k x(n - k) + \sum_{k=0}^{q} \beta_k w(n - k). \qquad (2.57)$$

Thus, the present value of the output $x(n)$ depends regressively upon its previous p sample values as well as the running (moving) average generated from $(q + 1)$ past samples of the input $w(n)$. Such systems are known as AutoRegressive Moving-Average (ARMA) models with denominator degree equal to p and the numerator degree equal to q, or in short, as ARMA(p, q) systems. If the input to such a system is white noise, then $x(n)$ is an ARMA(p, q) process. In particular, if $q = 0$, then $x(n)$ is purely autoregressive of order p $(x(n) \sim \mathrm{AR}(p))$ and if $p = 0$, then $x(n)$ is just a moving average process of order q $(x(n) \sim \mathrm{MA}(q))$. Clearly, if the input process is also wide-sense stationary with spectral density $S_w(\theta)$, then the output spectral density is given by

$$S_x(\theta) = \left| B(e^{j\theta}) \right|^2 S_w(\theta) \qquad (2.58)$$

and (2.58) reduces to (2.51) for any white noise input with unit spectral density.

The absence of zeros in $|z| < 1$ for the Wiener factor can be used to derive another useful relation involving its coefficients and the power spectrum.

To see this, note that $B(z)$ analytic and free of zeros in $|z| < 1$ implies that $\ln B(z)$ is also analytic in $|z| < 1$. Thus $\ln B(z)$ has a power series expansion in that region, and let [11]

$$\ln B(z) = \ln b_0 + \ln\left(1 + \sum_{k=1}^{\infty} b_k z^k / b_0\right) \triangleq d_0 + d_1 z + d_2 z^2 + \cdots , \quad (2.59)$$

where $d_0 = \ln b_0$. Using (2.52), this gives

$$\ln B(e^{j\theta}) = \sum_{k=0}^{\infty} d_k e^{jk\theta}$$

and

$$\ln B^*(e^{j\theta}) = \sum_{k=0}^{\infty} d_k^* e^{-jk\theta} .$$

Hence

$$\ln B(e^{j\theta}) + \ln B^*(e^{j\theta}) = \ln |B(e^{j\theta})|^2$$

or

$$\ln S(\theta) = \sum_{k=0}^{\infty} d_k e^{jk\theta} + \sum_{k=0}^{\infty} d_k^* e^{-jk\theta} .$$

As a result,

$$2d_0 = \frac{1}{2\pi} \int_{-\pi}^{\pi} \ln S(\theta) d\theta \qquad (2.60)$$

and, in general,

$$d_k = \frac{1}{2\pi} \int_{-\pi}^{\pi} \ln S(\theta) e^{-jk\theta} d\theta , \quad k \geq 1 . \qquad (2.61)$$

Substituting (2.60) and (2.61) into (2.59), we obtain

$$B(z) = exp\left(\sum_{k=0}^{\infty} d_k z^k\right) = e^{d_0} exp\left(\sum_{k=1}^{\infty} d_k z^k\right)$$

$$\triangleq e^{d_0}\left(1 + \alpha_1 z + \alpha_2 z^2 + \cdots\right) , \qquad (2.62)$$

where

$$1 + \alpha_1 z + \alpha_2 z^2 + \cdots = exp\left(\sum_{k=1}^{\infty} d_k z^k\right) . \qquad (2.63)$$

On comparing (2.62) and (2.49), we get

$$b_0 = e^{d_0} = exp\left(\frac{1}{4\pi} \int_{-\pi}^{\pi} \ln S(\theta) d\theta\right) \qquad (2.64)$$

and, in general,

$$b_k = \alpha_k e^{d_0}, \quad k = 1 \to \infty, \tag{2.65}$$

where α_k's are related to d_k's as in (2.63). Expanding the exponent in (2.63) and rearranging the terms gives the useful recursion formula [5, 12]

$$\alpha_0 = 1$$
$$\alpha_1 = d_1 \tag{2.66}$$
$$\alpha_2 = \frac{d_1^2}{2} + d_2 = \frac{\alpha_1 d_1}{2} + d_2$$

and, in general, for $k \geq 1$,

$$\alpha_k = \frac{1}{k} \sum_{i=1}^{k} i\, \alpha_{k-i} d_i. \tag{2.67}$$

The (Paley-Wiener) integrability condition in (2.47) is also known as the causality or physical-realizability criterion. To see this, let $A(\theta) = \sqrt{S(\theta)} > 0$. Then, from (2.51) and the above discussion, the condition

$$\frac{1}{2\pi} \int_{-\pi}^{\pi} \ln S(\theta) d\theta = \frac{1}{\pi} \int_{-\pi}^{\pi} \ln A(\theta) d\theta > -\infty \tag{2.68}$$

implies that, associated with the given Fourier transform magnitude function $A(\theta)$, there exists a phase function $\psi(\theta)$ such that $A(\theta) e^{j\psi(\theta)}$ can be represented in terms of a causal filter $B(z)$ with one-sided impulse response $\{b_k\}_{k=0}^{\infty}$ as

$$A(\theta) e^{j\psi(\theta)} = \sum_{k=0}^{\infty} b_k e^{jk\theta} = B(z)|_{z=e^{j\theta}}.$$

Clearly, for (2.68) to hold, $A(\theta)$ should not be zero over any finite interval.

The technique of obtaining the minimum-phase factor $B(z)$ from any given $S(\theta)$ as in (2.51) is known as the spectral factorization and several procedures have been developed in this context [6, 7]. In what follows, a constructive procedure [7] that is valid under (2.36) and (2.47) is illustrated.

From (2.48), since $\mathbf{T}_k > 0$ in (2.30) for every k, $\mathbf{T}_k$ can be factored using its unique lower triangular matrix factor $\mathbf{L}_k$ such that

$$\mathbf{T}_k = \mathbf{L}_k^* \mathbf{L}_k \tag{2.69}$$

through the Gauss factorization method [2, 3]. Then,

$$\mathbf{L}_k = \begin{bmatrix} L_{00}^{(k)} & 0 & \cdots & 0 \\ L_{10}^{(k)} & L_{11}^{(k)} & \cdots & 0 \\ \vdots & \vdots & \ddots & \vdots \\ L_{k0}^{(k)} & L_{k1}^{(k)} & \cdots & L_{kk}^{(k)} \end{bmatrix} \tag{2.70}$$

is lower triangular and it can be made unique in (2.69) by requiring all its diagonal elements to be positive. Let $\mathbf{L}_k$ represent such a unique matrix factor. Then, for every fixed $n \geq m \geq 0$ [7, 8],

$$\lim_{k \to \infty} L_{nm}^{(k)} = b_{n-m} \tag{2.71}$$

and in particular,

$$\lim_{k \to \infty} L_{n0}^{(k)} = b_n , \tag{2.72}$$

i.e., the entries in the first column of $\mathbf{L}_k$ tend to b_n, $n = 0 \to \infty$, the coefficients of the desired Wiener factor $B(z)$. By exploiting the Toeplitz structure, $\mathbf{L}_k$ in (2.70) can be computed efficiently through a recursive procedure. For example, letting

$$\mathbf{L}_{k+1} = \left[\begin{array}{c|c} c & 0 \\ \hline \mathbf{b} & \mathbf{A} \end{array} \right] , \tag{2.73}$$

where $\mathbf{A}$ is $(k+1) \times (k+1)$, $\mathbf{b}$ is $(k+1) \times 1$, and c is 1×1, we obtain

$$\mathbf{L}_{k+1}^* \mathbf{L}_{k+1} = \left[\begin{array}{c|c} d & \mathbf{b}^* \mathbf{A} \\ \hline \mathbf{A}^* \mathbf{b} & \mathbf{A}^* \mathbf{A} \end{array} \right] = \mathbf{T}_{k+1} = \left[\begin{array}{c|ccc} r_0 & r_1 & \cdots & r_{k+1} \\ \hline r_1^* & & & \\ \vdots & & \mathbf{T}_k & \\ r_{k+1}^* & & & \end{array} \right] ,$$

where

$$d = c^2 + \mathbf{b}^* \mathbf{b} = r_0 .$$

Using (2.69), this gives

$$\mathbf{A} = \mathbf{L}_k \tag{2.74}$$

$$\mathbf{b} = (\mathbf{L}_k^*)^{-1} \left[\begin{array}{c} r_1^* \\ r_2^* \\ \vdots \\ r_{k+1}^* \end{array} \right] = \left[\begin{array}{c} L_{10}^{(k+1)} \\ L_{20}^{(k+1)} \\ \vdots \\ L_{k+1,0}^{(k+1)} \end{array} \right] \tag{2.75}$$

and

$$c = L_{00}^{(k+1)} = (r_0 - \mathbf{b}^* \mathbf{b})^{1/2} = \left(r_0 - \sum_{i=1}^{k+1} |L_{i0}^{(k+1)}|^2 \right)^{1/2} > 0 . \tag{2.76}$$

Thus, at every stage, only the first column of the lower triangular matrix $\mathbf{L}_{k+1}$ needs to be computed and

$$\mathbf{L}_{k+1} = \left[\begin{array}{c|c} c & 0 \\ \hline \mathbf{b} & \mathbf{L}_k \end{array} \right] \tag{2.77}$$

with **b** and c as given in (2.75) and (2.76). Note that at stage $k + 1$, new information is brought in only through r_{k+1} and this is used together with $\mathbf{L}_k$ from the previous stage to evaluate **b** and c. From (2.72), for sufficiently large k, the first column of $\mathbf{L}_k$ tends to the coefficients b_m, $m = 0 \rightarrow \infty$, of the desired Wiener factor. Notice that r_k, $k = 0 \rightarrow \infty$, real also guarantees b_k, $k = 0 \rightarrow \infty$, to be real, i.e., Wiener factors for real processes can be chosen to be real and can be made unique by letting $b_0 > 0$. Though numerically this technique is extremely robust and in particular can tolerate zeros on the unit circle, nevertheless it does not give any information regarding the structure of $B(z)$ such as whether it is nonrational or rational and if so its degree, etc. This algorithm always generates the coefficients of the minimum-phase factor $B(z)$ and any other information must be taken into account separately. For example, if $B(z)$ is known to be a polynomial of degree n, then this information could be used to suppress b_m, $m > n$, at the final stage. However, for this algorithm, the best way to perform this in more complicated situations is not always clear.

Regular stochastic processes (those satisfying (2.47)) are inherently random, and this is best illustrated by considering the best linear predictor of such a process based on its past data.

2.1.1   LINEAR PREDICTION

Consider a finite power discrete-time regular stochastic process $x(nT)$ with power spectral density $S(\theta)$ that satisfies (2.36) and (2.47). Suppose observations on $x(nT)$ are available for the entire past and, based on that, its future values need to be predicted in some optimal manner. To be specific, let $x(-kT)$, $k = 0 \rightarrow \infty$, denote the past data (known) and let $x(T)$ denote the unknown sample at the next instant. Thus, the one-step (forward) linear predictor $\hat{x}(T)$ based on the entire past sample values is given by

$$\hat{x}(T) = \sum_{k=0}^{\infty} w_k x(-kT) \qquad (2.78)$$

and the problem is to find the best linear predictor that minimizes the mean-square error $E[|\epsilon(T)|^2]$ where the error $\epsilon(T)$ is

$$\epsilon(T) = x(T) - \hat{x}(T) = x(T) - \sum_{k=0}^{\infty} w_k x(-kT). \qquad (2.79)$$

Before solving this 'limiting case' that makes use of the entire set of past data, it is instructive to consider the same problem with only a finite number — say n — of its past samples. Thus with $x(-kT)$, $k = 0 \rightarrow n - 1$, known, the one-step forward linear predictor based on its n past samples is given by

$$\hat{x}_n(T) = \sum_{k=0}^{n-1} a_k x(-kT) . \tag{2.80}$$

The error associated with this predictor is

$$\epsilon_n(T) = x(T) - \hat{x}_n(T) = x(T) - \sum_{k=0}^{n-1} a_k x(-kT) . \tag{2.81}$$

Minimization of the mean-square error $E[|\epsilon_n(T)|^2]$ with respect to the unknowns results in the following standard set of linear equations

$$E[x(-iT)\, \epsilon_n^*(T)] = 0 , \quad i = 0 \to n - 1 .$$

Using (2.81) together with (2.29), the above equation reduces to

$$r_{i+1}^* - \sum_{k=0}^{n-1} a_k^* r_{k-i} = 0 , \quad i = 0 \to n - 1 , \tag{2.82}$$

and the residual mean-square error associated with the one-step predictor is given by

$$\sigma_n^2(1) = E[|\epsilon_n(T)|^2] = E[x(T)\, \epsilon_n^*(T)]$$

$$= r_0 - \sum_{k=0}^{n-1} a_k^* r_{k+1} . \tag{2.83}$$

Put together in matrix form, (2.82) reduces to

$$\begin{bmatrix} r_0 & r_1 & r_2 & \cdots & r_{n-1} \\ r_1^* & r_0 & r_1 & \cdots & r_{n-2} \\ \vdots & \vdots & \vdots & \ddots & \vdots \\ r_{n-1}^* & r_{n-2}^* & r_{n-3}^* & \cdots & r_0 \end{bmatrix} \begin{bmatrix} a_0^* \\ a_1^* \\ \vdots \\ a_{n-1}^* \end{bmatrix} = \begin{bmatrix} r_1^* \\ r_2^* \\ \vdots \\ r_n^* \end{bmatrix} = \gamma_1$$

or using (2.64) and (2.70), the best linear predictor is given by

$$\begin{bmatrix} a_0^* \\ a_1^* \\ \vdots \\ a_{n-1}^* \end{bmatrix} = \mathbf{T}_{n-1}^{-1} \begin{bmatrix} r_1^* \\ r_2^* \\ \vdots \\ r_n^* \end{bmatrix} = \mathbf{L}_{n-1}^{-1} (\mathbf{L}_{n-1}^*)^{-1} \begin{bmatrix} r_1^* \\ r_2^* \\ \vdots \\ r_n^* \end{bmatrix}$$

$$= \mathbf{L}_{n-1}^{-1} \begin{bmatrix} L_{10}^{(n)} \\ L_{20}^{(n)} \\ \vdots \\ L_{n0}^{(n)} \end{bmatrix} . \tag{2.84}$$

Substituting (2.84) into (2.83), the minimum mean-square error associated with the one-step predictor that makes use of n past samples simplifies to

$$\sigma_n^2(1) = r_0 - [r_1, r_2, \ldots, r_n] \begin{bmatrix} a_0^* \\ a_1^* \\ \vdots \\ a_{n-1}^* \end{bmatrix}$$

$$= r_0 - [r_1, r_2, \ldots, r_n] \mathbf{L}_{n-1}^{-1} \begin{bmatrix} L_{10}^{(n)} \\ L_{20}^{(n)} \\ \vdots \\ L_{n0}^{(n)} \end{bmatrix}$$

$$= r_0 - \sum_{k=1}^{n} |L_{k0}^{(n)}|^2 = |L_{00}^{(n)}|^2 > 0, \qquad (2.85)$$

where the last step follows from (2.76). Letting[8]

$$\Delta_n = det\, \mathbf{T}_n > 0, \quad n = 0 \to \infty, \qquad (2.86)$$

from (2.69), (2.77) and (2.76),

$$\Delta_n = |\mathbf{L}_n|^2 = |\mathbf{L}_{n-1}|^2 \cdot |L_{00}^{(n)}|^2 = \Delta_{n-1}|L_{00}^{(n)}|^2$$

or

$$\sigma_n^2(1) = |L_{00}^{(n)}|^2 = \frac{\Delta_n}{\Delta_{n-1}} > 0. \qquad (2.87)$$

As the number of known samples increases, the minimum mean-square error $\sigma_n^2(1)$ associated with the one-step predictor should not increase. This also follows from a fundamental matrix identity[9] [13] applied to Δ_{n+1}. In that case,

$$\Delta_{n+1} = \frac{1}{\Delta_{n-1}} \left(\Delta_n^2 - |\Delta_{n+1}^{(1)}|^2 \right), \qquad (2.88)$$

---

[8]Since $\mathbf{T}_n$ is Hermitian positive definite iff every principal minor is positive, we have $\mathbf{T}_n > 0 \Longleftrightarrow \Delta_i > 0, \ 0 \le i \le n$.

[9]Let $\mathbf{A}$ be an $n \times n$ matrix and $\Delta_{nw}, \Delta_{ne}, \Delta_{sw}, \Delta_{se}$ denote the $(n-1) \times (n-1)$ minors formed from consecutive rows and consecutive columns in the northwest, northeast, southwest and southeast corners. Further let Δ_c denote the central $(n-2) \times (n-2)$ minor of $\mathbf{A}$. Then from a special case of an identity due to Jacobi [14],

$$\Delta_c |\mathbf{A}| = \Delta_{nw}\Delta_{se} - \Delta_{ne}\Delta_{sw}.$$

where $\Delta_{n+1}^{(1)}$ represents the minor of $\mathbf{T}_{n+1}$ obtained by deleting its first column and last row. Thus,

$$|L_{00}^{(n+1)}|^2 = \frac{\Delta_{n+1}}{\Delta_n} = \frac{\Delta_n}{\Delta_{n-1}}\left(1 - \left|\frac{\Delta_{n+1}^{(1)}}{\Delta_n}\right|^2\right) \le \frac{\Delta_n}{\Delta_{n-1}} = |L_{00}^{(n)}|^2. \quad (2.89)$$

As a result, the minimum mean-square one-step prediction error $|L_{00}^{(n)}|^2 = \Delta_n/\Delta_{n-1}$ forms a monotonic nonincreasing sequence of positive numbers, and in the limit,

$$P_1 = \lim_{n\to\infty} \frac{\Delta_n}{\Delta_{n-1}} = \lim_{n\to\infty} |L_{00}^{(n)}|^2 = |b_0|^2, \quad (2.90)$$

from (2.72); i.e., the residual minimum mean-square error P_1 associated with the best one-step linear predictor in (2.78) is given by the square of the constant term in the Wiener factor. From (2.64), we also have

$$P_1 = \lim_{n\to\infty} \frac{\Delta_n}{\Delta_{n-1}} = |b_0|^2 = exp\left(\frac{1}{2\pi}\int_{-\pi}^{\pi} ln\, S(\theta)d\theta\right) > 0. \quad (2.91)$$

Thus, for a regular stochastic process, the minimum mean-square error associated with a one-step predictor is always positive. This irreducible quantity, in a sense, represents the inherent uncertainty contained in the process. From (2.87), if $\Delta_n = 0$ for some n, the prediction error is zero and at any stage, the process is completely predictable from n of its previous observations and it is essentially deterministic. From (2.45)–(2.46), the spectra associated with such processes are discrete. The Paley-Wiener criterion excludes such processes. The inherent uncertainty or prediction error in (2.91) is also closely related to the entropy for Gaussian processes. Since the joint probability density function of n zero-mean, complex Gaussian samples $\mathbf{x}_n = [x(0), x(T), \ldots, x((n-1)T)]^T$ is given by [15]

$$f_{\mathbf{x}_n}(x_n) = \frac{1}{\pi^n \Delta_{n-1}} exp\left[-\mathbf{x}_n^T \mathbf{T}_{n-1}^{-1} \mathbf{x}_n\right], \quad (2.92)$$

their entropy $\mathcal{H}(x_n)$ is given by

$$\mathcal{H}(x_n) = -E[ln\, f_{\mathbf{x}_n}(x_n)] = n\, ln\,(\pi e) + ln\, \Delta_{n-1}.$$

Thus, the differential entropy based on $n+1$ samples equals

$$\Delta\mathcal{H}_n = \mathcal{H}(x_{n+1}) - \mathcal{H}(x_n) = ln\,(\pi e) + ln\,\frac{\Delta_n}{\Delta_{n-1}}, \quad (2.93)$$

and from (2.91), as $n \to \infty$, the differential entropy $\mathcal{H}$ of the Gaussian process turns out to be (after subtracting the constant $ln\,\pi e$)

$$\mathcal{H} = \lim_{n\to\infty} \Delta\mathcal{H}_n = \lim_{n\to\infty} ln\,\frac{\Delta_n}{\Delta_{n-1}} = \frac{1}{2\pi}\int_{-\pi}^{\pi} ln\, S(\theta)d\theta. \quad (2.94)$$

Thus for Gaussian processes, the minimum mean-square error associated with the best one-step predictor is related to the differential entropy of the process as

$$P_1 = |b_0|^2 = exp(\mathcal{H}) . \qquad (2.95)$$

Following (2.94), in general, for an arbitrary process with power spectral density $S(\theta)$, the functional in (2.94) will be referred to as the entropy functional of that process. Referring back to (2.89)–(2.90),

$$|b_0|^2 \leq |L_{00}^{(k)}|^2 \qquad (2.96)$$

and from (2.84) with $n \to \infty$, the optimal coefficients in (2.78) can be recursively computed from the matrix equation

$$
\begin{bmatrix}
b_0 & 0 & 0 & \cdots \\
b_1 & b_0 & 0 & \cdots \\
b_2 & b_1 & b_0 & \ddots \\
\vdots & \vdots & \vdots & \ddots
\end{bmatrix}
\begin{bmatrix}
w_0^* \\
w_1^* \\
w_2^* \\
\vdots
\end{bmatrix}
=
\begin{bmatrix}
b_1 \\
b_2 \\
b_3 \\
\vdots
\end{bmatrix} . \qquad (2.97)
$$

The above arguments can be easily generalized in the case of a k-step predictor as well. In that case, $x(kT)$ is unknown and the k-step predictor based on the entire set of past samples is given by

$$\hat{x}(kT) = \sum_{i=0}^{\infty} w_i x(-iT) . \qquad (2.98)$$

As before, if only n past samples are known, in that case the k-step ahead predictor takes the form

$$\hat{x}_n(kT) = \sum_{i=0}^{n-1} \alpha_i x(-iT) \qquad (2.99)$$

and minimization of the mean-square error

$$E[|\epsilon_n(kT)|^2] = E[|x(kT) - \hat{x}_n(kT)|^2]$$

gives the n linear equations

$$E[x(-mT)\,\epsilon_n^*(kT)] = r_{m+k}^* - \sum_{i=0}^{n-1} \alpha_i^* r_{i-m} = 0 , \quad m = 0 \to n-1 . \qquad (2.100)$$

Rewriting in matrix form, this gives

$$
\begin{bmatrix}
r_0 & r_1 & r_2 & \cdots & r_{n-1} \\
r_1^* & r_0 & r_1 & \cdots & r_{n-2} \\
\vdots & \vdots & \vdots & \ddots & \vdots \\
r_{n-1}^* & r_{n-2}^* & r_{n-3}^* & \cdots & r_0
\end{bmatrix}
\begin{bmatrix}
\alpha_0^* \\
\alpha_1^* \\
\vdots \\
\alpha_{n-1}^*
\end{bmatrix}
=
\begin{bmatrix}
r_k^* \\
r_{k+1}^* \\
\vdots \\
r_{n+k-1}^*
\end{bmatrix}
= \gamma_k
$$

$$(2.101)$$

or

$$
\begin{bmatrix} \alpha_0^* \\ \alpha_1^* \\ \vdots \\ \alpha_{n-1}^* \end{bmatrix} = \mathbf{L}_{n-1}^{-1}(\mathbf{L}_{n-1}^*)^{-1} \begin{bmatrix} r_k^* \\ r_{k+1}^* \\ \vdots \\ r_{n+k-1}^* \end{bmatrix} = \mathbf{L}_{n-1}^{-1} \begin{bmatrix} L_{k0}^{(n+k-1)} \\ L_{k+1,0}^{(n+k-1)} \\ \vdots \\ L_{n+k-1,0}^{(n+k-1)} \end{bmatrix}.
$$

(2.102)

The last step follows from (2.75), since[10]

$$
(\mathbf{L}_{n+k-2}^*)^{-1} = \begin{bmatrix} \ddots & \vline & \times \\ & \vline & \times \\ \hline \mathbf{O} & \vline & \mathbf{L}_{n-1}^* \end{bmatrix}^{-1} = \begin{bmatrix} \ddots & \vline & \times \\ & \vline & \times \\ \hline \mathbf{O} & \vline & (\mathbf{L}_{n-1}^*)^{-1} \end{bmatrix}
$$

(2.103)

and hence using (2.101)

$$
(\mathbf{L}_{n+k-2}^*)^{-1} \begin{bmatrix} r_1^* \\ \vdots \\ \hline r_k^* \\ \vdots \\ r_{n+k-1}^* \end{bmatrix} = \begin{bmatrix} \times \\ \hline (\mathbf{L}_{n-1}^*)^{-1}\gamma_k \end{bmatrix} = \begin{bmatrix} L_{10}^{(n+k-1)} \\ \vdots \\ \hline L_{k0}^{(n+k-1)} \\ \vdots \\ L_{n+k-1,0}^{(n+k-1)} \end{bmatrix}
$$

or

$$
(\mathbf{L}_{n-1}^*)^{-1}\gamma_k = (\mathbf{L}_{n-1}^*)^{-1} \begin{bmatrix} r_k^* \\ r_{k+1}^* \\ \vdots \\ r_{n+k-1}^* \end{bmatrix} = \begin{bmatrix} L_{k0}^{(n+k-1)} \\ L_{k+1,0}^{(n+k-1)} \\ \vdots \\ L_{n+k-1,0}^{(n+k-1)} \end{bmatrix}.
$$

(2.104)

The minimum mean-square error corresponding to a k-step predictor that makes use of n past samples is given by

$$
\sigma_n^2(k) = E[x(kT)\,\epsilon_n^*(kT)] = r_0 - \sum_{i=0}^{n-1} \alpha_i^* r_{k+i}
$$

---

[10]The inverse of a partitioned matrix [16]:

$$
\begin{bmatrix} \mathbf{A} & \vline & \mathbf{B} \\ \hline \mathbf{C} & \vline & \mathbf{D} \end{bmatrix}^{-1} = \begin{bmatrix} \mathbf{E} & \vline & -\mathbf{EBD}^{-1} \\ \hline -\mathbf{F} & \vline & \mathbf{D}^{-1} + \mathbf{FBD}^{-1} \end{bmatrix},
$$

where $\mathbf{E} = (\mathbf{A} - \mathbf{BD}^{-1}\mathbf{C})^{-1}$ and $\mathbf{F} = \mathbf{D}^{-1}\mathbf{CE}$.

$$= r_0 - [\, r_k, \ r_{k+1}, \ \cdots \ r_{n+k-1} \,] \begin{bmatrix} \alpha_0^* \\ \alpha_1^* \\ \vdots \\ \alpha_{n-1}^* \end{bmatrix},$$

and using (2.102) this simplifies to

$$\sigma_n^2(k) = r_0 - \gamma_k^* \mathbf{L}_{n-1}^{-1} \begin{bmatrix} L_{k0}^{(n+k-1)} \\ L_{k+1,0}^{(n+k-1)} \\ \vdots \\ L_{n+k-1,0}^{(n+k-1)} \end{bmatrix}$$

$$= r_0 - \sum_{i=k}^{n+k-1} |L_{i,0}^{(n+k-1)}|^2 = \sum_{i=0}^{k-1} |L_{i,0}^{(n+k-1)}|^2 , \qquad (2.105)$$

where we have made use of (2.104) and (2.76). Once again, as n increases, since the error cannot increase, $\sigma_n^2(k)$ forms a monotone nonincreasing sequence in n and in the limit as $n \to \infty$, from (2.72), the minimum mean-square error P_k associated with the best k-step linear predictor is given by

$$P_k = \lim_{n \to \infty} \sigma_n^2(k) = \lim_{n \to \infty} \sum_{i=0}^{k-1} |L_{i,0}^{(n+k-1)}|^2$$

$$= \sum_{i=0}^{k-1} |b_i|^2 . \qquad (2.106)$$

Using (2.65), P_k can be expressed in terms of the entropy functional as

$$P_k = \left[1 + |\alpha_1|^2 + \cdots + |\alpha_{k-1}|^2 \right] exp\left(\frac{1}{2\pi} \int_{-\pi}^{\pi} ln\, S(\theta) d\theta \right) . \qquad (2.107)$$

From (2.107), since $P_k \geq P_{k-1}$ for any k, this implies that when the entire past is made use of in predicting, the residual prediction error P_k cannot decrease. This is not surprising since as k increases, the epoch of prediction gets farther and farther away from the known data. However, this is not always true if only a finite number of past samples are available. For example, suppose $x(T)$ and $x(2T)$ need to be predicted using $x(0)$. Then, from the above discussion, it follows that the best predictors are given by

$$\hat{x}(T) = \frac{r_1}{r_0} x(0)$$

and

$$\hat{x}(2T) = \frac{r_2}{r_0} x(0)$$

and the associated minimum mean-square errors for one- and two-step predictors are

$$\sigma_1^2(1) = (r_0^2 - |r_1|^2)/r_0$$

and

$$\sigma_1^2(2) = (r_0^2 - |r_2|^2)/r_0 ,$$

respectively. Since $|r_2|$ can be greater than $|r_1|$, the two-step prediction
error $\sigma_1^2(2)$ can be smaller than the one-step prediction error $\sigma_1^2(1)$ and in
general, $\sigma_n^2(k)$ need not be smaller than $\sigma_n^2(k-1)$ for finite values of n.

For AR processes, the present sample value $x(n)$ depends only on a
finite number of its past samples and hence the prediction error should not
decrease beyond a certain stage. For example, let $x(n)$ represent an AR(p)
process and consider the one-step prediction problem in (2.80)–(2.81) that
makes use of a finite number of its past samples. In that case, the prediction
error $\sigma_n^2(1) = \Delta_n/\Delta_{n-1}$ stays the same for $n \geq p$. To see this, from (2.57),
for an AR(p) process, we have

$$x(n) = -\sum_{i=1}^{p} \alpha_i x(n-i) + \beta_0 w(n) . \tag{2.108}$$

Since the present sample value of the white noise process $w(n)$ is uncorrelated with its own past sample values as well as the past samples of the
output $x(n-k)$, $k \geq 1$ (see (2.108)), we have

$$E[w(n)\, w^*(n-k)] = 0, \quad k \geq 1, \tag{2.109}$$

$$E[w(n)\, x^*(n-k)] = 0, \quad k \geq 1. \tag{2.110}$$

Using these equations together with (2.108), we obtain

$$E[x(n)\, x^*(n)] = r_0 = -\sum_{i=1}^{p} \alpha_i^* r_i + |\beta_0|^2 \tag{2.111}$$

and

$$E[x(n-k)\, x^*(n)] = r_k^* = -\sum_{i=1}^{p} \alpha_i^* r_{i-k}, \quad k \geq 1. \tag{2.112}$$

Rewriting the first p equations in (2.112) gives the unique nontrivial solution (see also (2.88))

$$\begin{bmatrix} \alpha_1^* \\ \alpha_2^* \\ \vdots \\ \alpha_p^* \end{bmatrix} = -\mathbf{T}_{p-1}^{-1} \begin{bmatrix} r_1^* \\ r_2^* \\ \vdots \\ r_p^* \end{bmatrix} \tag{2.113}$$

and using this together with (2.111), we also obtain

$$|\beta_o|^2 = \frac{\Delta_p}{\Delta_{p-1}} > 0.$$

However, the first $(p+1)$ equations in (2.112) can be rearranged as

$$\begin{bmatrix} r_1 & r_2 & \cdots & r_p & r_{p+1} \\ r_0 & r_1 & \cdots & r_{p-1} & r_p \\ \vdots & \vdots & \cdots & \vdots & \vdots \\ r_{p-1}^* & r_{p-2}^* & \cdots & r_0 & r_1 \end{bmatrix} \begin{bmatrix} \alpha_p \\ \alpha_{p-1} \\ \vdots \\ \alpha_1 \\ 1 \end{bmatrix} = \begin{bmatrix} 0 \\ 0 \\ \vdots \\ 0 \\ 0 \end{bmatrix}. \qquad (2.114)$$

Since α_k, $k = 1 \to p$, have a nontrivial solution given by (2.113), the coefficient matrix in (2.114) must be singular. (This also follows from (2.112), since the last column in the coefficient matrix of (2.114) is a linear combination of the first p columns.) But that matrix is obtained by deleting the first column and last row of $\mathbf{T}_{p+1}$, and using the definition in (2.88) we have

$$\Delta_{p+1}^{(1)} = 0$$

and hence $\Delta_{p+1} = \Delta_p^2 / \Delta_{p-1}$, or

$$\sigma_{p+1}^2(1) = \frac{\Delta_{p+1}}{\Delta_p} = \frac{\Delta_p}{\Delta_{p-1}} = \sigma_p^2(1).$$

More generally, for an AR(p) process

$$\Delta_{p+k}^{(1)} = \begin{vmatrix} r_1 & r_2 & \cdots & r_p & \cdots & r_{p+k} \\ r_0 & r_1 & \cdots & r_{p-1} & \cdots & r_{p+k-1} \\ r_1^* & r_0 & \cdots & r_{p-2} & \cdots & r_{p+k-2} \\ \vdots & \vdots & \vdots & \vdots & \vdots & \vdots \\ r_{p+k-2}^* & r_{p+k-3}^* & \cdots & \cdots & r_0 & r_1 \end{vmatrix} = 0, \quad k \geq 1,$$

$$(2.115)$$

since from (2.112), the last column (or the first row) can be expressed as a linear combination of the previous p columns (or rows) and hence

$$\sigma_{p+k}^2(1) = \sigma_p^2(1) = \frac{\Delta_p}{\Delta_{p-1}} > 0. \qquad (2.116)$$

i.e., for an AR(p) process, the performance of the one-step predictor that uses p of its past samples cannot be improved upon by making use of more samples from its past. It may be remarked that (2.115) is characteristic only to AR processes and hence for ARMA processes, improvement in performance can be obtained by incorporating more and more samples from the past.

Every power spectrum has a one-to-one correspondence with a positive function. Such functions have many remarkable properties and have been the subject of extensive study for over a century [17, 18]. They have found applications in several areas of engineering and, in particular, over the last several decades the theory of positive-real (p.r.) functions have been used in classical network theory to great advantage in establishing almost every major result in that field [19–22]. Before we examine the connection between power spectra and positive functions, we will describe some of the standard results from the theory of positive(-real) functions.

2.2 Positive Functions

Positive functions can be defined in the p-plane ($p = \sigma + j\omega$; continuous domain) or in the z-plane ($z = re^{j\theta}$; discrete domain). For applications involving analog signals and systems, the variable p is natural, and in the discrete case, the variable z is more appropriate. For the present applications, it is best to describe these functions in the z-domain. We begin with the standard definition of a positive function.

Definition: A function $Z(z)$ is said to be positive if

i) $Z(z)$ is analytic in $|z| < 1$ (the interior of the unit circle)
and
ii) $Re\, Z(z) \geq 0$ in $|z| < 1$.

Further if $Z(z)$ is also real for real z, then it is said to be positive-real (p.r.) [15–17]. Its physical meaning is best illustrated in classical network theory, where the impedance function $Z(p)$ associated with every passive network is shown to be a p.r. function.[11] Moreover a classical result due to Brune states that every rational p.r. function $Z(p)$ can be synthesized as the input impedance of a lumped passive network containing resistances and lossless elements such as inductors, capacitors and transformers [23, 24]. Thus the p.r. concept represents passivity in a strong fashion and plays a fundamental role in network analysis and synthesis.

Since a positive function $Z(z)$ is analytic in $|z| < 1$, it has a power series expansion with radius of convergence equal to unity; i.e.,

$$Z(z) = c_0 + 2\sum_{k=1}^{\infty} c_k z^k, \quad |z| < 1. \tag{2.117}$$

Without loss of generality, c_0 can be made positive and hence we will assume c_0 is always a positive number. However, in general c_k, $k = 1 \to \infty$,

---

[11]In the p-domain, $Z(p)$ is said to be positive if (i) $Z(p)$ is analytic in $Re\, p > 0$ and (ii) $Re\, Z(p) \geq 0$ in $Re\, p > 0$.

are complex quantities. Note that if $Z(z)$ is p.r., then every c_k, $k = 0 \rightarrow \infty$, is also real. Interestingly, the correspondence between positive functions and power spectra can be best seen from a result due to Schur associated with such functions.

Schur's Theorem: The power series[12]

$$Z(z) = c_0 + 2 \sum_{k=1}^{\infty} c_k z^k \tag{2.118}$$

defines a positive function iff every Hermitian Toeplitz matrix

$$\mathbf{C}_k = \begin{bmatrix} c_0 & c_1 & c_2 & \cdots & c_{k-1} & c_k \\ c_1^* & c_0 & c_1 & \cdots & c_{k-2} & c_{k-1} \\ \vdots & \vdots & \vdots & \vdots & \vdots & \vdots \\ c_{k-1}^* & c_{k-2}^* & \cdots & c_1^* & c_0 & c_1 \\ c_k^* & c_{k-1}^* & \cdots & \cdots & c_1^* & c_0 \end{bmatrix} \geq 0, \tag{2.119}$$

for every $k = 0 \rightarrow \infty$.

Proof [25]

Suppose $Z(z)$ is positive. Then, by definition, $Z(z)$ is analytic in $|z| < 1$ and

$$\frac{Z(z) + Z^*(z)}{2} \geq 0 \quad \text{in } |z| < 1. \tag{2.120}$$

Hence for every fixed $0 < \rho < 1$ and every trigonometric polynomial

$$h(\theta) = \sum_{m=0}^{k} h_m e^{-jm\theta}, \; \theta \text{ real}, \; k \geq 0,$$

we obtain

$$\frac{1}{2\pi} \int_{-\pi}^{\pi} \left(\frac{Z(\rho e^{j\theta}) + Z^*(\rho e^{j\theta})}{2} \right) |h(\theta)|^2 \, d\theta \geq 0.$$

Simplifying the left-hand side and using the convenient notation

$$c_{-n} = c_n^*, \quad n = 0 \rightarrow \infty, \tag{2.121}$$

we get

$$\frac{1}{2\pi} \int_{-\pi}^{\pi} \left(c_0 + \sum_{n=1}^{\infty} \rho^n \left(c_n e^{jn\theta} + c_{-n} e^{-jn\theta} \right) \right) \sum_{l=0}^{k} \sum_{m=0}^{k} h_l h_m^* e^{-j(l-m)\theta} d\theta$$

---

[12]The factor 2 in (2.117)–(2.118) is for later algebraic simplicity.

$$= \sum_{l=0}^{k} \sum_{m=0}^{k} h_l c_{l-m} \rho^{|l-m|} h_m^* \geq 0 \,.$$

In the limit as $\rho \to 1 - 0$, with k held fixed, the above expression reduces to

$$\sum_{l=0}^{k} \sum_{m=0}^{k} h_l c_{l-m} h_m^* = \mathbf{h}^* \mathbf{C}_k \mathbf{h} \geq 0 \,,$$

where $\mathbf{h} = [h_0, \; h_1, \; \ldots, \; h_k]^T$ and because of the arbitrary character of $\mathbf{h}$, we get

$$det \, \mathbf{C}_k \geq 0 \,, \quad k = 0 \to \infty \,.$$

As for sufficiency, let $det \, \mathbf{C}_k \geq 0$, $k = 0 \to \infty$. Thus, in particular, this implies that every

$$|c_k| \leq c_0 \,. \tag{2.122}$$

Thus,

$$|Z(z)| = \left| c_0 + 2 \sum_{k=1}^{\infty} c_k z^k \right| \leq c_0 + 2 \sum_{k=1}^{\infty} |c_k||z|^k$$

$$\leq c_0 \left[1 + 2 \sum_{k=1}^{\infty} |z|^k \right] = c_0 \left(\frac{1 + |z|}{1 - |z|} \right) < \infty \,, \quad \text{for } |z| < 1 \,.$$

Thus, $Z(z)$ is analytic in $|z| < 1$ and to show positivity, notice that

$$\frac{Z(z) + Z^*(z)}{2(1 - zz^*)} = \frac{\displaystyle\sum_{k=0}^{\infty} c_k z^k + \sum_{k=1}^{\infty} c_k^* z^{*k}}{(1 - zz^*)}$$

$$= \sum_{l=0}^{\infty} z^l z^{*l} \left(\sum_{k=0}^{\infty} c_k z^k + \sum_{k=1}^{\infty} c_k^* z^{*k} \right)$$

$$= \sum_{l=0}^{\infty} \sum_{k=0}^{\infty} c_k z^{l+k} z^{*l} + \sum_{l=0}^{\infty} \sum_{k=1}^{\infty} c_k^* z^l z^{*(l+k)} \,.$$

Let $l+k = m$, and $l = n$ in the first summation. Then $k = m-n \geq 0$ or $m \geq n$ and repeat this procedure in the second summation after interchanging the role of m and n respectively. Then, the above expression becomes

$$\frac{Z(z) + Z^*(z)}{2(1 - zz^*)} = \sum_{n=0}^{\infty} \sum_{m=n}^{\infty} c_{m-n} z^m z^{*n} + \sum_{m=0}^{\infty} \sum_{n=m+1}^{\infty} c_{n-m}^* z^m z^{*n}$$

$$= \left(\sum_{n=0}^{\infty} \sum_{m=n}^{\infty} + \sum_{m=0}^{\infty} \sum_{n=m+1}^{\infty} \right) (c_{m-n} z^m z^{*n})$$

$$= \sum_{m=0}^{\infty} \sum_{n=0}^{\infty} c_{m-n} z^m z^{*n} = \lim_{k \to \infty} \sum_{m=0}^{k} \sum_{n=0}^{k} c_{m-n} z^m z^{*n}$$

$$= \lim_{k \to \infty} \mathbf{d}^* \mathbf{C}_k \mathbf{d} \geq 0,$$

since $\mathbf{C}_k \geq 0$. Thus,

$$\frac{Z(z) + Z^*(z)}{2(1 - zz^*)} \geq 0, \quad |z| < 1,$$

and since $(1 - zz^*) < 1$ in $|z| < 1$, we have

$$\frac{Z(z) + Z^*(z)}{2} \geq 0, \quad |z| < 1,$$

or $Z(z)$ is a positive function, Q.E.D.

Because of the analyticity of $Z(z)$ in $|z| < 1$, it is free of poles in $|z| < 1$. Interestingly, $Z(z)$ is also free of zeros in $|z| < 1$.

To see this, notice that, $Z(z)$ (or any analytic function in $|z| < 1$) cannot have infinite number of zeros in $|z| < 1$, since in that case $Z(z)$ will have a limit point in $|z| \leq 1$. If such a limit point is inside the unit circle, then infinite number of zeros are inside, and a power series expansion $Z(z) = \sum_{k=0}^{\infty} a_k (z - z_0)^k$ about such a point z_0 implies that every $a_k = 0$, $k = 0 \to \infty$, and hence $Z(z) \equiv 0$. Thus, there are no limit points in $|z| < 1$. However, there can be limit points on the boundary of the unit circle. In both cases, the zeros in $|z| < 1$ are finite and isolated. Hence in a small neighborhood around each of these zeros, a power series expansion is possible. These zeros also must be of finite order; otherwise, using the above power series, we can again show that $Z(z) \equiv 0$. Thus zeros of $Z(z)$ in $|z| < 1$ are isolated, finite in number and of finite order. (All this is true for any function analytic in $|z| < 1$.)

Further, at a point z in $|z| < 1$, which is not a zero of $Z(z)$, we have

$$Re\left[\frac{1}{Z(z)}\right] = \frac{Re\, Z(z)}{|Z(z)|^2} \geq 0, \quad |z| < 1, \qquad (2.123)$$

i.e., if $Z(z)$ is positive, then at every point in $|z| < 1$ that is not a zero of $Z(z)$

$$Re\left[\frac{1}{Z(z)}\right] \geq 0.$$

Let z_0 be a zero of $Z(z)$ of order m in $|z| < 1$ (see Fig. 2.3). Since it must be an isolated zero, it is an isolated pole of $1/Z(z)$ also of order m. Thus, $1/Z(z)$ is analytic in the neighborhood of $z = z_0$. Hence by

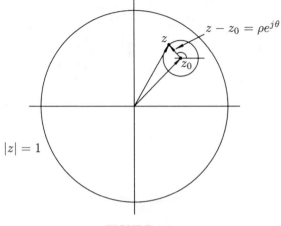

FIGURE 2.3.

Cauchy's theorem [26] for small neighborhoods around the pole, $1/Z(z)$ can be expanded in a Laurent series about that pole, i.e.,

$$\frac{1}{Z(z)} = a_{-m}(z - z_0)^{-m} + a_{-(m-1)}(z - z_0)^{-(m-1)} + \cdots$$

$$+ a_{-1}(z - z_0)^{-1} + a_0 + a_1(z - z_0) + \cdots. \qquad (2.124)$$

However, from (2.123), the real part of this function is nonnegative at every point in the neighborhood of z_0 and on the circle $z - z_0 = \rho e^{j\theta}$, $0 < \theta < 2\pi$. Thus, in the neighborhood of z_0 we have [20]

$$\frac{1}{Z(z)} = a_{-m}\rho^{-m}e^{-jm\theta} + o(1/\rho^m) \qquad (2.125)$$

and hence the most dominant term gives

$$Re\left[\frac{1}{Z(z)}\right] = \rho^{-m}\left(a_{-m}e^{-jm\theta} + a^*_{-m}e^{jm\theta}\right) \geq 0. \qquad (2.126)$$

Letting $\theta = 0$, this reduces to

$$a_{-m} + a^*_{-m} \geq 0 \qquad (2.127)$$

and for $\theta = \pi/2m$ and $\theta = -\pi/2m$, we get

$$-j(a_{-m} - a^*_{-m}) \geq 0$$

$$j(a_{-m} - a^*_{-m}) \geq 0.$$

Thus, $a_{-m} = a^*_{-m}$ and substituting this in (2.127), we get

$$a_{-m} > 0 \qquad (2.128)$$

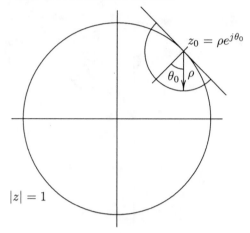

FIGURE 2.4.

and (2.126) becomes

$$Re\left(\frac{1}{Z(z)}\right) = 2\rho^{-m}a_{-m}\cos(m\theta) \geq 0, \quad -\pi < \theta < \pi. \tag{2.129}$$

However, for $m \geq 1$, $\cos(m\theta)$ changes sign in $|\theta| \leq \pi$, and we will have a contradiction in (2.129), unless $m = 0$; i.e., $Z(z)$ has no zeros in $|z| < 1$. In short, $Z(z)$ is free of poles and zeros in $|z| < 1$. Thus, $1/Z(z)$ is also free of poles and zeros in $|z| < 1$ and using (2.123), $1/Z(z)$ is also a positive function.

On the boundary of the unit circle ($|z| = 1$), a positive function can have poles and zeros. However, such poles and zeros must be simple, and further, the residues of such poles must be strictly positive [20].

To see this, let $z_0 = e^{j\theta_0}$, $-\pi < \theta_0 < \pi$, be a pole of $Z(z)$ of order $m \geq 1$ (see Fig. 2.4). Once again, proceeding as in (2.124) for $Z(z)$, in a small neighborhood of $z_0 = e^{j\theta_0}$, the most significant term gives

$$Z(z) \approx \frac{a_{-m}}{(z - z_0)^m}, \tag{2.130}$$

where

$$a_{-m} = |a_{-m}|e^{j\beta} \neq 0. \tag{2.131}$$

For any z in $|z| < 1$ and on the semicircle of radius ρ with center at $z_0 = e^{j\theta_0}$, we have $z - z_0 = \rho e^{j\theta}$, $-\frac{\pi}{2} < \theta < \frac{\pi}{2}$. For any such z, (2.130) simplifies into

$$Re[Z(z)] \approx \frac{|a_{-m}|}{\rho^m}\cos(m\theta + \beta) \geq 0. \tag{2.132}$$

For $-\frac{\pi}{2} < \theta < \frac{\pi}{2}$, since $\cos(m\theta + \beta)$ goes negative if $m > 1$, from (2.132), we must have, at most, $m = 1$. Even then, $Re[Z(z)]$ remains nonnegative only if $\beta = 0$. Thus, from (2.131), as in (2.128)

$$a_{-m} = |a_{-m}| > 0,$$

a positive number, and in the vicinity of the pole $z_0 = e^{j\theta_0}$,

$$Z(z) \approx \frac{a}{(z - z_0)}.$$

The constant a is the residue of the pole at $z = e^{j\theta_0}$, which is a positive number and the pole is simple. Lastly, every pole of $Z(z)$ is a zero of $1/Z(z)$ and since $1/Z(z)$ is also a positive function, every zero of a positive function on the unit circle is also simple.

Thus, a positive function is free of poles and zeros in $|z| < 1$, and on the unit circle its poles and zeros are simple. Further the residue of every such pole is positive [19, 20].

The boundary values of $Z(z)$ are defined as the interior ray limit [27]

$$Z(e^{j\theta}) = \lim_{r \to 1-0} Z(re^{j\theta}) \qquad (2.133)$$

and they exist for almost all θ. Hence,

$$Re\, Z(e^{j\theta}) = \lim_{r \to 1-0} Re\, Z(re^{j\theta}) \geq 0. \qquad (2.134)$$

Thus, a positive function has well defined ray limits almost everywhere on the unit circle, and at every such point its real part is also nonnegative and uniformly bounded.

It may be remarked that the actual boundary behavior of $Z(z)$ can be quite complicated and, in general, $Z(z)$ need not be analytic or continuous on $|z| = 1$. For example, it can have essential singularities on the boundary. To illustrate this, consider

$$Z(z) = 1 \pm e^{-(1+z)/(1-z)}. \qquad (2.135)$$

This is a positive function, since it is analytic in $|z| < 1$ and for every $z = \rho e^{j\theta}$, $\rho < 1$,

$$Re\,[Z(\rho e^{j\theta})] = 1 \pm exp\left(\frac{-(1-\rho^2)}{1+\rho^2 - 2\rho\cos\theta}\right) \cdot \cos\left(\frac{2\rho\sin\theta}{1+\rho^2 - 2\rho\cos\theta}\right) > 0.$$

(In fact, it is a positive-real function). However, it has an essential singularity at $z = 1$ since $Z(z)$ can take any value at $z = 1$ as that point is approached from the outside of the unit circle. Moreover, its tangential limit and interior ray limit at $z = 1$ do not agree, and the function is not even continuous at $z = 1$ (see (2.53)–(2.54)). Naturally, such functions with essential singularities by definition cannot be rational, and if we restrict our attention to rational positive functions, they are analytic in $|z| < 1$, and all of their poles and zeros on $|z| = 1$ are simple with positive residues for the poles.

Moreover, for rational $Z(z)$, since the only kind of singularities that can exist are poles, $Z(z)$ has at most simple poles (and zeros) on the boundary of the unit circle. Also, the analyticity of $Z(z)$ and $1/Z(z)$ in $|z| < 1$ implies that the numerator and denominator are Hurwitz polynomials. In network theory, every *rational* positive-real (p.r.) function $Z(p)$ corresponds to a lumped circuit containing positive resistances, and lossless elements such as inductors, capacitors, gyrators and transformers [19, 20]. When the network contains distributed elements, its input impedance and other characterizations become transcendental functions, and, depending upon the network configuration, some of these functions can exhibit essential singularities [22, 28].

Now, we are in a position to exhibit the one-to-one relationship between a power spectrum and a positive function. From (2.34)–(2.44),

$$S(\theta) = \sum_{k=-\infty}^{+\infty} r_k e^{jk\theta} \tag{2.136}$$

represents a power spectrum if and only if in (2.30), every $\mathbf{T}_k \geq 0$. On the other hand, from (2.118)–(2.119),

$$Z(z) = c_0 + 2\sum_{k=1}^{\infty} c_k z^k \tag{2.137}$$

represents a positive function if and only if in (2.119), every $\mathbf{C}_k \geq 0$. Thus by associating $c_k = r_k$, $k = 0 \rightarrow \infty$, and $r_{-k} = c_{-k} = c_k^* = r_k^*$, corresponding to every power spectrum $S(\theta)$, from Schur's theorem the function

$$Z(z) = r_0 + 2\sum_{k=1}^{\infty} r_k z^k \tag{2.138}$$

represents a positive function. Alternatively, since (2.138) can be rewritten as

$$Z(z) = \frac{1}{2\pi} \int_{-\pi}^{\pi} \left(\frac{1 + ze^{-j\theta}}{1 - ze^{-j\theta}} \right) S(\theta) d\theta, \quad |z| < 1,$$

it is analytic in $|z| < 1$ and further, since

$$Re\, Z(z) = \frac{1}{2\pi} \int_{-\pi}^{\pi} \left(\frac{1 - |z|^2}{|1 - ze^{-j\theta}|} \right) S(\theta) d\theta \geq 0, \quad |z| < 1,$$

by definition, $Z(z)$ in (2.138) represents a positive function. Clearly if $S(\theta)$ is even, then r_k's are real and $Z(z)$ becomes a p.r. function. Conversely, the real part of every positive function evaluated on the unit circle represents a power spectrum, since

$$S(\theta) \overset{\triangle}{=} Re\ [Z(e^{j\theta})] = \frac{Z(e^{j\theta}) + Z^*(e^{j\theta})}{2}$$

$$= \sum_{k=-\infty}^{+\infty} c_k e^{jk\theta} \geq 0\,. \tag{2.139}$$

The one-to-one relationship in (2.136)–(2.138) between power spectra and positive functions turns out to be an extremely useful tool in our analysis and will be used later on to exhibit solutions to various problems of interest. Before undertaking such a study, it is important to define a closely related useful concept, the concept of bounded functions.

2.2.1   BOUNDED FUNCTIONS

A function $d(z)$ is said to be bounded if

i) $d(z)$ is analytic in $|z| < 1$

and   (2.140)

ii) $|d(z)| \leq 1$ in $|z| < 1$.

If in addition, $d(z)$ is also real for real z, then it is said to be bounded-real (b.r.). For example, the function z^n, $n \geq 1$, is b.r., since $|z^n| = |z|^n \leq 1$ in $|z| \leq 1$. Similarly, the function $e^{-(1-z)}$ is also b.r., since it is analytic in $|z| < 1$, and inside the unit circle, $z = re^{j\theta}$, $r < 1$, we have $|e^{-(1-z)}| = e^{-(1-r\cos\theta)} < 1$.

Since a bounded function $d(z)$ is analytic in $|z| < 1$, it also has a power series expansion with radius of convergence equal to unity, i.e.,

$$d(z) = \sum_{k=0}^{\infty} d_k z^k\,, \quad |z| < 1\,. \tag{2.141}$$

Interestingly, every bounded function has a one-to-one correspondence with a positive function $Z(z)$ [20]. This follows by noticing that for every R_0 with $Re\,(R_0) > 0$, we have

$$d(z) \overset{\triangle}{=} \frac{Z(z) - R_0}{Z(z) + R_0^*}\,, \quad Re\,R_0 > 0 \tag{2.142}$$

is bounded, since

$$1 - |d(z)|^2 = \frac{|Z(z) + R_0^*|^2 - |Z(z) - R_0|^2}{|Z(z) + R_0^*|^2}$$

$$= \frac{2Re\,[Z(z)R_0^*] + 2Re\,[Z(z)R_0]}{|Z(z) + R_0^*|^2}$$

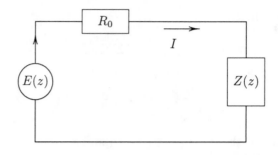

FIGURE 2.5.

$$= \frac{4Re\,[Z(z) \cdot Re\,(R_0)]}{|Z(z) + R_0^*|^2}$$

$$= \frac{4Re\,[Z(z)]\,Re\,(R_0)}{|Z(z) + R_0^*|^2} \geq 0, \quad |z| < 1, \qquad (2.143)$$

since $Z(z)$ is positive and $Re\,(R_0) > 0$. Thus,

$$|d(z)| \leq 1 \quad \text{in } |z| < 1$$

and since it is also analytic in $|z| < 1$, $d(z)$ in (2.142) represents a bounded function. If $Z(z)$ is p.r., by choosing R_0 to be any positive number,

$$d(z) = \frac{Z(z) - R_0}{Z(z) + R_0}, \quad R_0 > 0 \qquad (2.144)$$

represents a b.r. function. The function $d(z)$ so generated is known as the reflection coefficient of the p.r. function $Z(z)$ normalized to R_0. Its physical meaning is not difficult to grasp [20]. Consider the circuit in Fig. 2.5 formally, where a source $E(z)$ with constant internal resistance R_0 supplies power to a frequency dependent passive (p.r.) load $Z(z)$. The current $I(z)$ through the load is

$$I(z) = \frac{E(z)}{Z(z) + R_0}$$

and hence the power $P_d(\theta)$ delivered to the load at frequency θ is given by

$$P_d(\theta) = Re\,[Z(e^{j\theta})] \cdot |I|^2$$

$$= \frac{|E|^2}{R_0} \cdot \frac{Re\,[Z(e^{j\theta})] \cdot R_0}{|Z(e^{j\theta}) + R_0|^2}.$$

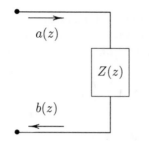

FIGURE 2.6. Incident and reflected waves.

With $d(z)$ as defined in (2.144) and using (2.143), the above expression reduces to

$$P_d(\theta) = \frac{|E|^2}{4R_0} \cdot \left(1 - |d(e^{j\theta})|^2\right) . \qquad (2.145)$$

Since $|E|^2/4R_0$ represents the total power available at the source, from (2.145), the quantity $|d(e^{j\theta})|^2$ (see also (2.157)) represents the fraction of the available power that is reflected back to the source because of impedance mismatch. Clearly if $Z(z) = R_0$, there is no mismatch, $d(z) \equiv 0$, and no power is reflected back to the source. Thus $d(z)$ in (2.144) represents the reflection coefficient of the positive function $Z(z)$ normalized to R_0, and if we introduce an incident wave $a(z)$ and a reflected wave $b(z)$ at the impedance terminals to symbolically represent the incoming and outgoing power carriers (Fig. 2.6), then from the above argument [20],

$$\frac{b(z)}{a(z)} = d(z) . \qquad (2.146)$$

Notice that the above incident wave and reflected wave depend on the normalization R_0 and do not exist independently. Thus, the reflection coefficient of $Z(z)$ normalized to any constant R_0 can be made to represent the ratio of an appropriate reflected wave to an incident wave across $Z(z)$. From (2.144), we also have the useful relation

$$Z(z) = R_0 \cdot \frac{1 + d(z)}{1 - d(z)}, \qquad R_0 > 0 \qquad (2.147)$$

for the p.r. function $Z(z)$ in terms of a b.r. function. Since $Z(z)$ is p.r. implies $1/Z(z)$ is also p.r., we have

$$Z_1(z) = \frac{1 - d(z)}{1 + d(z)} \qquad (2.148)$$

is p.r., and it corresponds to the b.r. function $-d(z)$.

The boundedness character of $d(z)$ can be parametrized in terms of the boundedness of certain lower triangular Toeplitz matrices. To see this, substituting (2.118) and (2.141) into (2.142), we have

$$\left(c_0 - R_0 + 2\sum_{k=1}^{\infty} c_k z^k\right) = \left(c_0 + R_0^* + 2\sum_{k=1}^{\infty} c_k z^k\right)\left(\sum_{k=0}^{\infty} d_k z^k\right).$$

Collating terms of equal power, we obtain

$$c_0 - R_0 = (c_0 + R_0^*)d_0$$
$$2c_1 = (c_0 + R_0^*)d_1 + 2c_1 d_0$$
$$\vdots$$
$$2c_k = (c_0 + R_0^*)d_k + 2\sum_{m=1}^{k} c_m d_{k-m}.$$

Put together in matrix form, these equations reduce to

$$\begin{bmatrix} (c_0 - R_0) & 0 & \cdots & 0 & 0 \\ 2c_1 & (c_0 - R_0) & 0 & \cdots & 0 \\ \vdots & 2c_1 & \ddots & \ddots & \vdots \\ \vdots & \vdots & \ddots & \ddots & \vdots \\ 2c_k & 2c_{k-1} & \cdots & 2c_1 & (c_0 - R_0) \end{bmatrix}$$

$$= \begin{bmatrix} c_0 + R_0^* & 0 & \cdots & 0 & 0 \\ 2c_1 & c_0 + R_0^* & 0 & \cdots & 0 \\ \vdots & 2c_1 & \ddots & \ddots & \vdots \\ \vdots & \vdots & \ddots & \ddots & \vdots \\ 2c_k & 2c_{k-1} & \cdots & 2c_1 & c_0 + R_0^* \end{bmatrix} \begin{bmatrix} d_0 & 0 & \cdots & 0 & 0 \\ d_1 & d_0 & 0 & \cdots & 0 \\ d_2 & d_1 & \ddots & \ddots & \vdots \\ \vdots & \vdots & \ddots & \ddots & \vdots \\ d_k & d_{k-1} & \cdots & d_1 & d_0 \end{bmatrix}.$$

$$\tag{2.149}$$

Let

$$\mathbf{C}_{l,k} = \begin{bmatrix} c_0 & 0 & \cdots & 0 & 0 \\ 2c_1 & c_0 & 0 & \cdots & 0 \\ 2c_2 & 2c_1 & \ddots & \ddots & \vdots \\ \vdots & \vdots & \ddots & \ddots & \vdots \\ 2c_k & 2c_{k-1} & \cdots & 2c_1 & c_0 \end{bmatrix} \tag{2.150}$$

and

$$\mathbf{D}_k = \begin{bmatrix} d_0 & 0 & \cdots & 0 & 0 \\ d_1 & d_0 & 0 & \cdots & 0 \\ d_2 & d_1 & \ddots & \ddots & \vdots \\ \vdots & \vdots & \ddots & \ddots & \vdots \\ d_k & d_{k-1} & \cdots & d_1 & d_0 \end{bmatrix} \tag{2.151}$$

represent two lower triangular Toeplitz matrices generated from c_k's and d_k's, respectively. Then (2.149) simplifies into

$$\mathbf{C}_{l,k} - R_0 \mathbf{I} = \mathbf{C}_{l,k} \mathbf{D}_k + R_0^* \mathbf{D}_k \,.$$

Let $R_0 = x + jy$. Then $x > 0$ and we get

$$\mathbf{C}_{l,k}(\mathbf{I} - \mathbf{D}_k) = x(\mathbf{I} + \mathbf{D}_k) + jy(\mathbf{I} - \mathbf{D}_k)$$

or

$$(\mathbf{C}_{l,k} - jy\mathbf{I})(\mathbf{I} - \mathbf{D}_k) = x(\mathbf{I} + \mathbf{D}_k)$$

and hence

$$\mathbf{C}_{l,k} = jy\mathbf{I} + x(\mathbf{I} + \mathbf{D}_k)(\mathbf{I} - \mathbf{D}_k)^{-1} \,. \tag{2.152}$$

With

$$\mathbf{J} = \begin{bmatrix} 0 & \cdots & 0 & 1 \\ 0 & \cdots & 1 & 0 \\ \vdots & \cdots & \vdots & \vdots \\ 1 & \cdots & 0 & 0 \end{bmatrix},$$

since from (2.119)

$$\mathbf{J}\mathbf{C}_k\mathbf{J} = \frac{\mathbf{C}_{l,k} + \mathbf{C}_{l,k}^*}{2} \geq 0, \quad k = 0 \to \infty,$$

using (2.152), we obtain

$$\mathbf{J}\mathbf{C}_k\mathbf{J} = \frac{x}{2}\left[(\mathbf{I} + \mathbf{D}_k)(\mathbf{I} - \mathbf{D}_k)^{-1} + (\mathbf{I} - \mathbf{D}_k^*)^{-1}(\mathbf{I} + \mathbf{D}_k^*)\right] \geq 0, \tag{2.153}$$

or

$$\mathbf{J}\mathbf{C}_k\mathbf{J} = x(\mathbf{I} - \mathbf{D}_k^*)^{-1}(\mathbf{I} - \mathbf{D}_k^*\mathbf{D}_k)(\mathbf{I} - \mathbf{D}_k)^{-1} \geq 0 \,.$$

As a result, for bounded functions, Schur's theorem translates into the familiar form

$$\mathbf{I} - \mathbf{D}_k^*\mathbf{D}_k \geq 0, \quad k = 0 \to \infty. \tag{2.154}$$

Since $(\mathbf{I} + \mathbf{D}_k)(\mathbf{I} - \mathbf{D}_k)^{-1} = (\mathbf{I} - \mathbf{D}_k)^{-1}(\mathbf{I} + \mathbf{D}_k)$, (2.153) can also be rewritten as

$$\mathbf{I} - \mathbf{D}_k\mathbf{D}_k^* \geq 0, \quad k = 0 \to \infty. \tag{2.155}$$

Thus, in particular,

$$|d_0|^2 \le 1 \quad \text{and} \quad \sum_{i=0}^{k} |d_i|^2 \le 1, \quad k = 0 \to \infty, \qquad (2.156)$$

i.e., every d_k in (2.141) is bounded by unity and, moreover, if for any bounded function we have $d_0 = 1$, then from (2.156) it follows that $d_k = 0$, $k = 1 \to \infty$, and hence $d(z) \equiv 1$.

In general, the boundary values of a bounded function are defined as the interior ray (radial) limit [27]

$$d(e^{j\theta}) = \lim_{r \to 1-0} d(re^{j\theta}) \qquad (2.157)$$

and they exist for almost all θ. Hence

$$|d(e^{j\theta})| = \left| \lim_{r \to 1-0} d(re^{j\theta}) \right| \le 1 \qquad (2.158)$$

whenever the above limit exists. Thus a bounded function has well defined interior ray limit almost everywhere on the unit circle, and at every such point, it is bounded by unity. However, as (2.135) shows, $Z(z)$ and hence $d(z)$ can exhibit essential singularities on the boundary and the behavior of these functions at those points on the boundary are not well defined. Clearly, for (2.140) to hold, a bounded function (rational or nonrational) cannot exhibit any poles on the boundary of the unit circle. Thus if $d(z)$ is given to be rational, then since it is free of poles in $|z| \le 1$, $d(z)$ is analytic and uniformly bounded by unity in $|z| \le 1$. Thus, for rational bounded functions, we also have

$$\sum_{k=0}^{\infty} |d_k| < \infty. \qquad (2.159)$$

To summarize, if $d(z)$ represents a rational bounded function, then

$$d(z) = \frac{h(z)}{g(z)}, \qquad (2.160)$$

where $g(z)$ is a strict Hurwitz polynomial[2] and $|h(z)/g(z)| \le 1$ in $|z| \le 1$. Thus unlike the positive function, bounded functions are free of poles in the closed unit circle. However, they can possess zeros in the interior/boundary of the unit circle (e.g., $d(z) = z^n$).

From the maximum modulus theorem [26], since a function that is analytic in any closed region attains its maximum absolute value only on the boundary and *not* inside that region, in $|z| \le 1$, bounded functions must possess their maximum value on the unit circle and not in its interior. Thus it follows that a *rational* function $d(z)$ is bounded if and only if

i) its denominator polynomial is strict Hurwitz

and   (2.161)

ii) $|d(e^{j\theta})| \le 1$, in $|\theta| \le \pi$.

Compared to (2.140), this is a simpler test since the boundedness on the
unit circle, together with the absence of poles in $|z| \le 1$ for $d(z)$ makes it
a bounded function.

In particular, any rational function of degree n that is free of poles in
$|z| \le 1$ satisfying

$$\left|d(e^{j\theta})\right| = 1, \quad |\theta| \le \pi,$$

also represents a bounded function. Such functions are known as regular
all-pass functions and up to a constant multiplier of unit magnitude, they
must have the form

$$d(z) = z^n \frac{g^*(1/z^*)}{g(z)} = \frac{\tilde{g}(z)}{g(z)}, \tag{2.162}$$

where $g(z)$ is a strict Hurwitz polynomial and

$$\tilde{g}(z) = z^n g^*(1/z^*) = z^n g_*(z) \tag{2.163}$$

is known as the polynomial reciprocal to $g(z)$. More generally, for any
rational function $f(z)$, let

$$f_*(z) = f^*(1/z^*) \tag{2.164}$$

denote its para-conjugate form as in (1.9). If $f(z)$ is real (for real z), then
$f_*(z) = f(1/z)$ and on the unit circle, (2.164) reduces to $f_*(e^{j\theta}) = f^*(e^{j\theta})$.
Thus, the para-conjugate form in (2.164) can be considered as an extension
of the familiar complex conjugate operation on the unit circle. For an all-
pass function, from (2.162)–(2.163) we have

$$d_*(z) = \frac{g(z)}{z^n g_*(z)} = \frac{1}{d(z)}. \tag{2.165}$$

The positive functions associated with bounded functions that are regular all-pass have certain interesting special properties. In network theory,
they are known as Foster functions and represent the input impedances of
lossless networks [19, 20, 29]. Thus, from (2.147) and (2.148),

$$Z_F(z) = \frac{1 + d(z)}{1 - d(z)} \tag{2.166}$$

is a Foster function whenever $d(z)$ is as given in (2.162). Using (2.164)–
(2.165), in that case

$$Z_{F*}(z) = \frac{1 + d_*(z)}{1 - d_*(z)} = \frac{d(z) + 1}{d(z) - 1} = -Z_F(z)$$

or

$$Z_F(z) + Z_{F*}(z) \equiv 0. \tag{2.167}$$

The above 'even part' condition is characteristic of Foster functions and it represents 'losslessness' or the all-pass character of the underlying bounded functions. Conversely, starting with the definition

$$Z(z) + Z_*(z) \equiv 0 \qquad (2.168)$$

for Foster functions, more generally with $d(z)$ as defined in (2.142), we have

$$1 - d(z)d_*(z) = \frac{2Re\,(R_0) \cdot (Z(z) + Z_*(z))}{(Z(z) + R_0^*)(Z_*(z) + R_0)} \equiv 0$$

or

$$d(z)d_*(z) \equiv 1 \quad \text{or} \quad |d(e^{j\theta})|^2 = 1\,;$$

i.e., Foster functions correspond to bounded functions that are regular all-pass. In the rational case from (2.162)–(2.166) with $g(z) = \sum_{k=0}^{n} g_k z^k$, since $|g_0/g_n| > 1$, $Z_F(z)$ must have the *same* degree n for its numerator and denominator polynomials. $Z_F(z)$'s being positive functions are free of poles in $|z| < 1$, and their zeros and poles on the unit circle are simple and equal in number. Moreover, from (2.167)–(2.168), it is also free of poles and zeros in $|z| > 1$, since every such pole/zero will correspond to a pole/zero in $|z| < 1$. Thus, a Foster function has all its poles and zeros on the unit circle. Since the all-pass function $d(z) = ze^{-j\theta_k}$ generates the Foster function

$$Z_F(z) = \frac{1 + ze^{-j\theta_k}}{1 - ze^{-j\theta_k}} = \frac{e^{j\theta_k} + z}{e^{j\theta_k} - z} \qquad (2.169)$$

with a simple pole at $z = e^{j\theta_k}$, as a result a degree n Foster function has the form

$$Z_F(z) = \sum_{m=1}^{n} P_m \left(\frac{1 + ze^{-j\theta_m}}{1 - ze^{-j\theta_m}}\right). \qquad (2.170)$$

Here $2P_m > 0$ represents the residue of the pole at $z = e^{j\theta_m}$, since

$$\lim_{z \to e^{j\theta_m}} \left(1 - ze^{-j\theta_m}\right) Z_F(z) = 2P_m > 0\,. \qquad (2.171)$$

Notice that if $Z_F(z)$ is also positive-real, then the θ_m's will occur in (plus / minus) pairs. Moreover, for $|z| < 1$, (2.170) can be expanded to give

$$Z_F(z) = \sum_{m=1}^{n} P_m \left(1 + 2\sum_{k=1}^{\infty} e^{-jk\theta_m} z^k\right)$$

$$= \sum_{m=1}^{n} P_m + 2\sum_{k=1}^{\infty} \left(\sum_{m=1}^{n} P_m e^{-jk\theta_m}\right) z^k$$

$$\stackrel{\triangle}{=} c_0 + 2\sum_{k=1}^{\infty} c_k z^k\,. \qquad (2.172)$$

Thus,

$$c_k = \sum_{m=1}^{n} P_m e^{-jk\theta_m}, \quad k = 0 \to \infty. \tag{2.173}$$

This result can be established more generally by starting with any degree n Foster function given by

$$Z_F(z) = \frac{B(z)}{A(z)} = \frac{(1 - ze^{-j\phi_1})(1 - ze^{-j\phi_2}) \cdots (1 - ze^{-j\phi_n})}{(1 - ze^{-j\theta_1})(1 - ze^{-j\theta_2}) \cdots (1 - ze^{-j\theta_n})}. \tag{2.174}$$

Since the numerator and denominator of any Foster function has the same degree, we can rewrite (2.174) as

$$Z_F(z) = K + \frac{\alpha_0 + \alpha_1 z + \cdots + \alpha_{\hat{n}-1} z^{n-1}}{(1 - ze^{-j\theta_1})(1 - ze^{-j\theta_2}) \cdots (1 - ze^{-j\theta_n})}$$

$$= K + \sum_{m=1}^{n} \frac{2P_m}{1 - ze^{-j\theta_m}}, \tag{2.175}$$

where, as in (2.171), $2P_m > 0$ represents the residue of the pole at $z = e^{j\theta_m}$. Further

$$Z_F(0) = c_0 = K + 2\sum_{m=1}^{n} P_m$$

gives

$$K = c_0 - 2\sum_{m=1}^{n} P_m \tag{2.176}$$

and hence (2.175) becomes

$$Z_F(z) = c_0 + 2\sum_{m=1}^{n} P_m \left(-1 + \frac{1}{1 - ze^{-j\theta_m}} \right)$$

$$= c_0 + 2\sum_{m=1}^{n} P_m \left(\frac{ze^{-j\theta_m}}{1 - ze^{-j\theta_m}} \right)$$

$$= c_0 + 2\sum_{m=1}^{n} P_m ze^{-j\theta_m} \left(\sum_{i=0}^{\infty} z^i e^{-ji\theta_m} \right)$$

$$= c_0 + 2\sum_{k=1}^{\infty} \left(\sum_{m=1}^{n} P_m e^{-jk\theta_m} \right) z^k \triangleq c_0 + 2\sum_{k=1}^{\infty} c_k z^k, \quad |z| < 1.$$

Thus

$$c_k = \sum_{m=1}^{n} P_m e^{-jk\theta_m}, \quad k \geq 1.$$

Further from (2.167) and (2.174)–(2.175)

$$Z_F(z) = -Z_{F*}(z) = -\frac{B_*(z)}{A_*(z)} = -\frac{\tilde{B}(z)}{\tilde{A}(z)}$$

$$= -K^* - \frac{z(\alpha_0^* z^{n-1} + \alpha_1^* z^{n-2} + \cdots + \alpha_{n-1}^*)}{(z - e^{j\theta_1})(z - e^{j\theta_2})\cdots(z - e^{j\theta_n})}$$

or

$$Z_F(0) = c_0 = -K^* > 0 \quad \Longrightarrow \quad K = -c_0$$

and together with (2.176) we have

$$c_0 = \sum_{m=1}^{n} P_m .$$

Thus

$$c_k = \sum_{m=1}^{n} P_m e^{-jk\theta_m} , \quad k \geq 0 , \tag{2.177}$$

for any Foster positive function and

$$Z_F(z) = \sum_{m=1}^{n} P_m \left(\frac{2}{1 - ze^{-j\theta_m}} - 1 \right) = \sum_{m=1}^{n} P_m \frac{1 + ze^{-j\theta_m}}{1 - ze^{-j\theta_m}} . \tag{2.178}$$

A quick examination shows that with $\mathbf{C}_k$'s as defined in (2.119) and using (2.177), we obtain

$$\mathbf{C}_k = \mathbf{A}_k \mathbf{P}_n \mathbf{A}_k^* ,$$

where

$$\mathbf{A}_k \triangleq \begin{bmatrix} 1 & 1 & \cdots & 1 \\ \lambda_1 & \lambda_2 & \cdots & \lambda_n \\ \lambda_1^2 & \lambda_2^2 & \cdots & \lambda_n^2 \\ \vdots & \vdots & \cdots & \vdots \\ \lambda_1^k & \lambda_2^k & \cdots & \lambda_n^k \end{bmatrix} , \quad \mathbf{P}_n = \begin{bmatrix} P_1 & 0 & \cdots & 0 \\ 0 & P_2 & \cdots & 0 \\ \vdots & \vdots & \ddots & \vdots \\ 0 & 0 & \cdots & P_n \end{bmatrix} ,$$

$$\lambda_m = e^{j\theta_m} , \quad m = 1 \to n$$

and since $\mathbf{C}_k$ is $(k+1) \times (k+1)$, we have

$$\mathbf{C}_k > 0 , \quad k = 0 \to n-1 \quad \text{and} \quad \det \mathbf{C}_k = 0 , \quad k \geq n . \tag{2.179}$$

From (2.136)–(2.138), for the power spectrum associated with a positive function since $r_k = c_k$, $k = 0 \to \infty$, using (2.177) in (2.136) the associated power spectrum is given by

$$S(\theta) = \sum_{m=1}^{n} P_m \delta(\theta - \theta_m) .$$

Thus there exists a one-to-one relationship between Foster functions and discrete spectra.

In the rational case, the degree[13] of a function is a measure of its complexity and the question of generating 'simpler' functions starting with the given function is an important one. Such a procedure, if possible, allows the complexity to be reduced at successive stages and ultimately generates a new rational function which is further 'irreducible'. For rational positive functions, measured in terms of their degree, Richards' theorem gives a complete answer to the problem of degree reduction and their simplification. In this context, an algorithm developed by Schur turns out to be extremely useful and it plays a fundamental role in formulating many of the ideas discussed here onwards.

2.2.2 SCHUR ALGORITHM AND LINE EXTRACTION

Let $d(z)$ represent the bounded function associated with a positive function $Z(z)$ normalized to some R_0 with positive real part. Thus

$$d(z) = \frac{Z(z) - R_0}{Z(z) + R_0^*}. \tag{2.180}$$

Consider the special case when $R_0 = Z(0)$. Then $d(z)$ has at least a simple zero at $z = 0$ and consequently the function

$$d_1(z) = \frac{1}{z} d(z) = \frac{1}{z} \cdot \frac{Z(z) - Z(0)}{Z(z) + Z^*(z)} \tag{2.181}$$

is analytic in $|z| < 1$. If $d(z)$ is also analytic on $z = e^{j\theta}$, then from (2.158)

$$\left| d_1(e^{j\theta}) \right| = \left| d(e^{j\theta}) \right| \leq 1$$

and hence from the maximum modulus theorem [26], $|d_1(z)| \leq 1$ in $|z| \leq 1$ and it is also analytic there. Of course, $d(z)$ need not be analytic on $z = e^{j\theta}$ and, more generally, from (2.181), $|d_1(z)| \leq 1/r$ for $|z| = r < 1$ and hence using Schwarz's lemma [26], $|d_1(z)| \leq 1$, in $|z| < 1$; i.e., $d_1(z)$ is always a bounded function. Let $Z_1(z)$ represent the positive function associated with $d_1(z)$ normalized to $Z(0)$. Thus,

$$d_1(z) = \frac{Z_1(z) - Z(0)}{Z_1(z) + Z^*(0)}. \tag{2.182}$$

Combining (2.181) and (2.182), we have

$$\frac{1}{z} \cdot \frac{Z(z) - Z(0)}{Z(z) + Z^*(0)} = \frac{Z_1(z) - Z(0)}{Z_1(z) + Z^*(0)} \tag{2.183}$$

---

[13]The degree of a rational function is its totality of poles or zeros including those at infinity with multiplicities counted.

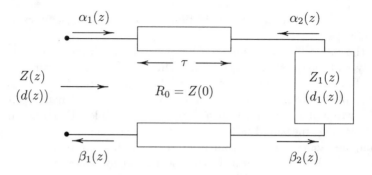

FIGURE 2.7. An ideal delay line.

or

$$d_1(z) = \frac{1}{z} d(z), \qquad (2.184)$$

i.e., the reflection coefficients associated with the two positive functions $Z(z)$ and $Z_1(z)$ so generated (both normalized to $Z(0)$) are related through the delay operator z. Thus, starting with any positive function $Z(z)$, the above procedure will always give rise to a new positive function $Z_1(z)$. This general result is due to Schur [18]. Equation (2.184) can be realized by extracting a delay operator from the reflection coefficient of $Z(z)$ normalized to $Z(0)$. Equations (2.183)–(2.184) can be given an interesting physical interpretation by making use of an ideal delay line and the incident and reflected waves introduced in (2.146).

Consider an ideal delay line as in Fig. 2.7 with one-way delay τ that has been terminated on a positive function $Z_1(z)$ and let $Z(z)$ represent the input positive function. Then, the reflection coefficient $d(z)$ of the input $Z(z)$ normalized to $R_0 = Z(0)$ is given by (2.180) and that of the termination $Z_1(z)$ is given by (2.182). Following the discussion in (2.146), let $\alpha_1(z)$, $\beta_1(z)$ represent the transforms of the input incident and reflected waves and let $\alpha_2(z)$, $\beta_2(z)$ represent those at the output terminals of the line. Under the assumption that all these waves are normalized to $Z(0)$, referring to Fig. 2.6 and making use of (2.146) and the remarks there, we have

$$\frac{\beta_1(z)}{\alpha_1(z)} = d(z) \qquad (2.185)$$

and

$$\frac{\alpha_2(z)}{\beta_2(z)} = d_1(z). \qquad (2.186)$$

(Note that $\tilde{\beta}_2(t)$ is the incident wave to the load.) Since the incident waves $\tilde{\alpha}_1(t)$ and $\tilde{\alpha}_2(t)$ only undergo a one-way delay of τ in the line, we have

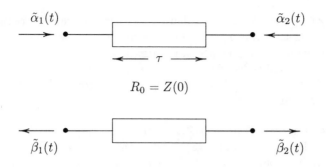

FIGURE 2.8. Incident and reflected wave pairs on a line.

(refer to Fig. 2.8),

$$\tilde{\beta}_2(t) = \tilde{\alpha}_1(t - \tau)$$

and

$$\tilde{\beta}_1(t) = \tilde{\alpha}_2(t - \tau).$$

Under zero initial conditions, the Laplace transform of the above equations give

$$\hat{\beta}_2(p) = e^{-p\tau}\,\hat{\alpha}_1(p)$$

and

$$\hat{\beta}_1(p) = e^{-p\tau}\,\hat{\alpha}_2(p).$$

Thus,

$$\frac{\hat{\alpha}_2(p)}{\hat{\beta}_2(p)} = e^{2p\tau}\frac{\hat{\beta}_1(p)}{\hat{\alpha}_1(p)}. \tag{2.187}$$

By identifying $e^{-2p\tau}$ to be the delay operator z, i.e.,

$$z = e^{-2p\tau} \tag{2.188}$$

and letting $\alpha_1(z)$ represent the incident wave under the above transformation, etc., (2.187) translates into

$$\frac{\alpha_2(z)}{\beta_2(z)} = \frac{1}{z}\cdot\frac{\beta_1(z)}{\alpha_1(z)},$$

and using (2.185)–(2.187), this reduces to

$$d_1(z) = \frac{1}{z}\,d(z).$$

Referring to Fig. 2.7, a line extraction from the input positive function $Z(z)$ has resulted in a new positive function $Z_1(z)$. Notice that the above line extraction is possible only if the normalization is carried out with respect

$$KEN\ KOSNiK$$

to $R_0 = Z(0)$, and clearly the quantity $Z(0)$ is also characteristic to the line. Following the transmission line terminology, we will denote $Z(0)$ to be the characteristic impedance of the first line [20, 30].

Under what conditions will the new positive function $Z_1(z)$ be 'simpler' than $Z(z)$? In the rational case, this question is completely answered by Richards' theorem.

Richards' Theorem [30]: Let $Z(z)$ represent a rational positive function of degree n. Define $d(z)$ and $d_1(z) = \frac{1}{z}d(z)$ as in (2.180) and (2.181). Then $d_1(z)$ is always bounded and if $Z_1(z)$ represents the associated positive function normalized to $Z(0)$, then the degree $\delta(Z_1(z)) \leq \delta(Z(z))$. Moreover, $\delta(Z_1(z)) = \delta(Z(z)) - 1$ if and only if

$$Z(z) + Z_*(z)|_{z=0} = 0. \qquad (2.189)$$

Proof: Let

$$Z(z) = \frac{b(z)}{a(z)} = \frac{b_0 + b_1 z + \cdots + b_n z^n}{a_0 + a_1 z + \cdots + a_n z^n}$$

represent a rational positive function of degree n. From (2.180),

$$d(z) = \frac{a_0^*}{a_0} \cdot \frac{b(z)a_0 - a(z)b_0}{b(z)a_0^* + a(z)b_0^*}$$

$$= \frac{a_0^*}{a_0} \cdot \frac{z\left[(a_0 b_1 - b_0 a_1) + (a_0 b_2 - b_0 a_2)z + \cdots + (a_0 b_n - b_0 a_n)z^{n-1}\right]}{(a_0^* b_0 + b_0^* a_0) + (a_0^* b_1 + b_0^* a_1)z + \cdots + (a_0^* b_n + b_0^* a_n)z^n}$$

and hence

$$d_1(z) = \frac{1}{z}d(z)$$

$$= \frac{a_0^*}{a_0} \cdot \frac{(a_0 b_1 - b_0 a_1) + (a_0 b_2 - b_0 a_2)z + \cdots + (a_0 b_n - b_0 a_n)z^{n-1}}{(a_0^* b_0 + b_0^* a_0) + (a_0^* b_1 + b_0^* a_1)z + \cdots + (a_0^* b_n + b_0^* a_n)z^n}.$$

$$(2.190)$$

From (2.182), $d_1(z)$ is bounded and from (2.190), $\delta(d_1(z)) \leq n = \delta(Z(z))$. Moreover, the associated positive function $Z_1(z)$ has the same degree as $d_1(z)$ and hence $\delta(Z_1(z)) \leq n$. Further, from (2.190), the degree of $d_1(z)$ becomes equal to $(n-1)$ if and only if the coefficient of the highest term in its denominator goes to zero, i.e.,

$$\delta(d_1(z)) = n - 1 \quad \Longleftrightarrow \quad a_0^* b_n + b_0^* a_n = 0,$$

i.e.,

$$\frac{b_0}{a_0} = -\frac{b_n^*}{a_n^*}.$$

But $b_0/a_0 = Z(0)$ and using (2.164), we have $b_n^*/a_n^* = Z^*(1/z^*)|_{z=0} = Z_*(0)$ and hence $\delta(Z_1) = \delta(Z) - 1$, iff

$$Z(0) + Z_*(0) = 0,$$

i.e., the function $Z(z) + Z_*(z) = 0$ at $z = 0$, Q.E.D.

Thus, degree reduction occurs in $Z_1(z)$ iff the 'even part' $(Z(z) + Z_*(z))$ of the input positive function $Z(z)$ has a zero at $z = 0$. Of course, the above line extraction procedure can be repeated on $Z_1(z)$ to generate a new positive function $Z_2(z)$ by carrying out the normalization with respect to $Z_1(0)$. From Richards' theorem, the degree of $Z_2(z)$ will be one less than that of $Z_1(z)$, iff $Z_1(z) + Z_{1*}(z) = 0$ at $z = 0$. If this condition is not satisfied, repeated application of Richards' theorem will not produce any more degree reductions and the positive functions so generated all will have the same degree as the input function $Z_1(z)$.

More generally (see Fig. 2.9), let $Z_r(z)$ represent the positive function at the rth stage resulting from r successive line extractions (with/without degree reduction). Then

$$R_{r-1} = Z_{r-1}(0), \quad r \geq 1, \tag{2.191}$$

represents the characteristic impedance of the rth line and as in (2.182), let

$$d_r(z) = \frac{Z_r(z) - R_{r-1}}{Z_r(z) + R_{r-1}^*}, \quad r \geq 1, \tag{2.192}$$

represent the reflection coefficient of $Z_r(z)$ normalized to R_{r-1}, the characteristic impedance of the previous line. To extract the next line, once again since the reflection coefficient of $Z_r(z)$ normalized to R_r has a zero at $z = 0$, proceeding as before

$$\frac{1}{z} \cdot \frac{Z_r(z) - R_r}{Z_r(z) + R_r^*}$$

is bounded, and letting $Z_{r+1}(z)$ denote the positive function associated with this bounded function, we have

$$\frac{1}{z} \cdot \frac{Z_r(z) - R_r}{Z_r(z) + R_r^*} = \frac{Z_{r+1}(z) - R_r}{Z_{r+1}(z) + R_r^*} = d_{r+1}(z). \tag{2.193}$$

Physically, $Z_{r+1}(z)$ represents the new positive function after $(r + 1)$ line extractions and $d_{r+1}(z)$ the associated bounded function normalized to the characteristic impedance of the last line (Fig. 2.9). Using (2.192) and (2.193), we have

$$Z_r(z) = \frac{R_{r-1} + R_{r-1}^* d_r(z)}{1 - d_r(z)} = \frac{R_r + R_r^* z d_{r+1}(z)}{1 - z d_{r+1}(z)}. \tag{2.194}$$

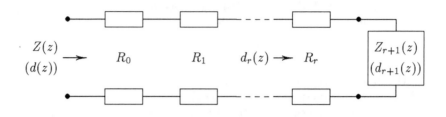

FIGURE 2.9. Schur algorithm and Richards' transformation. Here $d(z)$ is normalized with respect to R_0 and $d_k(z)$ is normalized with respect to R_{k-1}, $k = 1 \rightarrow r + 1$.

Let

$$d_r(0) = \frac{R_r - R_{r-1}}{R_r + R_{r-1}^*} \triangleq S_r \qquad (2.195)$$

represent the rth junction mismatch reflection coefficient. Clearly,

$$|S_r| = |d_r(0)| \leq 1$$

and using this, (2.194) simplifies to

$$
\begin{aligned}
d_r(z) &= \frac{(R_r - R_{r-1}) + z(R_r^* + R_{r-1})d_{r+1}(z)}{(R_r + R_{r-1}^*) + z(R_r^* - R_{r-1}^*)d_{r+1}(z)} \\
&= \frac{(R_r + R_{r-1}^*)S_r + z(R_r^* + R_{r-1})d_{r+1}(z)}{(R_r + R_{r-1}^*) + z(R_r^* + R_{r-1})S_r^* d_{r+1}(z)} \\
&= \frac{\lambda_r S_r + z\lambda_r^* d_{r+1}(z)}{\lambda_r + z\lambda_r^* S_r^* d_{r+1}(z)},
\end{aligned}
\qquad (2.196)
$$

where

$$\lambda_r \triangleq R_r + R_{r-1}^* . \qquad (2.197)$$

Notice that $d_r(z)$ and $d_{r+1}(z)$ represent the reflection coefficients of the positive functions $Z_r(z)$ and $Z_{r+1}(z)$ at the two terminals of the $(r+1)^{st}$ line that are normalized with respect to R_{r-1} and R_r, respectively. In the general case, the above expression that relates these two reflection coefficients is known as the Schur algorithm [18] and in the rational case we will refer to it as the Richards' transformation rule [31]. If $Z(z)$ is positive-real, then all R_k, $k \geq 0$, are positive and hence (2.196) simplifies into the more familiar form

$$d_r(z) = \frac{S_r + z d_{r+1}(z)}{1 + z S_r d_{r+1}(z)} . \qquad (2.198)$$

Although the above transformation rule appears to be different in the real and complex cases, this apparent difference can be avoided by redefining a modified bounded function in (2.192). Towards this, with $d_r(z)$ representing the reflection coefficient in (2.192), define the modified bounded functions

$$\rho_r(z) = d_r(z)\, e^{-j2\phi_{r-1}}, \quad r \geq 0, \tag{2.199}$$

where ϕ_r represents the sum of the phases of λ_i, $i = 0 \to r$, in (2.197), i.e.,

$$\lambda_i = R_i + R_{i-1}^* = |\lambda_i| e^{j\theta_i}, \quad i \geq 0, \tag{2.200}$$

and

$$\phi_r = \theta_1 + \theta_2 + \cdots + \theta_r. \tag{2.201}$$

(Notice that for $r = 0$, $\lambda_0 = R_0 > 0$ and hence $\theta_0 = 0$.) Clearly $\rho_r(z)$, $r \geq 0$, are bounded functions and a direct substitution into (2.196) yields

$$e^{j2\phi_{r-1}}\rho_r(z) = \frac{S_r \lambda_r + z\lambda_r^* \rho_{r+1}(z)\, e^{j2\phi_r}}{\lambda_r + z\lambda_r^* S_r^* \rho_{r+1}(z)\, e^{j2\phi_r}} = \frac{S_r + z\rho_{r+1}(z)\, e^{j2\phi_{r-1}}}{1 + zS_r^* \rho_{r+1}(z)\, e^{j2\phi_{r-1}}}$$

$$= e^{j2\phi_{r-1}} \cdot \frac{S_r e^{-j2\phi_{r-1}} + z\rho_{r+1}(z)}{1 + z\left(S_r e^{-j2\phi_{r-1}}\right)^* \rho_{r+1}(z)}.$$

Thus, in the general case, the Schur algorithm simplifies to

$$\rho_r(z) = \frac{s_r + z\rho_{r+1}(z)}{1 + zs_r^* \rho_{r+1}(z)}, \quad r \geq 0, \tag{2.202}$$

where

$$s_r \triangleq S_r\, e^{-j2\phi_{r-1}} = \frac{R_r - R_{r-1}}{R_r + R_{r-1}^*}\, e^{-j2\phi_{r-1}}$$

$$= d_r(0)\, e^{-j2\phi_{r-1}} = \rho_r(0), \quad r \geq 0, \tag{2.203}$$

represent the modified junction reflection coefficients. For $r = 0$, since $s_0 = d_0(0) = 0$, (2.202) gives

$$\rho_0(z) = z\rho_1(z).$$

But $\rho_0(z) = d(z)$ and $\theta_0 = 0$ gives $\rho_1(z) = d_1(z)e^{-j2\theta_0} = d_1(z)$, and the above expression reads

$$d(z) = zd_1(z) \tag{2.204}$$

which is the same as (2.184).

When $Z(z)$ is positive-real, every $R_r > 0$ implying that $\theta_i = 0$, $i \geq 0$, and hence

$$d_r(z) = \rho_r(z) \tag{2.205}$$

and

$$s_r = S_r = s_r^* .\tag{2.206}$$

Thus the transformation in (2.202) that uses the modified bounded functions and junction reflection coefficients represents both (2.196) and (2.198).

In the following chapters, we will discuss in some detail the applications of these ideas to problems involving spectrum extension as well as identification of linear time-invariant systems.

Problems

1. For any discrete-time W.S.S. stochastic process, let $\rho_k = r_k/r_0$ with r_k as in (2.29). Show that, for any k,

$$\rho_{2k} \geq 2\rho_k^2 - 1 .$$

   Thus, if $|r_k| > r_0/\sqrt{2}$, then r_{2k} is positive. Thus, the magnitude of r_k beyond a certain threshold determines the sign of r_{2k} to be positive.

2. Show that the optimum weight vector $[\beta_0, \ \beta_1, \ \ldots, \ \beta_{n-1}]^T$ for the best one-step backward predictor

$$\hat{x}(-T) = \sum_{k=0}^{n-1} \beta_k \, x(kT)$$

   that minimizes the mean-square error is given by the complex conjugate of (2.84). Thus, in the scalar case, the forward weight vector and the backward weight vector are complex conjugates of each other and the associated minimum mean-square errors remain the same. Prove that similar results hold also for a k-step predictor.

3. Show that the k-step prediction error $\sigma_n^2(k)$ in (2.105) using n samples forms a monotone nonincreasing sequence in n.

4. For an AR(p) process, show that the k-step predictor in (2.99) achieves its best performance at stage $n = p$, i.e., for all $n \geq p$, show that

$$\sigma_n^2(k) = \sigma_p^2(k) = \frac{1}{\Delta_{p-1}} \begin{vmatrix} r_0 & r_1 & \cdots & r_{p-1} & r_{p+k-1} \\ r_1^* & r_0 & \cdots & r_{p-2} & r_{p+k-2} \\ \vdots & \vdots & \ddots & \vdots & \vdots \\ r_{p-1}^* & r_{p-2}^* & \cdots & r_0 & r_k \\ r_{p+k-1}^* & \cdots & \cdots & r_k^* & r_0 \end{vmatrix} > 0 ,\tag{2.P.1}$$

   where Δ_p is as defined in (2.86). Thus, for AR(p) processes, the k-step ahead predictor using n samples achieves its best performance whenever $n \geq p$.

5. For an AR(p) process, show that

$$\Delta_{p+k} = \left(\frac{\Delta_p}{\Delta_{p-1}}\right)^k \Delta_p , \quad k \geq 1 .$$

Here Δ_p is as defined in (2.86).

6. Show that the sum of positive (p.r.) functions is a positive (p.r.) function. Similarly the product of bounded (b.r.) functions is a bounded (b.r.) function.

7. If $b(z)$ is a bounded function, then show that (a) $1 \pm b(z)$ and (b) $1 + exp[(b(z) + 1)/(b(z) - 1)]$ are positive functions.

8. Let $b(z)$ represent a regular all-pass function. Show that $(1 - b(z))/(1 + b(z))$ represents a positive Foster function.

9. Let $Z_1(z)$, $Z_2(z)$ and $Z(z)$ represent the positive functions (normalized to unity) associated with the bounded functions $b_1(z)$, $b_2(z)$ and $b(z) = b_1(z)b_2(z)$, respectively. Express $Z(z)$ in terms of $Z_1(z)$ and $Z_2(z)$.

10. **Schur-Cohn Test** [32] Let

$$g(z) = a_0 + a_1 z + \cdots + a_n z^n$$

represent a degree n strict Hurwitz polynomial. Then, its reciprocal polynomial

$$\tilde{g}(z) = z^n g^*(1/z^*) = a_n^* + a_{n-1}^* z + \cdots + a_0^* z^n$$

has all its zeros in $|z| < 1$ and $d(z) = \tilde{g}(z)/g(z)$ is a regular rational all-pass function as in (2.162). Since $d(z)$ is also bounded, its power series expansion $d(z) = \sum_{k=0}^{\infty} d_k z^k$, $|z| < 1$, satisfies (2.154)–(2.155).

(a) Show that in the case of rational all-pass functions, (2.154)–(2.155) reduces to the positivity of the Hermitian matrix

$$\mathbf{A} = \mathbf{A}_n^* \mathbf{A}_n - \mathbf{B}_n^* \mathbf{B}_n , \tag{2.P.2}$$

or equivalently to that of

$$\mathbf{B} = \mathbf{A}_n \mathbf{A}_n^* - \mathbf{B}_n \mathbf{B}_n^* , \tag{2.P.3}$$

where

$$\mathbf{A}_k = \begin{bmatrix} a_0 & 0 & \cdots & 0 & 0 \\ a_1 & a_0 & \cdots & 0 & 0 \\ \vdots & \vdots & \ddots & \vdots & \vdots \\ a_{k-2} & a_{k-3} & \cdots & a_0 & 0 \\ a_{k-1} & a_{k-2} & \cdots & a_1 & a_0 \end{bmatrix} , \quad k = 1 \to n ,$$

and

$$\mathbf{B}_k = \begin{bmatrix} a_n^* & 0 & \cdots & 0 & 0 \\ a_{n-1}^* & a_n^* & \cdots & 0 & 0 \\ \vdots & \vdots & \ddots & \vdots & \vdots \\ a_{n-k+2}^* & a_{n-k+3}^* & \cdots & a_n^* & 0 \\ a_{n-k+1}^* & a_{n-k+2}^* & \cdots & a_{n-1}^* & a_n^* \end{bmatrix}, \quad k = 1 \to n.$$

Thus, rationality together with the all-pass character of the above $d(z)$ reduces (2.154) to a finite test.

(b) Show that the principal minors of the Hermitian matrix $\mathbf{B}$ in (2.P.3) are given by

$$\Delta_k = |\mathbf{A}_k \mathbf{A}_k^* - \mathbf{B}_k \mathbf{B}_k^*| = \left| \begin{array}{c|c} \mathbf{A}_k & \mathbf{B}_k^* \\ \hline \mathbf{B}_k & \mathbf{A}_k^* \end{array} \right|, \qquad (2.P.4)$$

for $k = 0 \to n$. Since the positivity of a Hermitian matrix is equivalent to the positivity of its *every* principal minor, the polynomial $g(z) = a_0 + a_1 z + \cdots + a_n z^n$ is strict Hurwitz if and only if

$$\Delta_k > 0, \quad k = 0 \to n. \qquad (2.P.5)$$

The positivity of the later form in (2.P.4) for $k = 0 \to n$ represents the Schur-Cohn test [32].

(c) **Jury's Test** [33] In the case of real polynomials, show that the determinant in (2.P.4) simplifies into

$$\Delta_k = |\mathbf{X}_k + \mathbf{Y}_k| \cdot |\mathbf{X}_k - \mathbf{Y}_k|, \quad k = 0 \to n,$$

where

$$\mathbf{X}_k = \begin{bmatrix} a_0 & a_1 & \cdots & a_{k-1} & a_k \\ 0 & a_0 & \cdots & a_{k-2} & a_{k-1} \\ \vdots & \vdots & \cdots & \vdots & \vdots \\ 0 & 0 & \cdots & 0 & a_0 \end{bmatrix}$$

and

$$\mathbf{Y}_k = \begin{bmatrix} a_{n-k+1} & a_{n-k+2} & \cdots & a_{n-1} & a_n \\ a_{n-k+2} & a_{n-k+3} & \cdots & a_n & 0 \\ \vdots & \vdots & \cdots & \vdots & \vdots \\ a_n & 0 & \cdots & 0 & 0 \end{bmatrix}.$$

(*Hint:* $d(z)$ regular rational all-pass of degree n implies $\mathbf{T}_{n-1} > 0$ and $\mathbf{T}_k$ singular for $k \geq n$.)

11. **Foster's Theorem** [29] Let $Z_F(z)$ represent a rational (positive) Foster function. Then show that

(a) $Z_F(e^{j\theta}) = jX(\theta)$ where $X(\theta)$ is real and has the form

$$X(\theta) = \sum_{m=1}^{n} P_m \cot (\theta - \theta_m)/2, \quad P_m > 0,$$

and

(b) $\qquad -\dfrac{dX(\theta)}{d\theta} \geq |X(\theta)| > 0.$ $\qquad\qquad$ (2.P.6)

Since the poles and zeros of a rational Foster function are simple, equal in number and located on the unit circle, $\dfrac{dX(\theta)}{d\theta} < 0$ implies that ($X(\theta)$ is monotonic and hence) the above poles and zeros must alternate. In particular, if $g(z)$ is a strict Hurwitz polynomial and $\tilde{g}(z)$ its reciprocal, then the zeros of $g(z) + \tilde{g}(z)$ and $g(z) - \tilde{g}(z)$ are simple, interlace each other and lie on the unit circle (Routh-Hurwitz' Theorem [34]).
(*Hint:* Start from (2.178).)

(c) If $Z_F(z)$ is a rational positive-real Foster function, then more strongly

$$-\frac{dX(\theta)}{d\theta} \geq \left| \frac{X(\theta)}{\sin \theta} \right| > 0.$$ $\qquad\qquad$ (2.P.7)

Since the poles and zeros of a real function must occur in complex conjugate pairs, from the discussion in part (b) it follows that $z = \pm 1$ must be simple poles and/or zeros for a p.r. Foster function and hence there are only four possible generic plots for $X(\theta)$ [29].

(d) Conversely (Foster's Theorem), consider any real rational function $Z(z)$ with $Z(0) > 0$, whose poles and zeros are on the unit circle and simple. Further, the poles and zeros alternate. Show that $Z(z)$ is a Foster p.r. function [29].

12. (a) Let $b(z)$ represent a regular all-pass function with $b(e^{j\theta}) = e^{j\psi(\theta)}$. Show that

$$\frac{d\psi(\theta)}{d\theta} \geq |\sin \psi(\theta)|.$$ $\qquad\qquad$ (2.P.8)

(b) Let $H(z) = h_0 + h_1 z + \cdots + h_n z^n$ represent a degree n polynomial with all its zeros in $|z| < 1$. Write $H(e^{j\theta}) = A(\theta)e^{j\phi(\theta)}$, $A(\theta) > 0$. Show that (use (2.P.8))

$$\frac{d\phi(\theta)}{d\theta} \geq \frac{n}{2} + \frac{1}{2} |\sin (2\phi(\theta) - n\theta)|,$$ $\qquad\qquad$ (2.P.9)

i.e., the phase function of any polynomial with all its zeros in the interior of a unit circle is a monotone increasing function.

(c) Let $u(\theta)$ represent the real part of $H(z)$ on the unit circle. Thus $u(\theta) = A(\theta) \cos \phi(\theta)$. Use (2.P.9) to show that all zeros of $u(\theta)$ are real and simple. As a result, the zero crossings of $u(\theta)$ can be used to reconstruct $H(z)$ up to a multiplicative constant.

13. **Minimum Real Part Lemma [26]** Let $f(z)$ represent a function (nonconstant) that is analytic in $|z| \leq 1$. Then, the minimum value of the real part of $f(z)$ in $|z| \leq 1$ is achieved at some point on the boundary of the unit circle and *not* inside the unit circle.

    In particular, if $f(z) = Z(z)$ is a positive function that is analytic in $|z| \leq 1$, then its minimum real part in $|z| \leq 1$ is attained on the unit circle and, further, $Re\,(Z(z)) > 0$ if $|z| < 1$, unless $Re\,(Z(z)) \equiv 0$.

14. **Maximum Modulus Theorem [26]** Let $g(z)$ represent a function (nonconstant) that is analytic in $|z| \leq 1$. Then, the maximum of $|g(z)|$ in $|z| \leq 1$ is attained at some point on the unit circle and *not* inside the unit circle.

    In particular, if $g(z) = d(z)$ is a bounded function that is analytic in $|z| \leq 1$, then the maximum of $|d(z)|$ in $|z| \leq 1$ is attained on the unit circle, and $|d(z)| < 1$ in $|z| < 1$.

3

Admissible Spectral Extensions

3.1 Introduction

An interesting problem in the study of autocorrelation functions and their associated power spectral densities is that of estimating the spectrum from a finite extent of its autocorrelation function. Known as the trigonometric moment problem in the discrete case, it has been the subject of extensive study for a long time [7–8, 14, 35–36]. In view of the considerable mathematical interest as well as the practical significance of the moment problem in interpolation theory, system identification, power gain approximation theory and rational approximation of nonrational systems, it is best to review this problem in some detail. Towards this, consider a discrete-time zero mean wide-sense stationary stochastic process $x(nT)$ with autocorrelation function $\{r_k\}_{k=-\infty}^{\infty}$ as in (2.29). Its power spectral density function $S(\theta)$ is given by (2.34) and assume that the process has finite power and satisfies the causality criterion in (2.47). Thus

$$S(\theta) = \sum_{k=-\infty}^{\infty} r_k e^{jk\theta} \tag{3.1}$$

$$\frac{1}{2\pi} \int_{-\pi}^{\pi} S(\theta)\, d\theta = r_0 < \infty \tag{3.2}$$

and

$$\frac{1}{2\pi} \int_{-\pi}^{\pi} \ln S(\theta)\, d\theta > -\infty. \tag{3.3}$$

In that case, as in (2.48), every Hermitian Toeplitz matrix $\mathbf{T}_k$, formed from the autocorrelations $r_0 \to r_k$, $k = 0 \to \infty$, is positive definite, and moreover from (2.49)–(2.52) there exists a unique function

$$B(z) = \sum_{k=0}^{\infty} b_k z^k \tag{3.4}$$

that is analytic together with its inverse in $|z| < 1$ such that

$$S(\theta) = \left| B(e^{j\theta}) \right|^2, \quad a.e. \tag{3.5}$$

In addition, $B(z)$ is free of poles on $|z| = 1$ and it represents the minimum-phase Wiener factor associated with the power spectrum $S(\theta)$.

The trigonometric moment problem, also known as the interpolation problem, can be stated as follows: Given r_0, r_1, ..., r_n from a stationary regular stochastic process that satisfies (3.2) and (3.3), determine all solutions for the power spectral density $S(\theta)$ that are compatible with the given data; i.e., such an admissible solution $S(\theta)$ must satisfy

$$S(\theta) \geq 0 \tag{3.6}$$

and

$$\frac{1}{2\pi} \int_{-\pi}^{\pi} S(\theta)e^{-jk\theta}\, d\theta = r_k, \quad k = 0 \to n, \tag{3.7}$$

in addition to satisfying (3.2) and (3.3). From (2.44) and (2.48) for the existence of such a solution, the positivity of $\mathbf{T}_n$ is necessary, and interestingly that condition is also sufficient! It is well known that an infinite number of solutions exist to the above problem and various approaches can be used to obtain these solutions [4]. Among them, the geometrical interpretation to the class of all admissible extensions [11, 35] as well as the functional parametrization of all such solutions using the theory of positive and bounded functions [12, 18] are particularly interesting. In addition to a leisurely discussion of these solutions, an interesting generalization to the maximum entropy concept is also discussed here together with some of its consequences.

3.2   Geometrical Solution

The geometrical interpretation to the class of all extensions exploits the positivity of every Hermitian Toeplitz matrix generated from the auto-correlations [11, 12, 35]. In particular, using (2.86), since $\Delta_n > 0$, every admissible extension r_k, $k = n+1 \to \infty$, must satisfy

$$\Delta_k > 0, \quad k = n+1 \to \infty. \tag{3.8}$$

Consequently at the first step, the unknown $r_{n+1} = x$ should be chosen so that

$$\Delta(x) = \begin{vmatrix} r_0 & r_1 & \cdots & r_n & x \\ r_1^* & r_0 & \cdots & r_{n-1} & r_n \\ \vdots & \vdots & \ddots & \vdots & \vdots \\ r_n^* & r_{n-1}^* & \cdots & r_0 & r_1 \\ x^* & r_n^* & \cdots & r_1^* & r_0 \end{vmatrix} > 0. \tag{3.9}$$

Using the matrix identity in footnote 9 (chapter 2), $\Delta(x)$ can be expanded to give

$$\Delta(x) = \frac{\Delta_n^2 - |\mathbf{M}_n(x)|^2}{\Delta_{n-1}},$$

where

$$\mathbf{M}_n(x) \triangleq \begin{bmatrix} r_1 & r_2 & \cdots & r_n & x \\ r_0 & r_1 & \cdots & r_{n-1} & r_n \\ r_1^* & r_0 & \cdots & r_{n-2} & r_{n-1} \\ \vdots & \vdots & \cdots & \vdots & \vdots \\ r_{n-1}^* & r_{n-2}^* & \cdots & r_0 & r_1 \end{bmatrix}. \tag{3.10}$$

However, by making use of another determinantal expansion rule,[1] we have (after n interchanges of the first row)

$$|\mathbf{M}_n(x)| = (-1)^n \Delta_{n-1} |x - \xi_n|, \tag{3.11}$$

where

$$\xi_n = \mathbf{f}_n^T \mathbf{T}_{n-1}^{-1} \mathbf{b}_n, \tag{3.12}$$

$$\mathbf{f}_n = [r_1, \ r_2, \ \ldots, \ r_n]^T$$

and

$$\mathbf{b}_n = [r_n, \ r_{n-1}, \ \ldots, \ r_1]^T.$$

From (3.9)–(3.11) we obtain the well known relation

$$\Delta(x) = \frac{\Delta_n^2 - \Delta_{n-1}^2 |x - \xi_n|^2}{\Delta_{n-1}} > 0$$

and with $x = r_{n+1}$, this reads

$$\Delta_{n+1} = \frac{\Delta_n^2 - \Delta_{n-1}^2 |r_{n+1} - \xi_n|^2}{\Delta_{n-1}} > 0 \tag{3.13}$$

or

$$|r_{n+1} - \xi_n|^2 < \left(\frac{\Delta_n}{\Delta_{n-1}} \right)^2. \tag{3.14}$$

Equation (3.14) represents the interior of a circle with center ξ_n and radius Δ_n/Δ_{n-1}, and this implies that every admissible extension must have the unknown correlation r_{n+1} inside the above admissible circle.[2] Repeating

---

[1] If $|\mathbf{A}| \neq 0$, then

$$\begin{vmatrix} \mathbf{A} & \mathbf{B} \\ \mathbf{C} & \mathbf{D} \end{vmatrix} = |\mathbf{A}||\mathbf{D} - \mathbf{C}\mathbf{A}^{-1}\mathbf{B}|.$$

[2] If r_{n+1} is chosen to be on the boundary of the circle, then $\Delta_{n+1} = 0$ and the corresponding spectrum becomes discrete as in (2.46). In that case, as in (3.90)–(3.91), the solution is unique as the remaining r_k's are completely known.

this procedure (3.9)–(3.14), for $k > n$, we obtain

$$\Delta_{k+1} = \frac{\Delta_k^2 - |\mathbf{M}_k(r_{k+1})|^2}{\Delta_{k-1}}$$

$$= \frac{\Delta_k^2 - \Delta_{k-1}^2 |r_{k+1} - \xi_k|^2}{\Delta_{k-1}} > 0 \qquad (3.15)$$

or

$$|r_{k+1} - \xi_k|^2 < \gamma_k^2, \qquad (3.16)$$

where

$$\gamma_k = \frac{\Delta_k}{\Delta_{k-1}}, \qquad k = n \to \infty,$$

and ξ_k is as defined in (3.12) with n replaced by k. Thus at every stage the unknown correlation can be chosen in an infinite number of ways satisfying (3.16). Although at the first stage the center ξ_n and radius γ_n are completely determined from the given data, beyond that for every $k > n$, these parameters depend on the choice of r_k at the previous stages and because of the infinite choices available at each step, there are an infinite number of admissible solutions to the spectrum extension problem.

Using (2.89) the radius γ_k is also seen to be equal to the one-step minimum mean-square predictor error $|L_{00}^{(k)}|^2$, and from there as well as (3.15)

$$\gamma_{k+1} = \frac{\Delta_{k+1}}{\Delta_k} = \frac{\Delta_k}{\Delta_{k-1}} \left(1 - \left| \frac{\mathbf{M}_k(r_{k+1})}{\Delta_k} \right|^2 \right)$$

$$= \frac{\Delta_k}{\Delta_{k-1}} \left(1 - \left| \frac{\Delta_{k-1}|r_{k+1} - \xi_k|}{\Delta_k} \right|^2 \right) \leq \frac{\Delta_k}{\Delta_{k-1}} = \gamma_k, \ k > n, \ (3.17)$$

i.e., these radii of admissible circles form a monotone nonincreasing sequence of positive numbers that are bounded by the first entry γ_n in that sequence and hence

$$\gamma_k \leq \frac{\Delta_n}{\Delta_{n-1}}, \qquad k = n + 1 \to \infty. \qquad (3.18)$$

In the limit, from (2.90)–(2.91)

$$\gamma_\infty = \lim_{k \to \infty} \gamma_k = P_1 = |b_0|^2 = exp \left[\frac{1}{2\pi} \int_{-\pi}^{\pi} \ln S(\theta) \, d\theta \right]. \qquad (3.19)$$

Thus the 'final' radius γ_∞ also coincides with the one-step minimum mean-square prediction error $|b_0|^2$ with b_0 as in (3.4) representing the constant term of the Wiener factor generated by this procedure. Equation (3.19) also

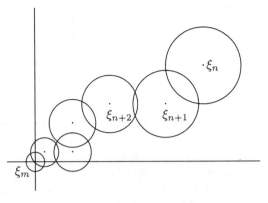

FIGURE 3.1. Migration of nonexpanding admissible circles.

exhibits the relationship between the final radius and the entropy functional $\mathcal{H}$ of the process defined in (2.94). From (3.18)–(3.19)

$$\gamma_\infty = exp\,(\mathcal{H}) \le \gamma_k \le \frac{\Delta_n}{\Delta_{n-1}} \tag{3.20}$$

for any $k > n$ and hence maximization of entropy is equivalent to choosing the extension in (3.15)–(3.16) such that the radius of the 'final circle' has its maximum possible value Δ_n/Δ_{n-1}. Using (3.18)–(3.19), clearly γ_∞ takes its maximum possible value Δ_n/Δ_{n-1} iff

$$\gamma_k = \frac{\Delta_n}{\Delta_{n-1}}, \quad k = n+1 \to \infty\,.$$

In that case, from (3.17) we have

$$r_{k+1} = \xi_k\,, \quad k = n \to \infty\,, \tag{3.21}$$

and every choice r_{k+1} must be at the center of the respective admissible circle for the maximum entropy method [12]. Since the center of the circle is maximally away from the circumference of the circle, such an extension is also most robust and maximally noncommittal. For every finite power process, since $r_k \to 0$, as $k \to \infty$ [37], by identifying with the maximum entropy extension, from (3.21) we have

$$\lim_{k \to \infty} \xi_k = \lim_{k \to \infty} r_{k+1} = 0\,,$$

i.e., for every extension the centers of the admissible circles ultimately progress towards the origin (see Fig. 3.1).

It is not difficult to obtain explicit parametric form for the maximum entropy extension. Towards this, notice that using (3.21) in (3.15) gives

$$|\mathbf{M}_k(r_{k+1})| = 0\,, \quad k = n \to \infty\,, \tag{3.22}$$

where $\mathbf{M}_k(x)$ is as defined in (3.10). On comparing (3.22) with (2.115), from the remark there it follows that this particular extension that maximizes entropy leads to an all-pole model of order n [38, 39], i.e.,

$$S_{ME}(\theta) = \frac{1}{|a_0 + a_1 e^{j\theta} + \cdots + a_n e^{jn\theta}|^2}, \tag{3.23}$$

where $a_0 + a_1 z + \cdots + a_n z^n$ represents the strict Hurwitz polynomial associated with the AR(n) model. Thus parametrically the maximum entropy extension, that also possesses the maximum possible value for the radius of every admissible circle, is equivalent to an AR(n) model that interpolates the given data $r_0 \to r_n$. Moreover from (2.95), this particular extension also has the property that given $r_0 \to r_n$, it has the maximum possible value for the one-step minimum mean-square prediction error, and in that sense it represents the most robust extension.

This leads us to the natural questions: Given $r_0 \to r_n$, what is the parametric form of the extension that maximizes the k-step minimum mean-square prediction error? More generally, can the infinite set of admissible extensions be parametrized functionally?

The functional parametrization of the class of all admissible extensions that interpolate the given autocorrelations follows from Schur's theory on bounded functions [18]. Youla also has given a neat parametrization to the entire class of solutions in a network theoretic setting in terms of bounded-real functions [12]. Before examining the extension associated with maximization of the multistep minimum mean-square prediction error, we begin with the general parametrization of the class of all extensions.

3.3   Parametrization of Admissible Extensions

Given $r_0, r_1, \ldots, r_n$ with $\mathbf{T}_n > 0$, the problem here is to parametrize all admissible extensions that satisfy (3.2) and (3.3) in addition to (3.7). In this context, Schur essentially has shown that the class of all such extensions that interpolate the given data can be parametrized by an arbitrary bounded function [18]. By making contact with network theory, Youla has given an interesting physical characterization of this problem, and in this context he has derived explicit formulas that express the admissible spectra in terms of arbitrary bounded functions [12].

The one-to-one correspondence between the power spectra and positive functions established in (2.136)–(2.139) can be used to exhibit such a parametrization. From (2.138) the extension problem is equivalent to finding all positive functions given their first $(n + 1)$ terms. Thus

$$Z(z) = r_0 + 2 \sum_{k=1}^{n} r_k z^k + O(z^{n+1}), \tag{3.24}$$

where $O(z^{n+1})$ is to be completed arbitrarily subject to the condition that $Z(z)$ turns out to be a positive function. From (2.142), the one-to-one correspondence between positive functions and bounded functions translates the problem in (3.24) into that of completing a bounded-real function. Since $r_0 \rightarrow r_n$ are known, from (2.149) the first $(n+1)$ terms $d_0 \rightarrow d_n$ of a bounded function $\rho(z)$ are known and the extension problem is to determine

$$\rho(z) = \sum_{k=0}^{n} d_k z^k + O(z^{n+1}). \tag{3.25}$$

Here $O(z^{n+1})$ represents the remaining terms to be determined arbitrarily subject to the sole requirement that $\rho(z)$ turns out to be a bounded function. The answer to this problem is contained in Schur's theory on bounded functions [18] and Richards' theorem on line extraction that is described in section 2.2.2.

To see this, consider the situation described in Fig. 2.9 where starting with a given bounded function $d(z)$, $(r+1)$ lines have been extracted to end up with a bounded function $d_{r+1}(z)$. In this process the bounded functions $d_1(z)$, $d_2(z)$, ..., $d_r(z)$ have been generated, and their interrelationship is given by (2.194) where S_r, $r \geq 0$, represent the junction reflection coefficients and R_r, $r \geq 0$, represent the characteristic impedances of the lines. Equivalently, this situation can be described in terms of the new set of bounded functions $\rho_1(z)$, $\rho_2(z)$, ..., $\rho_{r+1}(z)$ defined in (2.197) and their associated junction reflection coefficients s_k, $k = 0 \rightarrow n$, that are related through (2.200). To make further progress it is best to express the input bounded function $d(z) \equiv \rho(z)$ in terms of the terminal bounded function $\rho_{r+1}(z)$ for $r = n$.

Towards this purpose, letting $r = 0$ in (2.200) gives

$$\rho(z) = \rho_0(z) = \frac{s_0 + z\rho_1(z)}{1 + zs_0^*\rho_1(z)} \tag{3.26}$$

and $r = 1$ in (2.200) gives

$$\rho_1(z) = \frac{s_1 + z\rho_2(z)}{1 + zs_1^*\rho_2(z)}. \tag{3.27}$$

Substituting (3.27) in (3.26), we obtain (since $s_0 = 0$)

$$\rho(z) = \frac{(s_0 + zs_1) + z(s_0s_1^* + z)\rho_2(z)}{(1 + zs_0^*s_1) + z(s_1^* + zs_0^*)\rho_2(z)} = \frac{zs_1 + z^2\rho_2(z)}{1 + zs_1^*\rho_2(z)}. \tag{3.28}$$

Continuing this iteration for $n - 2$ more steps, we get

$$\rho(z) = \frac{h_{n-1}(z) + z\tilde{g}_{n-1}(z)\rho_n(z)}{g_{n-1}(z) + z\tilde{h}_{n-1}(z)\rho_n(z)}, \tag{3.29}$$

where $\tilde{g}_{n-1}(z) = z^{n-1}g^*_{n-1}(1/z^*)$ and $\tilde{h}_{n-1}(z) = z^{n-1}h^*_{n-1}(1/z^*)$ represent the polynomials reciprocal to $g_{n-1}(z)$ and $h_{n-1}(z)$, respectively. Once again updating (3.29) with the help of (2.200) for $r = n$, we obtain

$$\rho(z) = \frac{\left(h_{n-1}(z) + zs_n\tilde{g}_{n-1}(z)\right) + z\left(z\tilde{g}_{n-1}(z) + s^*_n h_{n-1}(z)\right)\rho_{n+1}(z)}{\left(g_{n-1}(z) + zs_n\tilde{h}_{n-1}(z)\right) + z\left(z\tilde{h}_{n-1}(z) + s^*_n g_{n-1}(z)\right)\rho_{n+1}(z)}$$

$$\triangleq \frac{h_n(z) + z\tilde{g}_n(z)\rho_{n+1}(z)}{g_n(z) + z\tilde{h}_n(z)\rho_{n+1}(z)}, \tag{3.30}$$

where, for $n \geq 1$ we can identify

$$\alpha_n g_n(z) = g_{n-1}(z) + zs_n\tilde{h}_{n-1}(z) \tag{3.31}$$

and

$$\alpha_n h_n(z) = h_{n-1}(z) + zs_n\tilde{g}_{n-1}(z). \tag{3.32}$$

Here α_n is at most a suitable *real* normalization constant yet to be determined. From (3.26), the above iteration starts with

$$g_0(z) = 1, \quad h_0(z) = 0 \tag{3.33}$$

and hence (3.31)–(3.32) give $\alpha_1 g_1(z) = 1$, $\alpha_1 h_1(z) = zs_1$, $\alpha_1\tilde{g}_1(z) = z$ and $\alpha_1\tilde{h}_1(z) = s^*_1$, and these relations agree with (3.28). A direct computation also shows

$$\alpha_n^2\left(g_n(z)g_{n*}(z) - h_n(z)h_{n*}(z)\right)$$
$$= (1 - |s_n|^2)\left(g_{n-1}(z)g_{n-1*}(z) - h_{n-1}(z)h_{n-1*}(z)\right)$$

and hence by setting

$$\alpha_n = \sqrt{1 - |s_n|^2}, \tag{3.34}$$

we have, in view of (3.33),

$$g_n(z)g_{n*}(z) - h_n(z)h_{n*}(z) = 1, \tag{3.35}$$

the familiar Feltketter relationship[3] [19, 20].

As Fig. 3.2 shows, (3.30) can be interpreted as the input reflection coefficient of an $(n + 1)$-stage transmission line terminated upon an *arbitrary* load represented by the positive function $Z_{n+1}(z)$. The bounded function $\rho_{n+1}(z)$ in (3.30) is related to $Z_{n+1}(z)$ through the terminal reflection coefficient $d_{n+1}(z)$ defined in (2.190) and (2.197). Here the input reflection coefficient $\rho(z)$ is normalized with respect to the characteristic impedance

---

[3]From (3.32) and (3.33) it is easy to see that $h_n(0) = 0$ for every $n \geq 0$ and we must have $h_n(z) = h_1 z + h_2 z^2 + \cdots + h_n z^n$. However, for (3.35) to hold, the highest term generated by $g_n(z)g_{n*}(z)$ must be cancelled by $h_n(z)h_{n*}(z)$ and hence $g_n(z)$ is at most of degree $n - 1$. This also follows from (3.31) and (3.33).

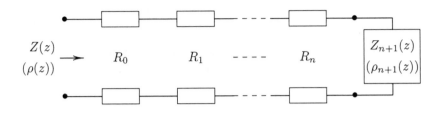

FIGURE 3.2. Positive functions and line extraction. Here $\rho(z)$ represents the input reflection coefficient and $\rho_{n+1}(z)$ represents the modified terminal reflection coefficient.

R_0 of the first line, and the terminal reflection coefficient is normalized with respect to the characteristic impedance R_n of the $(n+1)^{th}$ line. Note that (3.30) is yet another representation of $\rho(z)$ in terms of the transmission lines and the modified terminal reflection coefficient $\rho_{n+1}(z)$. Naturally, $g_n(z)$ and $h_n(z)$ characterize the $(n+1)$-stage transmission line and they are completely specified through the single step update rule,

$$\sqrt{1 - |s_n|^2}\, g_n(z) = g_{n-1}(z) + z s_n \tilde{h}_{n-1}(z), \quad n = 1 \to \infty, \qquad (3.36)$$

and

$$\sqrt{1 - |s_n|^2}\, h_n(z) = h_{n-1}(z) + z s_n \tilde{g}_{n-1}(z), \quad n = 1 \to \infty, \qquad (3.37)$$

that follows from (3.31)–(3.32) and (3.34). Using Rouché's theorem [32] together with an induction argument on (3.36), with the help of (3.33), it follows that $g_n(z)$, $n = 1, 2, \ldots$, are free of zeros in the closed unit circle (strict Hurwitz polynomials.) As a result, (3.35) together with the maximum modulus theorem [26] gives $h_n(z)/g_n(z)$ to be a bounded function. This also follows from (3.30) by setting $\rho_{n+1}(z) \equiv 0$ and hence $h_n(z)/g_n(z)$ represents the input reflection coefficient $\rho(z)$ with $\rho_{n+1}(z) \equiv 0$. Since $\rho(z)$ is normalized to R_0, from (2.147), the corresponding positive function $Z_0(z)$ in this case is given by

$$Z_0(z) = R_0 \left(\frac{1 + h_n(z)/g_n(z)}{1 - h_n(z)/g_n(z)} \right) = R_0 \left(\frac{g_n(z) + h_n(z)}{g_n(z) - h_n(z)} \right). \qquad (3.38)$$

Similarly in the case of an arbitrary $\rho_{n+1}(z)$ termination in (3.30), the corresponding input positive function $Z(z)$ is given by

$$Z(z) = R_0 \frac{1 + \rho(z)}{1 - \rho(z)} = 2 \frac{B_n(z) + z\rho_{n+1}(z)\tilde{B}_n(z)}{A_n(z) - z\rho_{n+1}(z)\tilde{A}_n(z)}, \qquad (3.39)$$

where

$$A_n(z) \overset{\triangle}{=} \frac{g_n(z) - h_n(z)}{\sqrt{R_0}} \qquad (3.40)$$

and

$$B_n(z) \overset{\triangle}{=} \sqrt{R_0} \, \frac{g_n(z) + h_n(z)}{2} \tag{3.41}$$

are known as the Levinson polynomials of the first and second kind, respectively [35]. Using (3.35), a direct calculation also shows that

$$A_n(z)B_{n*}(z) + A_{n*}(z)B_n(z) \equiv 1 . \tag{3.42}$$

Notice that $A_n(z)$ and $B_n(z)$ are always free of zeros in $|z| < 1$ since they represent the denominator and numerator polynomials of a positive function. Also from (3.35), $|g_n(e^{j\theta})| > |h_n(e^{j\theta})|$ and hence both $A_n(z)$ and $B_n(z)$ are strict Hurwitz polynomials. Similarly $A_n(z) - z\rho_{n+1}(z)\tilde{A}_n(z)$ is also free of zeros in $|z| < 1$, since it represents the denominator of a positive function.

With the help of (3.40)–(3.41), the positive function in (3.38) can be written as

$$Z_0(z) = \frac{2B_n(z)}{A_n(z)} \tag{3.43}$$

and together with (3.39)–(3.42), a direct calculation yields

$$Z(z) - Z_0(z) = Z(z) - \frac{2B_n(z)}{A_n(z)} = \frac{2z^{n+1}\rho_{n+1}(z)}{\left(A_n(z) - z\rho_{n+1}(z)\tilde{A}_n(z)\right)A_n(z)}$$

$$= 2 \sum_{k=n+1}^{\infty} \tilde{c}_k z^k , \quad |z| < 1; \tag{3.44}$$

i.e., the power series expansions about $z = 0$ of the positive functions $Z(z)$ and $2B_n(z)/A_n(z)$ agree for the first $(n+1)$ terms. Since $Z(z)$ contains an arbitrary bounded function $\rho_{n+1}(z)$, clearly that function does not affect the first $(n+1)$ coefficients in the expansion of (3.24). Equivalently, for any bounded $\rho_{n+1}(z)$, the first $(n + 1)$ coefficients of the driving-point positive function

$$Z(z) = R_0 \left(\frac{1 + \rho(z)}{1 - \rho(z)} \right) = r_0 + 2 \sum_{k=1}^{\infty} r_k z^k$$

match with those of $Z_0(z) = 2B_n(z)/A_n(z)$, and hence these coefficients are solely determined by $A_n(z)$ and $B_n(z)$. From (3.36)–(3.37) and (3.40)–(3.41) these functions only depend upon $s_1, s_2, \ldots, s_n$ (or equivalently R_0, $R_1, \ldots, R_n$), and the one-to-one relationship between the sets $\{r_k\}_{k=0}^{n}$ and $\{s_k\}_{k=1}^{n}$ that follows (see (3.69)) will complete this analysis regarding the class of all admissible extensions. To understand this from the transmission line point of view developed in [12], let the terminated structure in Fig. 3.2 be excited by an input current source $i(t) = \delta(t)$. This current impulse immediately launches a voltage impulse $r_0\delta(t)$ into the first line which reaches the junction separating the first and second line after τ seconds. At this point, part of the incident impulse is reflected back to the driving point and part of it is transmitted towards the next line. In the same way after τ more seconds, the transmitted part reaches the next junction where once again it is partly reflected and partly transmitted. This process continues till the transmitted part reaches the terminated passive load $Z_{n+1}(z)$. Interaction with this passive load can commence only after the lapse of $(n+1)\tau$ seconds, and even then it takes an additional $(n+1)\tau$ seconds for the returns from the load to reach the input. Thus the input voltage response $v(t)$ over the interval $0 \le t \le 2(n+1)\tau$ is determined entirely by the $(n+1)$ line cascade. Moreover, since 2τ represents the 2-way round trip delay common to all lines, as in (2.186) under the identification $z = e^{-2p\tau}$, it follows that the first $(n+1)$ coefficients of the driving-point impedance in the expansion (3.39) are determined by the characteristic impedances $R_0, R_1, \ldots, R_n$ of the first $(n+1)$ lines and *not* by the termination $Z_{r+1}(z)$ or equivalently $\rho_{r+1}(z)$. As Youla has shown [12], with the help of the even part condition in (3.42), the 'even part' of $Z(z)$ in (3.39) takes the simple form

$$\frac{Z(z) + Z_*(z)}{2} = \frac{1 - \rho_{n+1}(z)\rho_{n+1*}(z)}{\left(A_n(z) - z\rho_{n+1}(z)\tilde{A}_n(z)\right)\left(A_n(z) - z\rho_{n+1}(z)\tilde{A}_n(z)\right)_*}$$

$$= \frac{1 - \rho_{n+1}(z)\rho_{n+1*}(z)}{D_n(z)D_{n*}(z)}, \tag{3.45}$$

where

$$D_n(z) \triangleq A_n(z) - z\rho_{n+1}(z)\tilde{A}_n(z). \tag{3.46}$$

From (2.139), since the real part of every positive function on the unit circle corresponds to a power spectral density function, we have[4]

$$S(\theta) = Re\left(Z(e^{j\theta})\right) = \frac{1 - |\rho_{n+1}(e^{j\theta})|^2}{|D_n(e^{j\theta})|^2} \ge 0. \tag{3.47}$$

From (3.24)–(3.39), (3.45) and (2.139), $S(\theta)$ is parametrized by the bounded function $\rho_{n+1}(z)$ and for *every* such bounded function,

$$S(\theta) = Re\left(Z(e^{j\theta})\right) = \sum_{k=-n}^{n} r_k e^{jk\theta} + O(e^{j(n+1)\theta}).$$

Thus, (3.47) represents the class of all spectral extensions that interpolate the given autocorrelation sequence $r_0, r_1, \ldots, r_n$. Using (3.36)–(3.37) in (3.40), for the Levinson polynomial of the first kind we obtain the recursion

$$\sqrt{1 - |s_k|^2}\, A_k(z) = A_{k-1}(z) - zs_k\tilde{A}_{k-1}(z), \quad k = 1 \to \infty, \tag{3.48}$$

that begins under the initialization

$$A_0(z) = 1/\sqrt{r_0} \quad \text{and} \quad s_1 = r_1/r_0. \tag{3.49}$$

---

[4]On the unit circle, $[Z(z) + Z_*(z)]/2 = [Z(z) + Z^*(z)]/2 = Re\, Z(z)$.

Similarly from (3.41) for the Levinson polynomial of the second kind we also have the recursion

$$\sqrt{1 - |s_k|^2}\, B_k(z) = B_{k-1}(z) + z s_k \tilde{B}_{k-1}(z)\,, \quad k = 1 \to \infty\,, \tag{3.50}$$

that begins with $B_0(z) = \sqrt{r_0}/2$. Since $|s_k| < 1$ for all k (as pointed out before) it also follows that $A_k(z)$ (and $B_k(z)$), $k = 0, 1, \ldots, \infty$, are strict Hurwitz polynomials and this together with $|\rho_{n+1}(z)| \le 1$ in $|z| \le 1$ allows us to conclude from (3.46) that $D_k(z)$, $k = 0, 1, \ldots, \infty$, are analytic and free of zeros in $|z| < 1$. Clearly for (3.3) to hold, from (3.47) $\rho_{n+1}(z)$ must satisfy the causality criterion [3–5]

$$\frac{1}{2\pi} \int_{-\pi}^{\pi} ln\left(1 - |\rho_{n+1}(e^{j\theta})|^2\right) d\theta > -\infty \tag{3.51}$$

and in that case let $\Gamma_{n+1}(z)$ represent the solution of the spectral factorization

$$1 - \left|\rho_{n+1}(e^{j\theta})\right|^2 = \left|\Gamma_{n+1}(e^{j\theta})\right|^2 \tag{3.52}$$

that is analytic and free of zeros in $|z| < 1$. Hence

$$S(\theta) = \left|B_\rho(e^{j\theta})\right|^2 \tag{3.53}$$

and up to an arbitrary phase factor

$$B_\rho(z) = \frac{\Gamma_{n+1}(z)}{A_n(z) - z\rho_{n+1}(z)\tilde{A}_n(z)} = \frac{\Gamma_{n+1}(z)}{D_n(z)} \tag{3.54}$$

represents the Wiener factor associated with $S(\theta)$ in (3.47). By construction, $B_\rho(z)$ is analytic together with its inverse in $|z| < 1$. Notice that subject to (3.51) $\rho_{n+1}(z)$ can be chosen in a completely arbitrary manner. For example, the function $\rho_1(z) = e^{-(1-z)}$ is b.r., since in the closed unit circle $z = re^{j\theta}$, $r \le 1$, we have $|\rho_1(z)| = e^{-(1-r\cos\theta)} \le 1$. Further, since $\rho_1(z)$ also satisfies (3.51), it will give rise to an admissible nonrational extension. As remarked in section 2.2.1, neither the b.r. character of $\rho_{n+1}(z)$ nor the causality requirement in (3.51) preclude logarithmic or essential singularities on $|z| = 1$. (Note that $\rho_{n+1}(z)$ bounded implies that it is always free of poles in $|z| \le 1$.) For example, $\rho_2(z) = \frac{1}{2}(z+1) ln(z+1)$ has a logarithmic singularity at $z = -1$ and $\rho_3(z) = \frac{1}{2}(z-1)e^{1/(z-1)}$ has an essential singularity at $z = 1$. Clearly, $\rho_2(z)$ is analytic in $|z| < 1$ and since $1/(z-1)$ has a negative real part in $|z| < 1$, $\rho_3(z)$ is also analytic in $|z| < 1$. Moreover, both functions are bounded by unity in $|z| < 1$, continuous on $|z| \le 1$ (see also (2.53)–(2.54)) and satisfy (3.51). Hence they both give rise to admissible nonrational extensions in (3.54) with square summable impulse response. In particular, $B_\rho(z)$ is rational iff $\rho_{n+1}(z)$ is rational. In that case $\Gamma_{n+1}(z)$ is also rational and the factorization in (3.52) can be more compactly written as

$$1 - \rho_{n+1}(z)\rho_{n+1*}(z) = \Gamma_{n+1}(z)\Gamma_{n+1*}(z)\,, \tag{3.55}$$

where $\Gamma_{n+1}(z)$ is the unique factor that is analytic in $|z| \leq 1$ and free of zeros in $|z| < 1$. Referring to (3.52) and the above discussion, it follows that when $\rho_{n+1}(z)$ is rational, the numerator and denominator polynomials in (3.54) are always Hurwitz and their common factors, if any, must be in $|z| \geq 1$. However, as shown below, in the rational case, the numerator and denominator polynomials in the Wiener factor $B_\rho(z)$ are always relatively prime except possibly on the boundary of the unit circle. If not, let us assume $z = z_0$ is a common zero of $\Gamma_{n+1}(z)$ and $D_n(z)$ in (3.54). Then $|z_0| \geq 1$, since $D_n(z)$ and $\Gamma_{n+1}(z)$ are Hurwitz polynomials. From (3.46), this gives $D_n(z_0) = 0$, or

$$\frac{A_n(z_0)}{\tilde{A}_n(z_0)} = z_0\, \rho_{n+1}(z_0). \tag{3.56}$$

Also from (3.55), we obtain $\rho_{n+1}(z_0)\rho_{n+1}^*(1/z_0^*) = 1$, and using this, (3.56) reduces to

$$\rho_{n+1}^*(1/z_0^*) \frac{A_n(z_0)}{\tilde{A}_n(z_0)} = z_0. \tag{3.57}$$

Let us first consider the case $|z_0| > 1$. Then the boundedness of $\rho_{n+1}(z)$ gives $|\rho_{n+1}^*(1/z_0^*)| \leq 1$. Further since the ratio $A_n(z)/\tilde{A}_n(z)$ is analytic in $|z| > 1$ and has unit magnitude on $|z| = 1$, from the maximum modulus theorem [26], $|A_n(z)/\tilde{A}_n(z)| \leq 1$ in $|z| > 1$. Referring back to (3.57), this leads to a contradiction since the left-hand side

$$\left| \rho^*(1/z_0^*) \frac{A_n(z_0)}{\tilde{A}_n(z_0)} \right| \leq 1$$

and the right-hand side $|z_0| > 1$. This proves the relative primeness in $|z| > 1$ of the numerator and denominator polynomials in (3.54) for *every* rational function $\rho_{n+1}(z)$. With regard to the remaining case of common zeros on the unit circle ($z_0 = e^{j\theta_0}$), from (3.46), we have

$$D_n(z) = A_n(z) \left(1 - z\rho_{n+1}(z) \frac{\tilde{A}_n(z)}{A_n(z)} \right).$$

Since $\tilde{A}_n(z)/A_n(z)$ is bounded, $z\rho_{n+1}(z)\tilde{A}_n(z)/A_n(z)$ is also bounded and, hence, $(1 - z\rho_{n+1}(z)\tilde{A}_n(z)/A_n(z))$ is a positive function (see Problem 7(a), chapter 2). But from the discussion above (2.133), the zeros and poles of a positive function on $|z| = 1$ are simple, and thus every zero of $(1 - z\rho_{n+1}(z)\tilde{A}_n(z)/A_n(z))$ as well as that of $D_n(z)$ on $|z| = 1$ is simple. In that case, from (3.56), $|\rho_{n+1}(e^{j\theta_0})| = 1$ and using (3.52), (3.55), we obtain that $z = e^{j\theta_0}$ is also a zero of $\Gamma(z)$ of *at least* order one. Hence in the rational case, the ratio $B_\rho(z)$ is always analytic in $|z| \leq 1$ and except possibly by common (simple) zeros on the unit circle, there is no degree reduction between the numerator and denominator polynomials in (3.54).

To summarize, in the real case consider the b.r. function $\rho_{n+1}(z) = 1/(z-2)^2$. Then $D_n(1) = 0$ for any real $A_n(z)$. From the above discussion, such zeros of $D_n(z)$ on the unit circle will be simple and, moreover, they will be also zeros of $\Gamma_{n+1}(z)$. In fact, here $\Gamma_{n+1}(z) = 2(z-1)(a-bz)/(z-2)^2$, where $a = (\sqrt{5}+1)/2$ and $b = (\sqrt{5}-1)/2$. Thus, the Wiener factor $B_\rho(z)$ is free of poles in $|z| \leq 1$, and degree reduction between the numerator and denominator parts in (3.54) can happen only because of possible simple zeros on the unit circle. Incidentally, it may be remarked that this latter possibility does not arise when the numerator zeros in (3.54) are guaranteed to be outside the unit circle.

The maximum entropy extension can be identified easily by considering the expression for the entropy functional $\mathcal{H}_\rho$ associated with an otherwise arbitrary extension in (3.53). From (3.19)–(3.20), since for any process $\mathcal{H} = ln\,|B(0)|^2$, we have from (3.53)

$$\mathcal{H}_\rho = \frac{1}{2\pi}\int_{-\pi}^{\pi} ln\,S(\theta)d\theta = \frac{1}{2\pi}\int_{-\pi}^{\pi} ln\,|B_\rho(e^{j\theta})|^2 d\theta = ln\,|B_\rho(0)|^2$$

$$= ln\,|\Gamma_{n+1}(0)|^2 - ln\,|A_n(0)|^2$$

$$= ln\left(|1/A_n(0)|^2\right) - ln\left(|1/\Gamma_{n+1}(0)|^2\right). \tag{3.58}$$

Since $A_n(0)$ does not depend on $\rho_{n+1}(z)$ and since $|\Gamma_{n+1}(0)| \leq 1$, clearly the maximum value for H_ρ is achieved by selecting $\Gamma_{n+1}(0) = 1$. In that case, from the maximum modulus theorem [26], $\Gamma_{n+1}(z) \equiv 1$ and from (3.52), $\rho_{n+1}(z) \equiv 0$ which gives $d_{n+1}(z) \equiv 0$. Thus the maximum entropy extension [38, 39] corresponds to

$$B_{ME}(z) = \frac{1}{A_n(z)} \tag{3.59}$$

$$S_{ME}(\theta) = \frac{1}{|A_n(e^{j\theta})|^2} = \sum_{k=-n}^{n} r_k e^{jk\theta} + O(e^{j(n+1)\theta}) \tag{3.60}$$

and this agrees with the all-pole model in (3.23). Since $d_{n+1}(z)$ represents the reflection coefficient of the termination normalized with respect to R_n, we also have (see (2.192))

$$Z_{n+1}(z) = \left.\frac{R_n + R_n^* d_{n+1}(z)}{1 - d_{n+1}(z)}\right|_{d_{n+1}(z)\equiv 0} = R_n.$$

Moreover, $s_{n+1} = \rho_{n+1}(0) = 0$ implying that the terminating positive function $Z_{n+1}(z) = R_n$ is perfectly matched to the characteristic impedance R_n of the last line and hence there is no reflected wave from the termination (Fig. 3.2).

Referring back to the general case, an examination of (3.46)–(3.54) shows that the functions $g_n(z)$ and $h_n(z)$, containing the given information $\{r_k\}_{k=0}^n$, enter the final formulation for the power spectral density and its Wiener factor only through the strict Hurwitz polynomial $A_n(z)$. As (3.48)–(3.49) show, these Levinson polynomials of the first kind can be recursively generated, and to carry out that recursion it is necessary to exhibit the relationship between $\{r_k\}_{k=0}^n$ and the reflection coefficients $\{s_k\}_{k=1}^n$.

3.3.1 LEVINSON POLYNOMIALS AND REFLECTION COEFFICIENTS

From (3.40) and footnote 3, the Levinson polynomial $A_n(z)$ is of degree n and let[5]

$$A_n(z) = a_0^{(n)} + a_1^{(n)} z + \cdots + a_n^{(n)} z^n$$
$$= a_0 + a_1 z + a_2 z^2 + \cdots + a_n z^n, \tag{3.61}$$

where the coefficients depend on the given information $\{r_k\}_{k=0}^n$. To see their exact relationship, consider the maximum entropy extension described above for which $\rho_{n+1}(z) \equiv 0$. On comparing (3.59)–(3.60) with (3.47) and the discussions there, the maximum entropy extension as well as $A_n(z)$ are uniquely determined by $r_0 \to r_n$. From (3.39), the corresponding input positive function $Z_{ME}(z)$ has the form (see also (3.43))

$$Z_{ME}(z) = \frac{2B_n(z)}{A_n(z)}. \tag{3.62}$$

Using (3.45) or directly, this gives

$$\frac{Z_{ME}(z) + Z_{ME*}(z)}{2} = \frac{1}{A_n(z)A_{n*}(z)}. \tag{3.63}$$

However,

$$Z_{ME}(z) = r_0 + 2\sum_{k=1}^n r_k z^k + O(z^{n+1})$$

and together with (3.61), (3.63) simplifies to

$$\left(r_0 + \sum_{k=1}^\infty r_k^* z^{-k} + \sum_{k=1}^\infty r_k z^k\right)(a_n^* + a_{n-1}^* z + \cdots + a_0^* z^n) = \frac{z^n}{A_n(z)}. \tag{3.64}$$

Since $A_n(z) \neq 0$ in $|z| \leq 1$, $1/A_n(z)$ admits a power series expansion

$$\frac{1}{A_n(z)} = b_0 + b_1 z + b_2 z^2 + \cdots ; \quad b_0 = \frac{1}{a_0} \tag{3.65}$$

---

[5]Whenever there is no confusion, we will simply refer to $A_n(z)$ as the Levinson polynomial and use the later form in (3.61) without superscripts for its representation.

with radius of convergence greater than unity. Thus, by comparing coefficients of $1, z, z^2, \ldots, z^n$ on both sides of (3.64), we obtain

$$
\begin{bmatrix}
r_0 & r_1 & \cdots & r_n \\
r_1^* & r_0 & \cdots & r_{n-1} \\
\vdots & \vdots & \ddots & \vdots \\
r_n^* & r_{n-1}^* & \cdots & r_0
\end{bmatrix}
\begin{bmatrix}
a_n \\
a_{n-1} \\
\vdots \\
a_0
\end{bmatrix}
=
\begin{bmatrix}
0 \\
0 \\
\vdots \\
b_0
\end{bmatrix}.
\tag{3.66}
$$

However,

$$
A_n(z) = \begin{bmatrix} z^n & z^{n-1} & \cdots & z & 1 \end{bmatrix}
\begin{bmatrix}
a_n \\
a_{n-1} \\
\vdots \\
a_1 \\
a_0
\end{bmatrix}
= b_0 \begin{bmatrix} z^n & \cdots & z & 1 \end{bmatrix} \mathbf{T}_n^{-1}
\begin{bmatrix}
0 \\
0 \\
\vdots \\
0 \\
1
\end{bmatrix}
$$

$$
= \frac{-1}{a_0 \Delta_n}
\left|
\begin{array}{ccc|c}
 & & & 0 \\
 & \mathbf{T}_n & & \vdots \\
 & & & 1 \\
\hline
z^n & \cdots & 1 & 0
\end{array}
\right|
$$

$$
= \frac{1}{a_0 \Delta_n}
\left|
\begin{array}{ccccc}
r_0 & r_1 & \cdots & r_{n-1} & r_n \\
r_1^* & r_0 & \cdots & r_{n-2} & r_{n-1} \\
\vdots & \vdots & \ddots & \vdots & \vdots \\
r_{n-1}^* & r_{n-2}^* & \cdots & r_0 & r_1 \\
z^n & z^{n-1} & \cdots & z & 1
\end{array}
\right|.
$$

By applying Cramer's rule on (3.66) to solve for a_0 with the help of (3.65), we get

$$
a_0 = \sqrt{\frac{\Delta_{n-1}}{\Delta_n}}
$$

which finally gives the compact expression

$$
A_n(z) = \frac{1}{\sqrt{\Delta_n \Delta_{n-1}}}
\left|
\begin{array}{ccccc}
r_0 & r_1 & \cdots & r_{n-1} & r_n \\
r_1^* & r_0 & \cdots & r_{n-2} & r_{n-1} \\
\vdots & \vdots & \ddots & \vdots & \vdots \\
r_{n-1}^* & r_{n-2}^* & \cdots & r_0 & r_1 \\
z^n & z^{n-1} & \cdots & z & 1
\end{array}
\right|.
\tag{3.67}
$$

It may be remarked that an easy determinantal expansion (using footnote 9, chapter 2) on (3.67) also gives the recursion rule (3.48), and from there we can identify the reflection coefficient s_k as

$$s_k = \frac{(-1)^{k-1}}{\Delta_{k-1}} \begin{vmatrix} r_1 & r_2 & \cdots & r_{k-1} & r_k \\ r_0 & r_1 & \cdots & r_{k-2} & r_{k-1} \\ r_1^* & r_0 & \cdots & r_{k-3} & r_{k-2} \\ \vdots & \vdots & \cdots & \vdots & \vdots \\ r_{k-2}^* & r_{k-3}^* & \cdots & r_0 & r_1 \end{vmatrix} \triangleq (-1)^{k-1} \frac{\Delta_k^{(1)}}{\Delta_{k-1}}.$$

(3.68)

As in (2.88), $\Delta_k^{(1)}$ here represents the minor of Δ_k obtained by deleting its first column and last row. Alternatively, a direct expansion on (3.68) shows that

$$s_k = \frac{1}{\Delta_{k-1}} \begin{vmatrix} r_{k-1} & \cdots & r_1 & r_k \\ r_{k-2} & \cdots & r_0 & r_{k-1} \\ \vdots & \cdots & \vdots & \vdots \\ r_0 & \cdots & r_{k-2}^* & r_1 \end{vmatrix}$$

$$= \left(\sum_{i=1}^{k} r_i a_{k-i}^{(k-1)} \right) a_0^{(k-1)} = \left\{ A_{k-1}(z) \sum_{i=1}^{k} r_i z^i \right\}_k A_{k-1}(0), \quad (3.69)$$

where $a_i^{(k-1)}$ are as defined in (3.61) and $\{\ \}_k$ denotes the coefficients of z^k in $\{\ \}_k$. Since $A_n(0) = \sqrt{\Delta_{n-1}/\Delta_n}$, at times it is convenient to normalize the Levinson polynomials of the first kind by defining

$$a_n(z) \triangleq \sqrt{\frac{\Delta_n}{\Delta_{n-1}}} A_n(z) = \frac{1}{\Delta_{n-1}} \begin{vmatrix} r_0 & r_1 & \cdots & r_{n-1} & r_n \\ r_1^* & r_0 & \cdots & r_{n-2} & r_{n-1} \\ \vdots & \vdots & \cdots & \vdots & \vdots \\ r_{n-1}^* & r_{n-2}^* & \cdots & r_0 & r_1 \\ z^n & z^{n-1} & \cdots & z & 1 \end{vmatrix}$$

(3.70)

so that $a_n(0) = 1$. Clearly its recursion is given by

$$a_n(z) = a_{n-1}(z) - z s_n \tilde{a}_{n-1}(z), \quad n \geq 1, \quad (3.71)$$

with $a_0(z) = 1$. For uniformity, let

$$b_n(z) \triangleq \sqrt{\frac{\Delta_n}{\Delta_{n-1}}} B_n(z) \quad (3.72)$$

so that

$$b_n(z) = b_{n-1}(z) + z s_n \tilde{b}_{n-1}(z), \quad n \geq 1, \quad (3.73)$$

with $b_0(z) = r_0/2$. Notice that expressions (3.48)–(3.49) together with (3.69) can be easily implemented and together they constitute the Levinson recursion algorithm for the strict Hurwitz polynomials $A_k(z)$, $k = 1 \rightarrow n$, generated from the autocorrelations r_0, r_1, ..., r_n. Using (3.10) as well as (2.193)–(2.201), other equivalent forms for the reflection coefficients include

$$s_k = \frac{(-1)^{k-1}|\mathbf{M}_{k-1}(r_k)|}{\Delta_{k-1}} = S_k e^{-j2\phi_{k-1}},\tag{3.74}$$

where

$$S_k = \frac{R_k - R_{k-1}}{R_k + R_{k-1}^*}$$

and hence the characteristic impedances can be recursively expressed as

$$R_k = \frac{R_{k-1} + R_{k-1}^* S_k}{1 - S_k}, \quad k \geq 1,\tag{3.75}$$

starting with $R_0 = Z(0) = r_0$. From (3.17) and (3.74), we also have

$$\gamma_{k+1} = \frac{\Delta_{k+1}}{\Delta_k} = \gamma_k\left(1 - |s_{k+1}|^2\right) = r_0 \prod_{i=1}^{k+1}\left(1 - |s_i|^2\right), \quad k \geq 0,\tag{3.76}$$

and hence for every $k \geq 1$

$$\mathbf{T}_k > 0 \iff \Delta_i > 0, \; i = 0 \rightarrow k \iff |s_i| < 1, \; i = 1 \rightarrow k.\tag{3.77}$$

Clearly for any k_0 if $\Delta_{k_0-1} > 0$ and $\Delta_{k_0} = 0$, then from (3.76) $|s_{k_0}| = 1$. Thus

$$\Delta_{k_0-1} > 0 \text{ and } \Delta_{k_0} = 0 \iff |s_{k_0}| = 1.\tag{3.78}$$

Moreover, since

$$\gamma_\infty = \lim_{k \to \infty} \gamma_k = r_0 \prod_{i=1}^{\infty}\left(1 - |s_i|^2\right),\tag{3.79}$$

when the causality criterion (3.3) is satisfied, we have from (3.19), $\gamma_\infty = |b_0|^2 > 0$ and from the above expression every $|s_k| < 1$. Since $r_0 = \sum_{k=0}^{\infty}|b_k|^2$, from (3.79) we also obtain

$$\prod_{k=1}^{\infty}\left(1 - |s_k|^2\right) = \frac{|b_0|^2}{r_0} < 1.$$

From a classical result, the product $\prod_{k=1}^{\infty}(1-|s_k|^2)$ and the sum $\sum_{k=1}^{\infty}|s_k|^2$ converge or diverge together [40] and hence

$$\sum_{k=1}^{\infty}|s_k|^2 < \infty.$$

To summarize,

$$\int_{-\pi}^{\pi} \ln S(\theta)d\theta > -\infty \implies |s_k| < 1; \; \sum_{k=1}^{\infty}|s_k|^2 < \infty,\tag{3.80}$$

i.e., the causality criterion (3.3) translates into the square summability of the reflection coefficients, all of which are strictly bounded by unity. (For the converse, see Problem 3.2.) As remarked earlier (footnote 5, chapter 2), $S(\theta)$ that satisfies (3.80) can possess isolated poles on the unit circle and hence square summability of the reflection coefficients alone does not preclude poles on the unit circle and $S(\theta)$ need not be integrable (see Problem 3.7 for an example).

Alternatively, from (3.22) and (3.74) we also conclude that for the maximum entropy extension

$$s_{n+k} = 0, \quad k \geq 1, \tag{3.81}$$

and on comparing (2.115) and (3.68), this is seen to be true for every $\mathrm{AR}(n)$ process. Thus

$$x(nT) \sim \mathrm{AR}(n) \quad \Longleftrightarrow \quad s_{n+k} = 0, \quad k \geq 1. \tag{3.82}$$

As Geronimus has shown [41], the strict Hurwitz character of $A_n(z)$ can be exhibited more explicitly by starting with (3.48) and rewriting it as

$$A_n(z) = \left(\frac{1 - z s_n \tilde{A}_{n-1}(z)/A_{n-1}(z)}{\sqrt{1 - |s_n|^2}} \right) A_{n-1}(z)$$

$$= \frac{1}{\sqrt{r_0}} \prod_{k=1}^{n} \frac{1 - z s_k \tilde{A}_{k-1}(z)/A_{k-1}(z)}{\sqrt{1 - |s_k|^2}}.$$

Since $\tilde{A}_{k-1}(z)/A_{k-1}(z)$ is a regular all-pass function, we have

$$|\tilde{A}_{k-1}(z)/A_{k-1}(z)| \leq 1 \quad \text{in } |z| \leq 1$$

and hence

$$1 - |s_k| \leq |1 - z s_k \tilde{A}_{k-1}(z)/A_{k-1}(z)| \leq 1 + |s_k|$$

or in $|z| \leq 1$

$$\frac{1}{\sqrt{r_0}} \prod_{k=1}^{n} \sqrt{\frac{1 - |s_k|}{1 + |s_k|}} \leq |A_n(z)| \leq \frac{1}{\sqrt{r_0}} \prod_{k=1}^{n} \sqrt{\frac{1 + |s_k|}{1 - |s_k|}}. \tag{3.83}$$

Clearly, these upper and lower bounds for the Levinson polynomials $A_n(z)$ in $|z| \leq 1$ show that so long as $|s_k| < 1$, $k = 1 \rightarrow n$, $A_n(z)$ have no zeros in $|z| \leq 1$ and they are well behaved polynomials. From (3.80) such is the case under the causality criterion. Thus, in particular

$$\Delta_k > 0, \ k = 0 \rightarrow n \quad \Longrightarrow \quad A_k(z) \neq 0 \text{ in } |z| \leq 1, \ k = 0 \rightarrow n. \tag{3.84}$$

In general, since $\mathbf{T}_k \geq 0$, if at any stage Δ_{k_0} is singular, then there must exist an $n < k_0$ such that $\Delta_n > 0$ and $\Delta_{n+1} = 0$ and from (3.77)–(3.78),

$|s_k| < 1$, $k = 1 \to n$, and $s_{n+1} = e^{j\theta_0}$. Hence $A_n(z)$ and $a_n(z)$ are free of zeros in $|z| \le 1$, and let

$$a_n(e^{j\theta}) = R(\theta)e^{j\phi(\theta)}, \quad R(\theta) > 0.$$

Then from (3.71)

$$a_{n+1}(e^{j\theta}) = a_n(e^{j\theta}) - e^{j(\theta-\theta_0)}\tilde{a}_n(e^{j\theta})$$

$$= R(\theta)e^{j\phi(\theta)} - e^{j((n+1)\theta+\theta_0)}R(\theta)e^{-j\phi(\theta)}$$

$$= R(\theta)e^{j((n+1)\theta+\theta_0)/2}\sin\left(\phi(\theta) - \frac{(n+1)\theta}{2} - \frac{\theta_0}{2}\right). \quad (3.85)$$

Due to the strict Hurwitz nature of $a_n(z)$, as θ varies from 0 to 2π completing one revolution, there is no net increment in $\phi(\theta)$ [35], and the entire argument of the sine term in (3.85) increases by $(n+1)\pi$. Consequently $a_{n+1}(e^{j\theta})$ equals zero at least at $n+1$ distinct points θ_1, θ_2, ..., θ_{n+1}, $0 \le \theta_i < 2\pi$. However $a_{n+1}(z)$ is a polynomial of degree $n+1$ and can have at most $n+1$ zeros. Thus all the above $n+1$ zeros are simple. Arguing in a similar manner, from (3.73) it follows that all $n+1$ zeros of $b_n(z)$ are also simple and are located on the unit circle. From (3.39), (3.70) and (3.72), the input positive function is also given by

$$Z(z) = \frac{2\big(b_n(z) + z\rho_{n+1}(z)\tilde{b}_n(z)\big)}{a_n(z) - z\rho_{n+1}(z)\tilde{a}_n(z)}$$

and for the situation described above, since $s_{n+1} = \rho_{n+1}(0) = e^{j\theta_0}$, from the maximum modulus theorem [26], $\rho_{n+1}(z) = e^{j\theta_0}$ and the above expression reads

$$Z(z) = \frac{2\big(b_n(z) + zs_{n+1}\tilde{b}_n(z)\big)}{a_n(z) - zs_{n+1}\tilde{a}_n(z)}$$

$$= \frac{2b_{n+1}(z)}{a_{n+1}(z)}. \quad (3.86)$$

Thus the positive function has all its poles and zeros on the unit circle, and moreover from (3.42), (3.70), (3.72), in this case

$$\frac{Z(z) + Z_*(z)}{2} = \frac{a_{n+1}(z)b_{n+1*}(z) + a_{n+1*}(z)b_{n+1}(z)}{a_{n+1}(z)a_{n+1*}(z)}$$

$$= \frac{\Delta_{n+1}/\Delta_n}{a_{n+1}(z)a_{n+1*}(z)} = 0,$$

since $\Delta_{n+1} = 0$. Referring back to (2.166), we conclude that the positive function $Z(z)$ in (3.86) represents a Foster function with $n+1$ simple poles at $e^{j\theta_1}$, $e^{j\theta_2}$, ..., $e^{j\theta_{n+1}}$. Following the analysis in (2.171)–(2.175), we obtain

$$r_k = \sum_{m=1}^{n+1} P_m e^{-jk\theta_m}, \quad k \geq 0, \tag{3.87}$$

and

$$Z(z) = \sum_{m=1}^{n+1} P_m \frac{1 + ze^{-j\theta_m}}{1 - ze^{-j\theta_m}}, \quad P_m > 0. \tag{3.88}$$

In this case $\Delta_{n+k} = 0$ for every $k \geq 1$ and the corresponding power spectral density function is given by

$$S(\theta) = \sum_{k=-\infty}^{\infty} r_k e^{jk\theta} = \sum_{m=1}^{n+1} P_m \delta(\theta - \theta_m). \tag{3.89}$$

The above analysis shows that

$$\Delta_k > 0, \; k = 0 \to n, \; \Delta_{n+1} = 0 \iff r_k = \sum_{m=1}^{n+1} P_m e^{-jk\theta_m}, \; k \geq 0, \tag{3.90}$$

and

$$S(\theta) = \sum_{m=1}^{n+1} P_m \delta(\theta - \theta_m), \quad \Delta_{n+k} = 0, \; k \geq 1. \tag{3.91}$$

Conversely if the positive function $Z(z) = r_0 + 2 \sum_{k=1}^{\infty} r_k z^k$ is Foster, then from (2.176) for some $n \geq 0$, $Z(z)$ must have the form in (3.88) with

$$r_k = \sum_{m=1}^{n+1} P_m e^{-jk\theta_m}, \quad k \geq 0, \tag{3.92}$$

where $P_m > 0$ and from (2.177), $\Delta_k > 0$, $k = 0 \to n$, and $\Delta_{n+k} = 0$, $k \geq 1$. Thus, as pointed out before, Foster functions and discrete spectra have a one-to-one correspondence.

Toeplitz Inverses

Consider the Hermitian Toeplitz matrix $\mathbf{T}_n > 0$ generated from $r_0 \to r_n$. As Gohberg and Semuncul have shown explicitly [42, 43], it is possible to express the inverse of such Toeplitz matrices in terms of two lower triangular Toeplitz matrices generated from the coefficients of the Levinson polynomial $A_n(z)$.

To see this, it is best to make use of (3.42) together with (3.43). From (3.43), we have

$$2B_n^*(z) = A_n^*(z)Z_0^*(z) = Z_0^*(z)A_n^*(z). \tag{3.93}$$

Since $Z_0(z) = r_0 + 2\sum_{k=1}^{n} r_k z^k + O(z^{n+1})$, comparing coefficients of like powers of z^* on both sides of (3.93) and rearranging them in matrix form, we obtain

$$
2\begin{bmatrix} b_0^* & 0 & \cdots & 0 \\ b_1^* & b_0^* & \cdots & 0 \\ \vdots & \vdots & \cdots & \vdots \\ b_n^* & b_{n-1}^* & \cdots & b_0^* \end{bmatrix} = \begin{bmatrix} a_0^* & 0 & \cdots & 0 \\ a_1^* & a_0^* & \cdots & 0 \\ \vdots & \vdots & \cdots & \vdots \\ a_n^* & a_{n-1}^* & \cdots & a_0^* \end{bmatrix} \begin{bmatrix} r_0 & 0 & \cdots & 0 \\ 2r_1^* & r_0 & \cdots & 0 \\ \vdots & \vdots & \cdots & \vdots \\ 2r_n^* & 2r_{n-1}^* & \cdots & r_0 \end{bmatrix}.
\tag{3.94}
$$

Define the lower triangular Toeplitz matrices

$$
\mathbf{A}_0 \triangleq \begin{bmatrix} a_0^* & 0 & \cdots & 0 \\ a_1^* & a_0^* & \cdots & 0 \\ \vdots & \vdots & \ddots & \vdots \\ a_n^* & a_{n-1}^* & \cdots & a_0^* \end{bmatrix}
\tag{3.95}
$$

$$
\mathbf{B}_0 \triangleq \begin{bmatrix} b_0^* & 0 & \cdots & 0 \\ b_1^* & b_0^* & \cdots & 0 \\ \vdots & \vdots & \ddots & \vdots \\ b_n^* & b_{n-1}^* & \cdots & b_0^* \end{bmatrix}
\tag{3.96}
$$

and

$$
\mathbf{C}_n = \begin{bmatrix} r_0 & 0 & \cdots & 0 \\ 2r_1^* & r_0 & \cdots & 0 \\ \vdots & \vdots & \ddots & \vdots \\ 2r_n^* & 2r_{n-1}^* & \cdots & r_0 \end{bmatrix}.
\tag{3.97}
$$

Then (3.94) simplifies to

$$
2\mathbf{B}_0 = \mathbf{A}_0\mathbf{C}_n = \mathbf{C}_n\mathbf{A}_0,
\tag{3.98}
$$

where the last equality follows from the last equality in (3.93). Thus

$$
\mathbf{C}_n = 2\mathbf{A}_0^{-1}\mathbf{B}_0 = 2\mathbf{B}_0\mathbf{A}_0^{-1}.
\tag{3.99}
$$

But

$$
\mathbf{T}_n = \begin{bmatrix} r_0 & r_1 & \cdots & r_n \\ r_1^* & r_0 & \cdots & r_{n-1} \\ \vdots & \vdots & \ddots & \vdots \\ r_n^* & r_{n-1}^* & \cdots & r_0 \end{bmatrix} \triangleq \frac{\mathbf{C}_n + \mathbf{C}_n^*}{2}
\tag{3.100}
$$

and using (3.99), we obtain

$$
\mathbf{T}_n = \mathbf{A}_0^{-1}\mathbf{B}_0 + \mathbf{B}_0^*(\mathbf{A}_0^*)^{-1} = \mathbf{A}_0^{-1}\left(\mathbf{B}_0\mathbf{A}_0^* + \mathbf{A}_0\mathbf{B}_0^*\right)(\mathbf{A}_0^*)^{-1}
\tag{3.101}
$$

or equivalently

$$\mathbf{T}_n = \mathbf{B}_0\mathbf{A}_0^{-1} + (\mathbf{A}_0^*)^{-1}\mathbf{B}_0^* = (\mathbf{A}_0^*)^{-1}\Big(\mathbf{A}_0^*\mathbf{B}_0 + \mathbf{B}_0^*\mathbf{A}_0\Big)\mathbf{A}_0^{-1}. \quad (3.102)$$

To make further progress, we can rewrite (3.42) in terms of the reciprocal polynomials $\tilde{A}_n(z)$ and $\tilde{B}_n(z)$ as

$$A_n(z)\tilde{B}_n(z) + \tilde{A}_n(z)B_n(z) = z^n$$

or in a more convenient form

$$A_n(z)\big(z\tilde{B}_n(z)\big) + \big(z\tilde{A}_n(z)\big)B_n(z) = z^{n+1}. \quad (3.103)$$

Comparing the coefficients of z^k, $k = 0 \to 2n + 1$, on both sides of (3.103) as before, their complex conjugates can be expressed compactly in terms of lower triangular Toeplitz matrices as

$$\left[\begin{array}{c|c} \mathbf{A}_0 & \mathbf{0} \\ \hline \mathbf{A}_n^* & \mathbf{A}_0 \end{array}\right]\left[\begin{array}{c|c} \mathbf{B}_n & \mathbf{0} \\ \hline \mathbf{B}_0^* & \mathbf{B}_n \end{array}\right] + \left[\begin{array}{c|c} \mathbf{A}_n & \mathbf{0} \\ \hline \mathbf{A}_0^* & \mathbf{A}_n \end{array}\right]\left[\begin{array}{c|c} \mathbf{B}_0 & \mathbf{0} \\ \hline \mathbf{B}_n^* & \mathbf{B}_0 \end{array}\right] = \left[\begin{array}{c|c} \mathbf{0} & \mathbf{0} \\ \hline \mathbf{I}_{n+1} & \mathbf{0} \end{array}\right],$$

$$(3.104)$$

where $\mathbf{A}_0$ and $\mathbf{B}_0$ are as defined in (3.95)–(3.96),

$$\mathbf{A}_n \triangleq \begin{bmatrix} 0 & 0 & \cdots & 0 & 0 \\ a_n & 0 & \cdots & 0 & 0 \\ a_{n-1} & a_n & \cdots & 0 & 0 \\ \vdots & \vdots & \ddots & \vdots & \vdots \\ a_1 & a_2 & \cdots & a_n & 0 \end{bmatrix} \quad (3.105)$$

and

$$\mathbf{B}_n \triangleq \begin{bmatrix} 0 & 0 & \cdots & 0 & 0 \\ b_n & 0 & \cdots & 0 & 0 \\ b_{n-1} & b_n & \cdots & 0 & 0 \\ \vdots & \vdots & \ddots & \vdots & \vdots \\ b_1 & b_2 & \cdots & b_n & 0 \end{bmatrix}. \quad (3.106)$$

From (3.104), we get

$$\mathbf{A}_0\mathbf{B}_n + \mathbf{A}_n\mathbf{B}_0 = \mathbf{0} \quad (3.107)$$

and

$$\mathbf{A}_n^*\mathbf{B}_n + \mathbf{A}_0\mathbf{B}_0^* + \mathbf{A}_0^*\mathbf{B}_0 + \mathbf{A}_n\mathbf{B}_n^* = \mathbf{I}_{n+1}. \quad (3.108)$$

Once again invoking commutativity of lower triangular Toeplitz matrices, (3.107) gives

$$\mathbf{B}_n = -\mathbf{A}_0^{-1}\mathbf{A}_n\mathbf{B}_0 = -\mathbf{A}_0^{-1}\mathbf{B}_0\mathbf{A}_n = -\mathbf{A}_n\mathbf{B}_0\mathbf{A}_0^{-1} = -\mathbf{B}_0\mathbf{A}_n\mathbf{A}_0^{-1} \quad (3.109)$$

and using this, from (3.108)

$$\begin{aligned} \mathbf{A}_n^*\mathbf{B}_n + \mathbf{A}_0^*\mathbf{B}_0 &= -\mathbf{A}_n^*\mathbf{A}_n\mathbf{B}_0\mathbf{A}_0^{-1} + \mathbf{A}_0^*\mathbf{A}_0\mathbf{A}_0^{-1}\mathbf{B}_0 \\ &= \big(-\mathbf{A}_n^*\mathbf{A}_n + \mathbf{A}_0^*\mathbf{A}_0\big)\mathbf{B}_0\mathbf{A}_0^{-1}, \quad (3.110) \end{aligned}$$

where we have made use of (3.99). Similarly

$$\mathbf{A}_0\mathbf{B}_0^* + \mathbf{A}_n\mathbf{B}_n^* = \mathbf{A}_0\mathbf{A}_0^*(\mathbf{A}_0^*)^{-1}\mathbf{B}_0^* - \mathbf{A}_n\mathbf{A}_n^*\mathbf{B}_0^*(\mathbf{A}_0^*)^{-1}$$
$$= (\mathbf{A}_0\mathbf{A}_0^* - \mathbf{A}_n\mathbf{A}_n^*)(\mathbf{A}_0^*)^{-1}\mathbf{B}_0^*. \qquad (3.111)$$

Substituting (3.110) and (3.111) into (3.108), we have

$$(\mathbf{A}_0^*\mathbf{A}_0 - \mathbf{A}_n^*\mathbf{A}_n)\mathbf{B}_0\mathbf{A}_0^{-1} + (\mathbf{A}_0\mathbf{A}_0^* - \mathbf{A}_n\mathbf{A}_n^*)(\mathbf{A}_0^*)^{-1}\mathbf{B}_0^* = \mathbf{I}_{n+1}. \quad (3.112)$$

But for any two lower triangular Toeplitz matrices $\mathbf{A}_0$ and $\mathbf{A}_n$ as in (3.95), (3.105), by induction it is easy to show that

$$\mathbf{A}_0\mathbf{A}_0^* - \mathbf{A}_n\mathbf{A}_n^* = \mathbf{A}_0^*\mathbf{A}_0 - \mathbf{A}_n^*\mathbf{A}_n \qquad (3.113)$$

and hence (3.112) simplifies to

$$(\mathbf{A}_0\mathbf{A}_0^* - \mathbf{A}_n\mathbf{A}_n^*)(\mathbf{B}_0\mathbf{A}_0^{-1} + (\mathbf{A}_0^*)^{-1}\mathbf{B}_0^*) = \mathbf{I}_{n+1}$$

and using (3.102), this gives

$$(\mathbf{A}_0\mathbf{A}_0^* - \mathbf{A}_n\mathbf{A}_n^*)\mathbf{T}_n = \mathbf{I}_{n+1}$$

or

$$\mathbf{T}_n^{-1} = \mathbf{A}_0\mathbf{A}_0^* - \mathbf{A}_n\mathbf{A}_n^* = \mathbf{A}_0^*\mathbf{A}_0 - \mathbf{A}_n^*\mathbf{A}_n, \qquad (3.114)$$

the desired inverse formula for Hermitian Toeplitz matrices in terms of two lower/upper triangular Toeplitz matrices that involve the coefficients of the Levinson polynomial $A_n(z)$. Equation (3.114) once again exhibits the rich structure present in a Hermitian Toeplitz matrix.

3.3.2   REFLECTION COEFFICIENTS AND CHARACTERISTIC IMPEDANCES

So far we have seen that, when the power spectral density function $S(\theta)$ satisfies the integrability condition and the Paley-Wiener criterion, the reflection coefficients s_k are strictly bounded by unity and square summable (see (3.80)) and consequently $s_k \to 0$ as $k \to \infty$. What happens to the characteristic impedances $R_k = Z_k(0)$ in Fig. 3.2 as $k \to \infty$ under these conditions?

To investigate this in the case of positive-real functions, from (2.193), (2.203)–(2.204), the reflection coefficients and the characteristic impedances are related by

$$s_r = S_r = d_r(0) = \frac{R_r - R_{r-1}}{R_r + R_{r-1}}, \quad r \geq 1, \qquad (3.115)$$

and the above equation can be rewritten as[6]

$$R_r = R_{r-1} \frac{1 + s_r}{1 - s_r} = R_0 \prod_{k=1}^{r} \frac{1 + s_k}{1 - s_k}.$$ (3.116)

Clearly, the sequence $\{R_r\}_{r=0}^{\infty}$ and the product $\prod_{k=1}^{r}(1 + s_k)/(1 - s_k)$ converge or diverge together. Following Sublette [44] and using a method due to Knopp [40], from (3.116)

$$\begin{aligned} ln\,(R_r/R_0) &= \sum_{k=1}^{r} ln\,(1 + s_k) - ln\,(1 - s_k) \\ &= \sum_{k=1}^{r} \left(s_k + \frac{s_k^2}{2} + \frac{s_k^3}{3} + \cdots \right) - \left(-s_k + \frac{s_k^2}{2} - \frac{s_k^3}{3} + \cdots \right) \\ &= 2 \left(\sum_{k=1}^{r} s_k - \sum_{k=1}^{r} s_k^3 \left(\frac{1}{3} + \frac{s_k^2}{5} + \frac{s_k^4}{7} + \cdots \right) \right). \end{aligned}$$ (3.117)

Let

$$\alpha_k = \frac{1}{3} + \frac{s_k^2}{5} + \frac{s_k^4}{7} + \cdots \leq \frac{1}{1 - s_k^2}.$$

Since $|s_k| < 1$ and $s_k \to 0$ as $k \to \infty$, $\alpha_k \to \frac{1}{3}$ and $\{\alpha_k\}_{k=1}^{\infty}$ represent a bounded sequence. As a result, (3.117) becomes

$$ln\,(R_r/R_0) = 2 \left(\sum_{k=1}^{r} s_k - \sum_{k=1}^{r} \alpha_k s_k^3 \right)$$

or

$$R_r = R_0\,exp\left(2 \sum_{k=1}^{r} s_k \right) \cdot exp\left(-2 \sum_{k=1}^{r} \alpha_k s_k^3 \right).$$ (3.118)

Since $|s_k| < 1$, we have

$$\sum_{k=1}^{\infty} |s_k|^3 < \sum_{k=1}^{\infty} |s_k|^2 < \infty,$$

where the last step follows from the Paley-Wiener criterion. The absolute convergence of the series $\sum_{k=1}^{\infty} s_k^3$ together with the bounded character of $\{\alpha_k\}_{k=1}^{\infty}$ implies that $\sum_{k=1}^{\infty} \alpha_k s_k^3$ is convergent. Thus

$$R_r = K_0 \cdot exp\left(2 \sum_{k=1}^{r} s_k \right)$$ (3.119)

---

[6]From (3.74)–(3.75), in the complex case we have, $|R_r| \leq |R_{r-1}|(1+|s_k|)/(1-|s_k|)$.

and hence the sequence $\{R_r\}_{r=1}^{\infty}$ converges to a nonzero (positive) limit iff the sum $\sum_{k=1}^{\infty} s_k$ converges.[7]

In the rational case, a beautiful proof by Ikeno (in the continuous p-domain) shows that [45]

$$R_k \longrightarrow Z(1-0) = r_0 + 2\sum_{k=1}^{\infty} r_k = S(0), \qquad (3.120)$$

where the last two equalities follow provided $\sum_{k=1}^{\infty} r_k < \infty$. For rational p.r. functions, since $\sum_{k=1}^{\infty} s_k^2 < \infty$, together with (3.119) this implies $\sum_{k=1}^{\infty} s_k < \infty$, i.e., in the rational case the reflection coefficients always form a summable sequence.

Returning back to the general case, from (3.119), R_k converges to a definite limit iff $\sum_{k=1}^{\infty} s_k < \infty$ and to examine this limit, we can make use of (3.P.5). Thus, as $n \to \infty$

$$\frac{h_n(z)}{g_n(z)} \longrightarrow \rho(z) = \frac{Z(z) - R_0}{Z(z) + R_0}, \qquad |z| < 1.$$

However, under the additional condition of absolute summability of the reflection coefficients, we have

$$\sum_{k=1}^{\infty} |s_k| < \infty \qquad (3.121)$$

and in that case, it can be shown that [46]

$$\lim_{n\to\infty} \frac{h_n(z)}{g_n(z)} = \rho(z) = \frac{Z(z) - R_0}{Z(z) + R_0}, \qquad |z| \leq 1; \qquad (3.122)$$

i.e., the absolute convergence of the reflection coefficients guarantees that the sequence of functions $h_n(z)/g_n(z)$ converges uniformly to $\rho(z)$ in the closed unit circle, and moreover $\rho(z)$ is absolutely continuous in $|z| \leq 1$. In particular, $h_n(1)/g_n(1)$ is bounded by unity and using (3.36)–(3.37), we have

$$\frac{h_n(1)}{g_n(1)} = \frac{s_n + h_{n-1}(1)/g_{n-1}(1)}{1 + s_n h_{n-1}(1)/g_{n-1}(1)}.$$

Thus

$$\frac{h_1(1)}{g_1(1)} = s_1 = \frac{R_1 - R_0}{R_1 + R_0}.$$

---

[7]If $\sum_{k=1}^{\infty} s_k = -\infty$, then from (3.119), $R_\infty = 0$. As Schur has noted, the summability of the reflection coefficients does not follow from the integrability condition of the spectrum and the Paley-Wiener criterion (see Problem 3.8 for an example).

and using this together with (3.115) for $r = 2$, we have

$$\frac{h_2(1)}{g_2(1)} = \frac{R_2 - R_0}{R_2 + R_0}.$$

By induction, it quickly follow that

$$\frac{h_n(1)}{g_n(1)} = \frac{s_n(R_{n-1} + R_0) + (R_{n-1} - R_0)}{(R_{n-1} + R_0) + s_n(R_{n-1} - R_0)} = \frac{R_n - R_0}{R_n + R_0}$$

and from (3.121)–(3.122), since $Z(z)$ is continuous at $z = 1$,

$$\lim_{n \to \infty} \frac{h_n(1)}{g_n(1)} = \lim_{n \to \infty} \frac{R_n - R_0}{R_n + R_0} = \frac{Z(1-0) + R_0}{Z(1-0) + R_0}$$

or, under the absolute summability of the reflection coefficients, we have

$$R_n \longrightarrow Z(1-0) = r_0 + 2\sum_{k=1}^{\infty} r_k = S(0), \qquad (3.123)$$

where the last two equalities follow provided $\sum_{k=1}^{\infty} r_k < \infty$.

Next, we take up the second question raised in section 3.2 and examine the parametric form of the extension that maximizes the k-step minimum mean-square prediction error. Towards this, first the two-step predictor case is taken up in the next section. The general situation involving the k-step predictor is analyzed in Appendix 3.A.

3.4 Two-Step Predictor

Given the autocorrelations $r_0, r_1, \ldots, r_n$, with $\mathbf{T}_n > 0$, of all the admissible completions given by (3.53) and (3.54), the problem here is to find the particular extension that maximizes the k-step minimum mean-square prediction error. From section 3.2, since maximization of the one-step prediction error leads to an all-pole model, we examine next the two-step predictor case.

Using (2.106) and (2.107), the two-step minimum mean-square prediction error P_2 is given by

$$P_2 = |b_0|^2 + |b_1|^2 = (1 + |\alpha_1|^2) \exp\left(\frac{1}{2\pi}\int_{-\pi}^{\pi} \ln S(\theta)d\theta\right). \qquad (3.124)$$

Naturally, maximization of P_2 is with respect to the unknown autocorrelations $r_{n+1}, r_{n+2}, \ldots$ and using the relation

$$\alpha_1 = d_1 = \frac{1}{2\pi}\int_{-\pi}^{\pi} \ln S(\theta)e^{-j\theta}d\theta$$

given in (2.61)–(2.65), this leads to

$$\frac{\partial \alpha_1}{\partial r_k} = \frac{1}{2\pi} \int_{-\pi}^{\pi} \frac{e^{j(k-1)\theta}}{S(\theta)} d\theta \tag{3.125}$$

and hence

$$\frac{\partial P_2}{\partial r_k} = \frac{|b_0|^2}{2\pi} \int_{-\pi}^{\pi} \left(\frac{1 + |\alpha_1|^2 + \alpha_1 e^{j\theta} + \alpha_1^* e^{-j\theta}}{S(\theta)} \right) e^{jk\theta} d\theta$$

$$= \frac{|b_0|^2}{2\pi} \int_{-\pi}^{\pi} \frac{|1 + \alpha_1 e^{j\theta}|^2}{S(\theta)} e^{jk\theta} d\theta = 0, \quad |k| > n. \tag{3.126}$$

Since $|b_0|^2 > 0$, (3.126) implies that the Fourier series expansion for the periodic nonnegative function $(|1 + \alpha_1 e^{j\theta}|^2)/S(\theta)$ truncates after the nth term and hence it must have the form

$$\frac{|1 + \alpha_1 e^{j\theta}|^2}{S(\theta)} = \sum_{k=-n}^{n} c_k e^{jk\theta} = |g(e^{j\theta})|^2, \tag{3.127}$$

where

$$g(z) = g_0 + g_1 z + \cdots + g_n z^n \tag{3.128}$$

represents the strict Hurwitz polynomial associated with the factorization in (3.127). (Notice that since $S(\theta)$ is free of poles on the unit circle, $g(z)$ cannot have any zeros on $|z| = 1$.) Then,

$$S(\theta) = \frac{|1 + \alpha_1 e^{j\theta}|^2}{|g(e^{j\theta})|^2} = |B_2(e^{j\theta})|^2, \tag{3.129}$$

where

$$B_2(z) = \frac{a(z)}{g(z)} \tag{3.130}$$

represents the minimum-phase factor and $a(z)$ represents the strict Hurwitz polynomial of degree 1 associated with the factorization $|1 + \alpha_1 e^{j\theta}|^2 = |a(e^{j\theta})|^2$. Depending on the value of α_1, this gives rise to two choices, i.e.,

$$a(z) = (1 + \alpha_1 z) \quad \text{or} \quad (\alpha_1 + z). \tag{3.131}$$

Thus, subject to $r_0, r_1, \ldots, r_n$, the Wiener factor $B_2(z)$ in (3.130) that maximizes the two-step minimum mean-square prediction error, if it exists, is of the type ARMA$(n, 1)$. To complete the argument, we must demonstrate the existence of such a factor that is analytic together with its inverse in $|z| < 1$.

Towards this purpose, notice that in the case of real autocorrelations,[8] this specific extension, if admissible, should follow from (3.53)–(3.54) for a

---

[8]Real correlations are in a sense more complicated, since the solutions must be shown to be real.

certain choice of the rational bounded-real function $\rho(z) = \rho_{n+1}(z)$, and on comparing (3.130) and (3.54), because of degree restrictions, $\rho(z)$ must have the form[9]

$$\rho(z) = \frac{1}{a+bz}. \tag{3.132}$$

For (3.132) to be bounded-real, it is necessary that there exist no poles in $|z| \leq 1$, i.e.,

$$\left|\frac{a}{b}\right| > 1, \tag{3.133}$$

and

$$\left|\frac{1}{a+be^{j\theta}}\right| \leq 1 \iff (a \pm b)^2 \geq 1, \tag{3.134}$$

in addition to a and b being real. Conversely, when (3.133) and (3.134) are true, from the maximum modulus theorem [26], analyticity in $|z| \leq 1$, together with (3.134), implies boundedness for $\rho(z)$ in $|z| \leq 1$. In the present situation, the existence of such a bounded-real function as in (3.132) can be verified by solving for a, b using (3.46)–(3.55) and examining whether they satisfy the necessary and sufficient conditions (3.133) and (3.134). In that case, using (3.132) in (3.46)–(3.54) we obtain

$$D_n(z) \triangleq \frac{A_n(z)(a+bz) - z\tilde{A}_n(z)}{a+bz} = \frac{g_0 + g_1 z + \cdots + g_n z^n}{a+bz}, \tag{3.135}$$

where

$$g_k = a_{k-1}b + a_k a - a_{n-k+1}, \quad k = 0 \to n, \tag{3.136}$$

and the degree n requirement for $g(z)$ in (3.128) yields

$$b = \frac{a_0}{a_n}. \tag{3.137}$$

Here, a_k, $k = 0 \to n$, are the coefficients of the degree n Levinson polynomial $A_n(z)$ in (3.61). Similarly using (3.55),

$$\Gamma(z) = \frac{\alpha + \beta z}{a+bz}, \tag{3.138}$$

where α, β satisfy

$$\alpha^2 + \beta^2 = a^2 + b^2 - 1 \tag{3.139}$$

$$\alpha\beta = ab. \tag{3.140}$$

The minimum-phase character of $\Gamma(z)$ gives $|\alpha| > |\beta|$ and hence from (3.139)–(3.140), we have

$$\alpha = \pm\frac{1}{2}\left(\sqrt{(a+b)^2 - 1} + \sqrt{(a-b)^2 - 1}\right) \tag{3.141}$$

---

[9]It is shown in Appendix 3.B that $\rho(z) = (1+dz)/(a+bz+cz^2)$, etc., are not acceptable forms.

and

$$\beta = \pm\frac{1}{2}\left(\sqrt{(a+b)^2-1} - \sqrt{(a-b)^2-1}\right). \tag{3.142}$$

Clearly, the signs of α and β must be chosen so as to satisfy (3.140). Notice that the numerator and denominator polynomials of $\Gamma(z)$ in (3.138) must be always relatively prime to each other. If not, the b.r. function $\Gamma(z) = \mu$, a constant strictly less than unity, and using (3.55) this implies $\rho(z)$ is an all-pass function. This contradicts (3.132), the only acceptable form for $\rho(z)$. Moreover from (3.54), in $|z| < 1$ the Wiener factor $B_2(z)$ yields the power series expansion

$$B_2(z) = \frac{\Gamma(z)}{D_n(z)} = \frac{\alpha+\beta z}{\displaystyle\sum_{k=0}^{n} g_k z^k} = \frac{\alpha}{g_0} + \left(\frac{\beta}{g_0} - \frac{\alpha g_1}{g_0^2}\right)z + \cdots$$

$$\overset{\triangle}{=} b_0 + b_1 z + b_2 z^2 + \cdots \tag{3.143}$$

and hence from (2.65)–(2.66)

$$\alpha_1 = \frac{b_1}{b_0} = \frac{\beta}{\alpha} - \frac{g_1}{g_0}. \tag{3.144}$$

On the other hand, comparing the numerator parts of (3.130)–(3.131) and (3.143), we obtain

$$\alpha_1 = \frac{\beta}{\alpha} \quad\text{or}\quad \frac{\alpha}{\beta} \tag{3.145}$$

(since b_k, a_k, $k = 0 \to \infty$, are real in the case of real autocorrelations). It is easy to show that the first choice $\alpha_1 = \beta/\alpha$ does not *always* lead to a bounded-real solution for $\rho(z)$. In fact, by letting $\alpha_1 = \beta/\alpha$ and equating this to (3.144), we obtain $g_1/g_0 = 0$, which in turn implies $g_1 = 0$ (since $g_0 = a_0 a$ is a nonzero finite number) and, hence from (3.136) we obtain $a^{(0)} \overset{\triangle}{=} (a_n^2 - a_0^2)/a_0 a_n$ for a. From (3.133), for $\rho(z)$ to be bounded-real, the ratio

$$x_0 \overset{\triangle}{=} \frac{a^{(0)}}{b} = \frac{a_n^2 - a_0^2}{a_0 a_1} \tag{3.146}$$

generated by this choice must satisfy $|x_0| > 1$. However, as the strict Hurwitz polynomial $A_2(z) = (z+2)(2z+3) = 2z^2 + 7z + 6$ shows

$$|x_0| = \frac{6^2 - 2^2}{6 \times 7} = \frac{32}{42} < 1$$

and, hence, $\alpha_1 = \beta/\alpha$ does not always lead to an admissible solution. At times, however, the above choice can lead to admissible solutions. For example, in the case of the AR(2) process characterized by the strict Hurwitz polynomial $A_2(z) = 5 - 2z + z^2$, the above choice leads to $a = 12$

and $b = 5$, which results in a bounded-real $\rho(z)$ in (3.132) and hence an admissible Wiener factor. Since the above choice does not *always* lead to an admissible one, turning back to (3.145), this leads to the only other possibility,

$$\alpha_1 = \frac{\alpha}{\beta}. \tag{3.147}$$

In this case, equating (3.144) and (3.147), we obtain the key equation

$$\frac{\beta^2 - \alpha^2}{\alpha\beta} = \frac{g_1}{g_0}. \tag{3.148}$$

From (3.139)–(3.140)

$$(\alpha^2 - \beta^2)^2 = (\alpha + \beta)^2(\alpha - \beta)^2$$
$$= \{(a + b)^2 - 1\}\{(a - b)^2 - 1\} \tag{3.149}$$

and using this, (3.148) reduces to

$$\frac{(a^2 - b^2)^2 - \{(a + b)^2 + (a - b)^2\} + 1}{b^2} = \left(b - \frac{a_n}{a_0} - \frac{aa_1}{a_0}\right)^2$$

or

$$\frac{b^4\{(a/b)^2 - 1\}^2 - 2b^2\{(a/b)^2 + 1\} + 1}{b^2} = \left(b - \frac{1}{b} + \frac{aa_1}{a_0}\right)^2.$$

Let

$$x = \frac{a}{b}. \tag{3.150}$$

After some algebra, the above equation simplifies to

$$b^2x^4 - 2\left(b^2 + 1 + \frac{a_1^2}{2a_n^2}\right)x^2 - \frac{2a_1}{a_n}\left(b - \frac{1}{b}\right)x = 0$$

and since $x \neq 0$, this reduces to the cubic equation with real coefficients

$$x^3 + px + q = 0, \tag{3.151}$$

where

$$p = -2\left(1 + \frac{1}{b^2} + \frac{a_1^2}{2a_0^2}\right) < 0 \tag{3.152}$$

and

$$q = -\frac{2a_1}{a_0}\left(1 - \frac{1}{b^2}\right). \tag{3.153}$$

Clearly, the Wiener factor $B_2(z)$ in (3.143) that maximizes the two-step minimum mean-square prediction error represents an admissible solution, if the cubic equation (3.151) has at least one real solution with magnitude greater than unity and further that solution together with (3.137) satisfies (3.134). To examine this, notice that if the discriminant

$$D = \left(\frac{q}{2}\right)^2 + \left(\frac{p}{3}\right)^3 \tag{3.154}$$

is negative, then (3.151) has three real roots and if $D > 0$, it has one real root and two complex roots that form a conjugate pair [47]. However, as shown in Appendix 3.C, the above discriminant is always negative and the corresponding three real roots can be obtained explicitly by making use of Cardano's formula. In that case, let

$$R = sgn(q) \left(-\frac{p}{3}\right)^{1/2} = -sgn\left(\frac{a_1}{a_0}\right) \sqrt{\frac{2}{3}\left(1 + \frac{1}{b^2} + \frac{a_1^2}{2a_0^2}\right)}. \tag{3.155}$$

Then, the three roots are given by [47]

$$x_1 = -2R\cos(\phi/3) \tag{3.156}$$
$$x_2 = -2R\cos(\phi/3 + 2\pi/3) \tag{3.157}$$

and

$$x_3 = -2R\cos(\phi/3 + 4\pi/3) \tag{3.158}$$

where

$$\cos\phi = \frac{q}{2R^3} = \frac{\left|\frac{a_1}{a_0}\right|\left(1 - \frac{1}{b^2}\right)}{\left[\frac{2}{3}\left(1 + \frac{1}{b^2} + \frac{a_1^2}{2a_0^2}\right)\right]^{3/2}}. \tag{3.159}$$

It is easy to show that two of these roots always have magnitude greater than unity. This follows from a well known sufficiency condition due to Cohn,[10] which in the case of (3.151) reduces to $|p| > |q| + 1$, for two of its roots to have magnitude greater than unity. To verify this condition, notice that

$$|p| - |q| - 1 = \left(1 - \left|\frac{a_1}{a_0}\right|\right)^2 + \frac{2}{b^2}\left(1 + \left|\frac{a_1}{a_0}\right|\right) > 0 \tag{3.160}$$

and hence (3.151) has two roots with magnitude greater than unity and one with magnitude less than unity. Moreover, since a_0/a_n represents the product of the n roots of the strict Hurwitz polynomial $A_n(z)$, from (3.137), we have $|b| > 1$ and hence without exception, $1 > \cos\phi \geq 0$. Thus, without loss of generality we can choose $0 < \phi \leq \pi/2$ or

$$1 > \cos\left(\frac{\phi}{3}\right) \geq \cos\left(\frac{\pi}{6}\right) = \frac{\sqrt{3}}{2} \tag{3.161}$$

---

[10]For a polynomial $f(z) = a_0 + a_1 z + \cdots + a_p z^p + \cdots + a_n z^n$, $a_p \neq 0$, if $|a_p| > |a_0| + |a_1| + \cdots + |a_{p-1}| + |a_{p+1}| + \cdots + |a_n|$, then from Rouché's Theorem, $f(z)$ has exactly p zeros inside the unit circle [32].

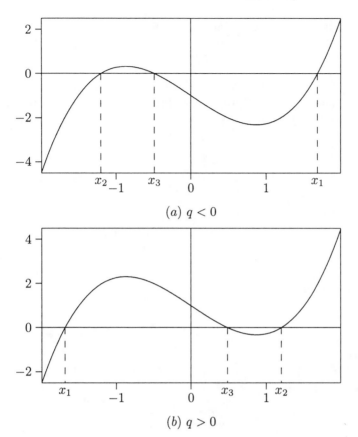

(a) $q < 0$

(b) $q > 0$

FIGURE 3.3. Structure of the cubic function $x^3 + px + q$, $p < 0$. (See (3.151)–(3.158).)

and

$$\frac{1}{2} < -cos\left(\frac{\phi}{3} + \frac{2\pi}{3}\right) \le \frac{\sqrt{3}}{2}, \quad 0 \le -cos\left(\frac{\phi}{3} + \frac{4\pi}{3}\right) < \frac{1}{2}. \qquad (3.162)$$

As a result, x_2 and x_3 have the same sign that is opposite to the sign of x_1. Clearly, the sign of x_2 and x_3 is the same as that of q. The structure of these roots is summarized in Fig. 3.3. Further, using (3.161) and (3.162), we also have $2|R| > |x_1| \ge \sqrt{3}|R|$, $\sqrt{3}|R| \ge |x_2| > |R|$ and $|x_3| < |R|$ which gives

$$|x_1| \ge |x_2| > |x_3|. \qquad (3.163)$$

Thus, using the Cohn criterion, *only* x_1 and x_2 have magnitudes greater than unity, i.e., $\rho(z)$ obtained using any one of these two roots is always analytic in $|z| \le 1$. For bounded reality, it remains to show that a, b so obtained from these roots also satisfy

$$(a \pm b)^2 \geq 1. \tag{3.164}$$

In what follows, we demonstrate (3.164) for the largest root x_1 given by (3.156). The existence of a second solution generated by x_2 in (3.157) is proved in Appendix 3.D. In the case of the largest root, using (3.150), (3.155), (3.156) and (3.161), x_1 gives rise to

$$|a| \overset{\triangle}{=} |x_1||b| \geq \sqrt{3}\,|R||b| = \sqrt{2\left(1 + b^2 + \frac{a_1^2}{2a_n^2}\right)} > 2.$$

Again, without loss of generality, we can assume a and b are of the same sign — say positive (i.e., $x_1 > 0$, $b > 1$) – and hence,

$$(a + b)^2 > 9 > 1. \tag{3.165}$$

To verify the second part, let

$$a - b \geq \sqrt{2\left(1 + b^2 + \frac{a_1^2}{2a_n^2}\right)} - b \overset{\triangle}{=} K. \tag{3.166}$$

To find its minimum, notice that by straightforward algebra, $\partial K/\partial b = 0$ yields the solution

$$\tilde{b}^2 = 1 + \frac{a_1^2}{2a_n^2}. \tag{3.167}$$

But, the second derivative

$$\frac{\partial^2 K}{\partial b^2} = \frac{\sqrt{2}\,\tilde{b}_0^2}{(b^2 + \tilde{b}_0^2)^{3/2}}$$

is always positive and in particular, at $b = \tilde{b}$, the quantity K in (3.166) achieves its minimum value given by

$$K_{min} = \sqrt{2(2\tilde{b}^2)} - \tilde{b} = \tilde{b}.$$

As a result, from (3.166)–(3.167),

$$(a - b)^2 \geq K_{min}^2 = \tilde{b}^2 > 1. $$

This completes the proof for the bounded reality of the first solution and together with Appendix 3.D, we have two bounded-real solutions for $\rho(z)$ under the second choice (3.147) [48].

However, interestingly as shown below, the first solution x_1 *never* satisfies the key equation (3.148) and hence it does not give rise to an admissible extension that maximizes the two-step minimum mean-square prediction error. To see this, from (3.141)–(3.142), since

$$\beta^2 - \alpha^2 < 0,$$

(3.148) gives

$$\alpha\beta\frac{g_1}{g_0} = \beta^2 - \alpha^2 < 0. \tag{3.168}$$

But

$$\alpha\beta\frac{g_1}{g_0} = ab\frac{g_1}{g_0} = ab\left(\frac{a_0 b + a_1 a - a_n}{a_0 a}\right)$$

$$= b^2\left(1 - \frac{1}{b^2} + \frac{a_1}{a_0}\cdot\frac{a}{b}\right)$$

and with $x = a/b$ as in (3.150), (3.168) reads

$$b^2\left(1 - \frac{1}{b^2} + \frac{a_1}{a_0}x\right) < 0.$$

Clearly, to maximize P_2 the solutions x_1 and x_2 in (3.156)–(3.157) must satisfy

$$1 - \frac{1}{b^2} + \frac{a_1}{a_0}x_i < 0, \quad i = 1,\ 2. \tag{3.169}$$

However, from (3.155)–(3.156) and (3.161)

$$x_1 = 2\,sgn\left(\frac{a_1}{a_0}\right)|R\cos(\phi/3)|$$

and thus $(a_1/a_0)x_1 > 0$, or

$$1 - \frac{1}{b^2} + \frac{a_1}{a_0}x_1 = 1 - \frac{1}{b^2} + \left|\frac{a_1}{a_0}x_1\right| > 0$$

which contradicts (3.169). Thus x_1 never gives rise to an admissible spectral extension[11] in terms of maximizing P_2 and hence x_2 together with x_0 in (3.146) are the *only* possible choices for a/b that could maximize P_2. Once again it is possible to show that if x_0 generates a b.r. solution for $\rho(z)$, then x_2 does not satisfy (3.169) and conversely if x_2 satisfies (3.169), then x_0 does not generate a bounded-real solution. As a result, the Wiener factor that maximizes the two-step minimum mean-square prediction error is always *unique* and is generated by either x_0 in (3.146) or x_2 in (3.157) (and not both, unless $x_2 = x_0$).

---

[11]Notice that whenever x_1 or x_2 does not satisfy (3.148), the violation is only because of a multiplier factor ± 1. This easily follows since they both satisfy the squared equation (3.149) that leads to the cubic equation (3.151).

To complete this proof, first consider the situation where $|x_0| > |x_2|$. Then from Appendix 3.D, since x_2 always generates a b.r. solution we have in this case, $|x_0| > |x_2| > 1$, and, as before with $a^{(0)} = x_0 b$,

$$|a^{(0)} \pm b| = |x_0 \pm 1||b| \geq |x_2 \pm 1||b| = |a^{(2)} \pm b| > 1 \qquad (3.170)$$

or x_0 gives rise to a b.r. solution that maximizes P_2 through the first choice in (3.145). However, from (3.146), (3.155)–(3.157), since x_0 and x_2 always have the same sign and $(a_1/a_0)x_2 < 0$, in this case ($|x_0| > |x_2|$) we have

$$\frac{a_1}{a_0} x_2 = -\left|\frac{a_1}{a_0} x_2\right| > -\left|\frac{a_1}{a_0} x_0\right| = \frac{a_1}{a_0} x_0$$

and this violates (3.169) since

$$1 - \frac{1}{b^2} + \frac{a_1}{a_0} x_2 > 1 - \frac{1}{b^2} + \frac{a_1}{a_0} x_0 = 1 - \frac{1}{b^2} + \frac{a_1}{a_0} \cdot \frac{a_n^2 - a_0^2}{a_0 a_1}$$

$$= 1 - \frac{1}{b^2} + \frac{a_n^2 - a_0^2}{a_0^2} = 1 - \frac{1}{b^2} + \frac{1}{b^2} - 1 = 0, \qquad (3.171)$$

i.e., x_2 *does not* satisfy (3.148) and consequently when $|x_0| > |x_2|$, x_0 is the *only* admissible solution that maximizes the minimum mean-square prediction error P_2.

On the other hand, when $|x_2| > |x_0|$,

$$\frac{a_1}{a_0} x_2 = -\left|\frac{a_1}{a_0} x_2\right| < -\left|\frac{a_1}{a_0} x_0\right| = \frac{a_1}{a_0} x_0$$

and (3.169) translates into

$$1 - \frac{1}{b^2} + \frac{a_1}{a_0} x_2 < 1 - \frac{1}{b^2} + \frac{a_1}{a_0} x_0$$

$$= 1 - \frac{1}{b^2} + \frac{a_1}{a_0} \cdot \frac{a_n^2 - a_0^2}{a_0 a_1} = 0 \qquad (3.172)$$

and hence x_2 satisfies (3.148) and generates an admissible solution that maximizes P_2. To establish the desired uniqueness, it remains to show that in this case, x_0 *does not* generate a bounded-real solution. Towards this, notice that the condition $|x_2| > |x_0|$ can be written as

$$x_2^2 = 4R^2 \cos^2(\phi/3 + 2\pi/3) > \frac{(a_n^2 - a_0^2)^2}{a_0^2 a_1^2} = x_0^2.$$

Using (3.155) and letting

$$\lambda = \cos^2(\phi/3 + 2\pi/3), \qquad (3.173)$$

the above equation simplifies into

$$\frac{8}{3}\left(1 + \frac{1}{b^2} + \frac{a_1^2}{2a_0^2}\right)\lambda > \frac{a_0^2}{a_1^2}\left(\frac{1}{b^2} - 1\right)^2$$

and with

$$u = \left(\frac{a_1}{a_0}\right)^2$$

this inequality ($|x_2| > |x_0|$) gives

$$f(\lambda, u) = 4\lambda b^4 u^2 + 8\lambda b^2 (b^2 + 1)u - 3(b^2 - 1)^2 > 0. \tag{3.174}$$

From (3.162), the variable λ in (3.173)–(3.174) always satisfies $\frac{1}{4} < \lambda \le \frac{3}{4}$ and it can be expressed in terms of u by observing that

$$\lambda^{3/2} = (1/2)^3 (2\cos\phi + 6\lambda^{1/2}), \tag{3.175}$$

where $\cos\phi$ is as given in (3.159), and further simplification of (3.175) yields

$$g(\lambda, u) = \lambda(4\lambda - 3)^2 (b^2 u + 2b^2 + 2)^3 - 27b^2 (b^2 - 1)^2 u = 0. \tag{3.176}$$

Since $|x_2| > |x_0|$ is equivalent to (3.174), if any triplet a_0, a_1, a_n satisfying (3.174) is also able to generate a b.r. solution in x_0, then we must have (i) $|x_0| > 1$ and (ii) $|a^{(0)} \pm b| \ge 1$, where $a^{(0)} = x_0 b$. Of these, $|x_0| > 1$ is equivalent to

$$|x_0| = \left|\frac{a_n^2 - a_0^2}{a_0 a_1}\right| \ge 1 \implies \left(\frac{a_0}{a_1}\right)^2 \left(\frac{b^2 - 1}{b^2}\right)^2 \ge 1 \implies u \le \left(\frac{b^2 - 1}{b^2}\right)^2 \tag{3.177}$$

and $|a^{(0)} \pm b| \ge 1$ gives

$$(a^{(0)} + b)^2 \ge 1 \implies b^2 \left(\frac{a_1}{a_0}\right)^2 - 2b^2 \left(\frac{a_1}{a_0}\right) + (b^2 - 1) \ge 0$$

or, for $b > 1$,

$$\frac{a_1}{a_0} \le \frac{b-1}{b} \quad \text{or} \quad \frac{a_1}{a_0} \ge \frac{b+1}{b}. \tag{3.178}$$

Similarly,[12]

$$(a^{(0)} - b)^2 \ge 1 \implies b^2 \left(\frac{a_1}{a_0}\right)^2 + 2b^2 \left(\frac{a_1}{a_0}\right) + (b^2 - 1) \ge 0$$

---

[12]Note that for $b < -1$, (3.178) reads

$$\frac{a_1}{a_0} \le \frac{b+1}{b} \quad \text{or} \quad \frac{a_1}{a_0} \ge \frac{b-1}{b}$$

and similarly (3.181) becomes

$$u \le \left(\frac{b+1}{b}\right)^2 = u_0.$$

or

$$\frac{a_1}{a_0} \le -\frac{b+1}{b} \quad \text{or} \quad \frac{a_1}{a_0} \ge -\frac{b-1}{b}. \tag{3.179}$$

Thus, in terms of u, (3.178) and (3.179) can be combined to give

$$|a^{(0)} \pm b| \ge 1 \implies u \le \left(\frac{b-1}{b}\right)^2 \quad \text{or} \quad u \ge \left(\frac{b+1}{b}\right)^2. \tag{3.180}$$

For any $b > 1$, since

$$\left(\frac{b-1}{b}\right)^2 < \left(\frac{b^2-1}{b^2}\right)^2 < \left(\frac{b+1}{b}\right)^2,$$

from (3.177) and (3.180) it follows that for x_0 to be bounded-real,

$$u \le \left(\frac{b-1}{b}\right)^2 \overset{\triangle}{=} u_0. \tag{3.181}$$

Clearly, the desired solution, if any, should satisfy both (3.174) and (3.181) and to examine whether any such u exists, it is enough to determine the sign of $f(\lambda, u)$ at $u = u_0$ and the positive zeros of $f(\lambda, u)$ together with their multiplicities. Towards this, elimination of λ using the pair of equations (3.174), (3.176), results in the resultant

$$R(u) = (b^2 u + 2b^2 + 2)^3 (b^2 u - (b-1)^2)(b^2 u - (b+1)^2) \\ \times (3b^4 u^2 + 2b^2(b^2+1)u - (b^2-1)^2) = 0$$

and this equation has three simple positive roots and two negative roots. In the present context, the positive roots given by

$$u_1 = \left(\frac{b-1}{b}\right)^2 = u_0 \tag{3.182}$$

$$u_2 = \left(\frac{b+1}{b}\right)^2 > u_0 \tag{3.183}$$

and

$$u_3 = \frac{-(b^2+1) + 2\sqrt{b^4 + b^2 + 1}}{3b^2} < u_0 \tag{3.184}$$

alone are admissible candidates as zeros of $f(\lambda, u)$, and when substituted into $f(\lambda, u) = 0$, they yield

$$\lambda_i \overset{\triangle}{=} \lambda(u_i) = \frac{3(b^2-1)^2}{4b^4 u_i^2 + 8b^2(b^2+1)u_i}, \quad i = 1, 2, 3. \tag{3.185}$$

A direct substitution of (3.182)–(3.184) into (3.185) shows that $\lambda_i(b^2)$, $i = 1 \to 3$, are monotonic functions of b^2 that for $|b| > 1$ satisfy

$$\frac{1}{4} < \lambda_1 \le \frac{3}{4}, \quad 0 \le \lambda_2 < \frac{1}{4} \quad \text{and} \quad \frac{3}{4} \le \lambda_3 < 1.$$

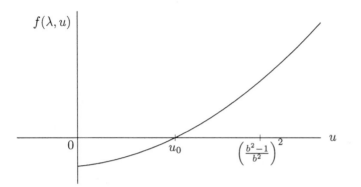

FIGURE 3.4. Structure of $f(\lambda, u)$ in (3.174). $|x_2| > |x_0|$ only if $f(\lambda, u) > 0$ and x_0 yields a b.r. solution iff $u \leq u_0$. See (3.174)–(3.181).

However, from the discussion following (3.174), the range for λ_1 alone is acceptable for λ, and hence the pair (u_1, λ_1) is the *only* admissible solution that satisfies $f(\lambda, u) = 0$ in (3.174). Further, a direct calculation shows that $f'(\lambda_1, u_1) = 3b^2(b+1)^2 \neq 0$ and hence $u = u_1 = u_0$ represents a simple zero of $f(\lambda, u)$. Similarly if $b < -1$, from footnote 12, u_2 in (3.183) equals u_0 and it represents the only positive (simple) zero of $f(\lambda, u)$. Finally, since at $u = 0$, $\lambda = 3/4$, $f(\lambda, 0) = -3(b^2 - 1)^2 < 0$, we have $f(\lambda, u) > 0$ only if $u > u_0$ and together with (3.181) it now follows that there exists no triplet a_0, a_1, a_n that satisfies $|x_2| > |x_0|$ and yields a bounded-real solution for[13] x_0. This situation is summarized in Fig. 3.4. This completes the proof for the uniqueness of the extension that maximizes the two-step minimum mean-square prediction error.

The general case involving complex autocorrelations can be dealt with in a similar manner. In that case, to exhibit an admissible solution, it is enough to demonstrate the existence of a bounded function (not necessarily a real function) that satisfies the key equations.

Figures 3.5–3.7 together with Table 3.1 show details of the maximum entropy extension and the two-step predictor discussed above in three different situations. In Fig. 3.5(a), the original spectrum (solid line) corresponds to an ARMA$(4, 1)$ model with transfer function

$$B(z) = \frac{1.5 + z}{5.3 - 2.6z + 4.3z^2 - 1.7z^3 + 2.5z^4} \tag{3.186}$$

and the number of known autocorrelations is taken to be 5 here. The max-

---

[13]If $|x_2| = |x_0|$, from previous remarks and (3.174), $x_2 = x_0$, $f(\lambda, u) = 0 \implies u = (|b|-1)^2/b^2 = (a_1/a_0)^2$ or $|a_0|-|a_n| = \pm a_1$. For example, $A_3(z) = 5+3z-2z^3$ gives $x_2 = x_0 = -1.4$.

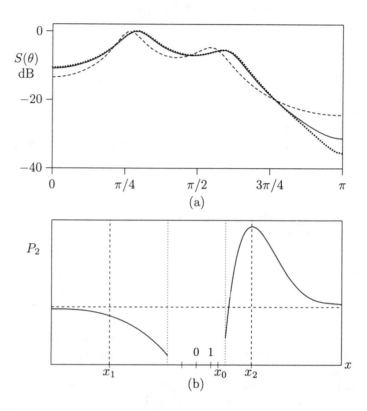

FIGURE 3.5. One- and two-step predictors. (a) Original spectrum (solid line) corresponds to an ARMA(4, 1) model given by (3.186). Dashed line corresponds to the AR(4) model in (3.187) and the dotted line corresponds to the maximally robust two-step predictor generated by x_2 in (3.188). (b) Two-step minimum mean-square prediction error P_2 vs. x.

imum entropy extension results in an AR(4) model given by

$$B_1(z) = \frac{1}{3.254 - 2.892z + 3.664z^2 - 1.989z^3 + 1.770z^4} \qquad (3.187)$$

and is represented by the dashed curve. In this case, since $|x_2| > |x_0|$, the maximally robust two-step predictor is generated by x_2. This is also evident from Fig. 3.5(b) that shows the variation of the two-step minimum prediction error P_2 for all admissible value of x and it is maximized by x_2. The associated minimum-phase ARMA(4, 1) transfer function is given by

$$B_2(z) = \frac{0.582 + 0.465z}{2.151 - 0.974z + 1.681z^2 - 0.631z^3 + z^4}, \qquad (3.188)$$

and it is represented by the dotted curve. Notice that compared to the AR-case, the spectrum corresponding to the two-step predictor in Fig. 3.5(a)

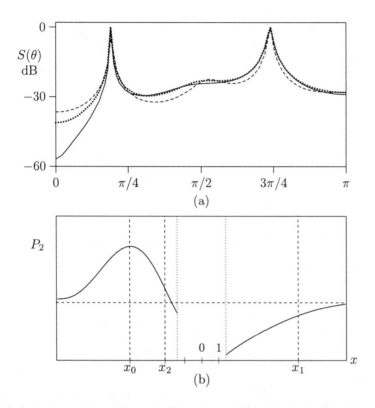

FIGURE 3.6. One- and two-step predictors. (a) Original spectrum (solid line) corresponds to an ARMA$(6,2)$ model given by (3.189). Dashed line corresponds to the AR(7) model in (3.190) and the dotted line corresponds to the maximally robust two-step predictor generated by x_0 in (3.191). (b) Two-step minimum mean-square prediction error P_2 vs. x.

is 'closer' to the original spectrum and, in a sense, this shows the improvement over the maximum entropy estimator. Figure 3.6 illustrates a situation where the maximally robust two-step predictor is generated by x_0. In Fig. 3.6(a), the original spectrum (solid line) corresponds to an ARMA$(6,2)$ model with transfer function

$$B(z) = \frac{2.1 - 3.0z + z^2}{4.84 - 2.3z + 1.5z^2 - 1.79z^3 + 4.2z^4 - 1.35z^5 + 2.3z^6} \quad (3.189)$$

and the number of autocorrelations is taken to be 8 here. In this case, the maximum entropy spectrum (dashed line) is given by the transfer function

$$\frac{1}{-1.87 - 0.86z - 0.63z^2 + 0.38z^3 - 0.82z^4 - 0.75z^5 - 1.17z^6 - 0.83z^7}$$
$$(3.190)$$

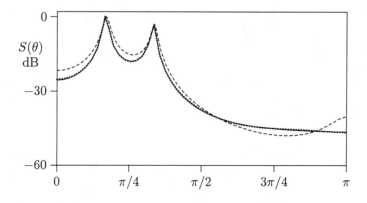

FIGURE 3.7. Effect of the number of known autocorrelations on the two-step predictor. Original spectrum (solid line) corresponds to the ARMA(4, 3) model in (3.192). Dashed line corresponds to $n = 5$ case and the dotted line corresponds to $n = 7$ case (see Table 3.1).

and the two-step predictor (dotted curve) is generated by

$$B_2(z) = \frac{2.489 - 1.584z}{4.904 + 1.146z^2 - 1.432z^3 + 3.279z^4 + 0.470z^5 + 2.377z^6 + z^7}. \tag{3.191}$$

Once again as Fig. 3.6(b) shows, $|x_0| > |x_2|$ and x_0 maximizes P_2 in this case. To illustrate the effect of the number of known Fourier coefficients on the extension, Fig. 3.7 considers an ARMA(4, 3) model with transfer function

$$B(z) = \frac{-8.616 + 5.662z - 2.762z^2 + z^3}{1.083 - 2.899z + 3.883z^2 - 2.787z^3 + z^4}. \tag{3.192}$$

As before, the solid line represents the true spectrum. From the theory of Padé approximations [49] (see also section 4.2), in the case of an ARMA(p, q) system, if the number of known autocorrelations $n+1$ is less than $p+q+1$, the resulting extensions are not unique and to illustrate the effect of the number of known autocorrelations on the two-step predictor under discussion, two cases are considered here. They correspond to $n = 5$ and $n = 7$, and in both cases the solutions are given by x_2 and they are exhibited in Fig. 3.7. Notice that in this case, $n = 5$ (dashed line) results in a stable ARMA(5, 1) form and $n = 7$ (dotted line) results in a stable ARMA(7, 1) form with details as in Table 3.1. Compared to the spectrum corresponding to the ARMA(5, 1) form, the spectrum associated with ARMA(7, 1) form is 'closer' to the original spectrum.

To summarize, given $r_0, r_1, \ldots, r_n$, there exists a *unique* Wiener factor that maximizes the minimum mean-square error associated with the two-step predictors that are compatible with the given autocorrelations. This

Wiener factor turns out to be an ARMA$(n, 1)$ filter given by

$$B_2(z) = \frac{\alpha + \beta z}{g_0 + g_1 z + \cdots + g_n z^n} , \qquad (3.193)$$

where

$$g_k = a_{k-1} b + a_k a - a_{n-k+1} , \quad k = 0 \to n$$

$$\alpha = \pm \frac{1}{2} \left(\sqrt{(a+b)^2 - 1} + \sqrt{(a-b)^2 - 1} \right)$$

$$\beta = \pm \frac{1}{2} \left(\sqrt{(a+b)^2 - 1} - \sqrt{(a-b)^2 - 1} \right) ,$$

with

$$a = \begin{cases} -2bR\cos(\phi/3 + 2\pi/3) \\[2mm] \dfrac{a_n^2 - a_0^2}{a_1 a_n} \end{cases} \quad \begin{pmatrix} \text{whichever is} \\ \text{larger in} \\ \text{magnitude} \end{pmatrix} \qquad (3.194)$$

$$b = \frac{a_0}{a_n} .$$

Here, a_k, $k = 0 \to n$, represent the coefficients of the degree n Levinson polynomial in (3.61). The signs of α and β should be selected so as to satisfy (3.140). Further, R and ϕ are as in (3.155) and (3.159), respectively. As remarked earlier, the largest quantity (in magnitude) between the two choices for the parameter a in (3.194) gives rise to a *unique* bounded-real function in (3.132) and an admissible Wiener factor that maximizes the two-step minimum mean-square prediction error. The corresponding spectrum satisfies (3.2), (3.3) and the interpolation property (3.7).

TABLE 3.1

TWO-STEP PREDICTORS THAT
MAXIMIZE THE MINIMUM MEAN-SQUARE ERROR (SEE FIG. 3.7)

	Original Coefficients		Two-Step (x_2) Coefficients $n = 5$		Two-Step (x_2) Coefficients $n = 7$	
	Num.	Den.	Num.	Den.	Num.	Den.
z^0	-8.616	1.083	0.677	1.322	0.714	27.56
z^1	5.662	-2.899	0.304	-2.352	0.353	-42.02
z^2	-2.762	3.883		1.668		25.92
z^3	1.000	-2.787		0.621		11.58
z^4		1.000		-1.604		-15.52
z^5				1.000		1.792
z^6						8.535
z^7						1.000

Thus, as an application to the class of all spectral extensions described in section 3.3, we have demonstrated the existence of a unique admissible power spectrum with the property that the corresponding Wiener factor maximizes the minimum mean-square error associated with two-step predictors and matches the given $(n + 1)$ autocorrelations of a discrete-time stationary stochastic process. Maximization of the two-step minimum mean-square prediction error is shown to result in a unique minimum-phase $ARMA(n, 1)$ solution. Since maximization of the entropy functional is equivalent to maximization of the minimum mean-square error associated with one-step predictors, this is a true generalization of Burg's maximum entropy extension which results in a stable $AR(n)$ filter [38, 39].

As remarked in section 3.2, from a geometrical point of view [12], the maximum entropy extension is also maximally robust because the new estimates so obtained are always at the centers of the admissible circles. As a result, in that case from (3.17)–(3.22), the radius $\gamma_\infty = b_0$ of the 'final circle' has the maximum possible value $\sqrt{\Delta_n/\Delta_{n-1}}$. Naturally, any other extension cannot have all its new estimates at the centers of the admissible circles, and from (3.20) it now follows that the radius of the 'final circle' for any other admissible extension will be strictly less than that corresponding to the maximum entropy extension. Thus, the radius of the 'final circle' may be taken as a measure of robustness for the extension under consideration. Given $r_0 \to r_n$, in the case of any arbitrary $ARMA(n, 1)$ admissible extension, from (3.139)–(3.140) and the discussion therein,

$$\alpha^2 = \frac{1}{4} \left((a+b)^2 - 1 + (a-b)^2 - 1 + 2\sqrt{((a+b)^2 - 1)((a-b)^2 - 1)} \right)$$

$$> \frac{a^2}{2} + \frac{2(b^2 - 1)}{2} > \frac{a^2}{2}$$

and from (3.143), the 'final radius' is given by

$$b_0 = \frac{\alpha}{g_0} = \frac{\alpha}{a_0 a} \geq \frac{1}{\sqrt{2}a_0} = \frac{1}{\sqrt{2}} \sqrt{\frac{\Delta_n}{\Delta_{n-1}}}, \qquad (3.195)$$

where $|a| > |b| + 1$ and $|b| > 1$. Since the radii of admissible circles form a monotone nonincreasing sequence, this implies that every $ARMA(n, 1)$ extension has a 'final radius,' which is at least 70% of the maximum possible value. Since the unique $ARMA(n, 1)$ extension discussed earlier maximizes the two-step minimum mean-square prediction error $(|b_0|^2 + |b_1|^2)$, the final radius in that case should possess a tighter lower bound.

The uniqueness of the above $ARMA(n, 1)$ extension should not be confused with other admissible $ARMA(n, 1)$ extensions. In fact, from (3.132), with b as given by (3.137) and, for example, letting $|a| > |b| + 1$ results in a bounded-real function $\rho(z)$ that generates an admissible $ARMA(n, 1)$ extension. Thus, there exist an infinite number of admissible $ARMA(n, 1)$

extensions that match the first $(n + 1)$ autocorrelations, and the particular one described above is distinguished by the fact that it maximizes the two-step minimum mean-square prediction error among all such admissible extensions.

The existence of a single AR(n) type of extension as well as the abundance of admissible ARMA$(n, 1)$ extensions that match the given $(n + 1)$ autocorrelations follow from a general result on Padé approximations [49]. More generally, given $(p + q + 1)$ autocorrelations, an ARMA(p, q) admissible extension (Wiener factor), if it exists, is unique, i.e., there can be only one admissible extension of the ARMA(p, q) type that matches the first $(p + q + 1)$ autocorrelations. Of course, if $n < p + q$, then, as in this case, the admissible extensions are not unique.

Given a set of $(n + 1)$ autocorrelations, the question of maximizing the entropy functional among all admissible ARMA$(n, 1)$-type extensions is also interesting by itself. The entropy functional $H(a)$ of an ARMA$(n, 1)$ admissible extension is given by $H(a) = ln\, b_0^2$, with b_0 as in (3.195) where a and b as a pair satisfy $|a| > |b| + 1$, $b = |a_0/a_n| > 1$. In that case, it is easy to show that $\partial H/\partial a > 0$ and $a = \infty$ is its only stationary point, which does not lead to a realizable bounded-real solution for $\rho(z)$ in (3.132). Thus, given $(n + 1)$ autocorrelations, an ARMA$(n, 1)$-type admissible extension that maximizes the entropy functional does not exist.

The natural generalization to k-step predictors $(k > 2)$ that maximize the corresponding minimum mean-square error is shown to result in a well-structured stable ARMA$(n, k - 1)$ filter in Appendix 3.A. In general, to identify an ARMA(p, q) system, $(p + q + 1)$ autocorrelations are needed. But if fewer autocorrelations are to be used to identify the same system, then this approach leads to a feasible solution. In that case, beginning with $(p+1)$ autocorrelations, maximization of $(q+1)$-step minimum mean-square prediction error leads to a stable ARMA(p, q) solution. The parametrization of such filters is, of course, much more complicated.

Appendix 3.A

Maximization of the k-Step
Minimum Mean-Square Prediction Error

Given a finite set of $(n + 1)$ autocorrelations r_0, r_1, ..., r_n, the problem here is to find the transfer function corresponding to the extension that maximizes the minimum mean-square prediction error associated with the k-step predictor. Using (2.65) and (2.66), the k-step minimum mean-square prediction error is given by

$$P_k = \sum_{i=0}^{k-1} |b_i|^2 = (1 + |\alpha_1|^2 + |\alpha_2|^2 + \cdots + |\alpha_{k-1}|^2)e^{2d_0}$$

$$= (1 + |\alpha_1|^2 + |\alpha_2|^2 + \cdots + |\alpha_{k-1}|^2) exp \left(\frac{1}{2\pi} \int_{-\pi}^{\pi} \ln S(\theta) d\theta \right) \quad (3.A.1)$$

with $2d_0$ as defined in (2.60). Maximization of P_k with respect to the free parameters $r_{n+1}, r_{n+2}, \ldots$ leads to

$$\frac{\partial P_k}{\partial r_m} = 0, \quad |m| > n,$$

and using (3.A.1), this is equivalent to

$$\frac{\partial (2d_0)}{\partial r_m} \left(\sum_{i=0}^{k-1} |\alpha_i|^2 \right) + \frac{\partial}{\partial r_m} \left(\sum_{i=0}^{k-1} |\alpha_i|^2 \right) = 0, \quad |m| > n. \quad (3.A.2)$$

We will show that the left-hand side of the above expression (3.A.2) reduces to

$$\frac{1}{2\pi} \int_{-\pi}^{\pi} \frac{1}{S(\theta)} \left| \sum_{i=0}^{k-1} \alpha_i e^{ji\theta} \right|^2 e^{jm\theta} d\theta \quad (3.A.3)$$

and this in turn implies that the periodic function given by the integrand of (3.A.3) is identically equal to zero for all $|m| > n$. As a result, the nonnegative periodic function $\left| \sum_{i=0}^{k-1} \alpha_i e^{ji\theta} \right|^2 / S(\theta)$ must truncate after n terms beyond the origin and hence

$$\frac{1}{S(\theta)} \left| \sum_{i=0}^{k-1} \alpha_i e^{ji\theta} \right|^2 = \sum_{i=-n}^{n} f_i e^{ji\theta} = \left| \sum_{m=0}^{n} g_m e^{jm\theta} \right|^2,$$

where $\sum_{m=0}^{n} g_m z^m$ represents a strict Hurwitz polynomial (see comments after (3.128)). Thus,

$$S(\theta) \overset{\triangle}{=} |H_k(e^{j\theta})|^2, \quad (3.A.4)$$

where

$$H_k(z) \overset{\triangle}{=} \frac{\alpha_0 + \alpha_1 z + \cdots + \alpha_{k-1} z^{k-1}}{g_0 + g_1 z + g_2 z^2 + \cdots + g_n z^n} \sim \text{ARMA}(n, k-1), \quad (3.A.5)$$

i.e., given $r_0, r_1, \ldots, r_n$, the transfer function that maximizes the k-step minimum mean-square prediction error is a stable ARMA$(n, k-1)$. Thus, to complete the proof, it is enough to establish the equality between the left-hand side of (3.A.2) and (3.A.3). Towards this, from (2.60), we obtain

$$\frac{\partial (2d_0)}{\partial r_m} = \frac{1}{2\pi} \int_{-\pi}^{\pi} \frac{1}{S(\theta)} e^{jm\theta} d\theta \quad (3.A.6)$$

and by direct expansion,

$$\left| \sum_{i=0}^{k-1} \alpha_i e^{ji\theta} \right|^2 = \sum_{i=0}^{k-1} |\alpha_i|^2 + \sum_{i=0}^{k-1} \sum_{l=0}^{i-1} 2 Re(\alpha_i^* e^{-ji\theta} \cdot \alpha_l e^{jl\theta}). \quad (3.A.7)$$

Substituting (3.A.6) in (3.A.2) and (3.A.7) in (3.A.3), the desired equality condition between (3.A.2) and (3.A.3) reduces to

$$\sum_{i=0}^{k-1} \frac{\partial}{\partial r_m} |\alpha_i|^2 = \frac{1}{2\pi} \int_{-\pi}^{\pi} \frac{e^{jm\theta}}{S(\theta)} \left[\sum_{i=0}^{k-1} \sum_{l=0}^{i-1} 2Re(\alpha_i^* e^{-ji\theta} \cdot \alpha_l e^{jl\theta}) \right] d\theta, \quad (3.A.8)$$

for $|m| > n$. However, the left-hand side of the above equation simplifies to

$$\sum_{i=0}^{k-1} \frac{\partial}{\partial r_m} |\alpha_i|^2 = \sum_{i=0}^{k-1} \frac{\partial}{\partial r_m} (\alpha_i \alpha_i^*) = \sum_{i=0}^{k-1} 2Re \left(\alpha_i^* \frac{\partial \alpha_i}{\partial r_m} \right) \quad (3.A.9)$$

and hence from (3.A.9), for (3.A.8) to hold, it is enough to establish that

$$\frac{\partial \alpha_i}{\partial r_m} = \frac{1}{2\pi} \int_{-\pi}^{\pi} \frac{e^{j(m-i)\theta}}{S(\theta)} \left(\sum_{l=0}^{i-1} \alpha_l e^{jl\theta} \right) d\theta, \quad |m| > n. \quad (3.A.10)$$

We will prove (3.A.10) through an induction argument on i by assuming it to be true up to $i = k - 1$ and extending that result to $i = k$. Clearly (3.A.10) is well known for $i = 0$ and is proved for $i = 1$ in section 3.4. To make further progress for $i \geq 2$, we can make use of (2.67). Thus

$$\alpha_i = \frac{1}{i}(d_1 \alpha_{i-1} + 2d_2 \alpha_{i-2} + \cdots + (i-1)d_{i-1}\alpha_1 + id_i\alpha_0), \quad i > 1, \quad (3.A.11)$$

and hence for $i = k$,

$$\frac{\partial \alpha_k}{\partial r_m} = \frac{1}{k} \left(\sum_{p=1}^{k} p \frac{\partial d_p}{\partial r_m} \alpha_{k-p} + \sum_{p=1}^{k} p d_p \frac{\partial \alpha_{k-p}}{\partial r_m} \right)$$

$$= \frac{1}{k} \left(\sum_{p=1}^{k} p \alpha_{k-p} \cdot \frac{1}{2\pi} \int_{-\pi}^{\pi} \frac{e^{j(m-p)\theta}}{S(\theta)} d\theta + \sum_{p=1}^{k-1} p d_p \frac{\partial \alpha_{k-p}}{\partial r_m} \right)$$

where we have made use of (2.61) and the identity $\alpha_0 = 1$. By assumption, since (3.A.10) is true up to $i = k - 1$, substituting that into the above expression, we obtain

$$\frac{\partial \alpha_k}{\partial r_m} = \frac{1}{2\pi} \int_{-\pi}^{\pi} \frac{a(\theta) e^{j(m-k)\theta}}{S(\theta)} d\theta,$$

where

$$a(\theta) = 1 + \frac{1}{k} \sum_{p=1}^{k-1} p \left(\alpha_{k-p} e^{j(k-p)\theta} + d_p e^{jp\theta} \sum_{l=0}^{k-p-1} \alpha_l e^{jl\theta} \right). \quad (3.A.12)$$

Expanding the terms in (3.A.12), we obtain

$$
\begin{aligned}
a(\theta) = & + \frac{1}{k}\left[\alpha_{k-1}e^{j(k-1)\theta} + d_1 e^{j\theta}\left(1 + \alpha_1 e^{j\theta} + \cdots + \alpha_{k-2}e^{j(k-2)\theta}\right)\right.\\
& + 2\alpha_{k-2}e^{j(k-2)\theta} + 2d_2 e^{j2\theta}\left(1 + \alpha_1 e^{j\theta} + \cdots + \alpha_{k-3}e^{j(k-3)\theta}\right)\\
& \vdots\\
& + (k-2)\alpha_2 e^{j2\theta} + (k-2)d_{k-2}e^{j(k-2)\theta}(1 + \alpha_1 e^{j\theta})\\
& \left. + (k-1)\alpha_1 e^{j\theta} + (k-1)d_{k-1}e^{j(k-1)\theta}\right].
\end{aligned}
$$

Rearranging the terms in the above expression with the help of (3.A.11), the above expression simplifies to

$$
a(\theta) = \sum_{l=0}^{k-1} \alpha_l e^{jl\theta}.
$$

This establishes the identity (3.A.10) for $i = k$, and completes the proof.

Thus, given $r_0, r_1, \ldots, r_n$, the spectral density that maximizes the k-step minimum mean-square prediction error has the form

$$
S(\theta) = |H_k(e^{j\theta})|^2
$$

where the associated stable transfer function is given by

$$
H_k(z) = \frac{\displaystyle\sum_{m=0}^{k-1} \alpha_m z^m}{\displaystyle\sum_{m=0}^{n} g_m z^m} \sim \mathrm{ARMA}(n, k-1).
$$

Notice that the coefficients in the numerator factor here are precisely those given by (2.65)–(2.66). In general, the numerator polynomial $\sum_{m=0}^{k-1} \alpha_m z^m$ need not be Hurwitz (see (3.133)) and hence the Wiener factor (minimum-phase) $B_k(z)$ associated with the above extension is given by $H_k(z)b(z)$ where $b(z)$ is an appropriate all-pass function that makes $(\sum_{m=0}^{k-1} \alpha_m z^m)b(z)$ a Hurwitz polynomial. Moreover, from (3.54)–(3.55), this minimum-phase factor $B_k(z) = H(z)b(z)$ should, if at all, follow for a specific choice of rational bounded-real function $\rho_{n+1}(z)$ with numerator degree $(k-2)$ and denominator degree $(k-1)$. The denominator degree requirement for $H_k(z)$ together with the above restrictions on its numerator factor completely specifies the b.r. function $\rho_{n+1}(z)$ and hence the Wiener factor. For the existence of such a Wiener factor, the set of equations so obtained must yield a b.r. solution. Whether this is always possible for $k \geq 3$ is still an open question. Needless to say, the possibility of the all-pass factor $b(z)$ in the Wiener factor complicates the situation.

Appendix 3.B

Uniqueness of $\rho(z)$

For uniqueness of $\rho(z)$ in (3.132), it remains to show that other feasible b.r. forms such as

$$\rho(z) = \frac{1 + d_1 z + d_2 z^2 + \cdots + d_{m-1} z^{m-1}}{a + bz + c_1 z^2 + c_2 z^3 + \cdots + c_{m-1} z^m}, \quad m > 1, \qquad (3.B.1)$$

cannot give rise to ARMA$(n, 1)$-type Wiener factors in (3.130), unless $c_i = d_i = 0$, $i = 1 \rightarrow m - 1$. To prove this, consider first

$$\rho(z) = \frac{1 + dz}{a + bz + cz^2}. \qquad (3.B.2)$$

From (3.54), since $B_2(z) = \Gamma(z)/D_n(z)$, the degree one requirement in the numerator of the Wiener factor $B_2(z)$ in (3.130) translates into the same requirement for the numerator of $\Gamma(z)$. Thus, with (3.B.2) in (3.55), we obtain

$$(a + bz + cz^2)(a + bz + cz^2)_* - (1 + dz)(1 + dz)_*$$
$$= (\alpha + \beta z)(\alpha + \beta z)_*$$

which is the same as

$$ac(z^2 + z^{-2}) + (ab + bc - d)(z + z^{-1}) + a^2 + b^2 + c^2 - d^2 - 1$$
$$= \alpha\beta(z + z^{-1}) + \alpha^2 + \beta^2.$$

Comparing the equal powers of z on both sides, we must have either $a = 0$ or $c = 0$. If $a = 0$, then

$$\rho(z) = \frac{1 + dz}{z(b + cz)}$$

and it is not a b.r. function because of the pole at the origin. On the other hand, if $c = 0$, then

$$\rho(z) = \frac{1 + dz}{a + bz}$$

and substituting this in (3.56), we obtain

$$D_n(z) = \frac{-da_0 z^{n+2} + (ba_n - a_0 + da_1)z^{n+1} + \cdots}{a + bz}.$$

Since $D_n(z)$ is of degree n, necessarily $da_0 = 0 \implies d = 0$, since $a_0 \neq 0$ and this leads to

$$\rho(z) = \frac{1}{a + bz}. \qquad (3.B.3)$$

Arguing in a similar manner, starting with any $m > 2$, (3.B.1) can be shown to reduce to (3.B.3).

Appendix 3.C

Negative Discriminant

In this Appendix, we will show that the discriminant D in (3.154) is always negative. To prove this, with (3.152)–(3.153) in (3.154) and letting $u = a_1^2/a_0^2$, the discriminant D simplifies to

$$D = \left(\frac{q}{2}\right)^2 + \left(\frac{p}{3}\right)^3$$

$$= \frac{5}{9}\left(1 + \frac{1}{b^4}\right)u - \frac{26}{9b^2}u - \frac{1}{27}\left(u^3 + 6\left(1 + \frac{1}{b^2}\right)u^2 + 8\left(1 + \frac{1}{b^2}\right)^3\right), \quad u > 0.$$

$$(3.C.1)$$

Since $|b| > 1$, we have $(1/|b|) < 1$, which gives $(u/b^4) < (u/b^2)$ and with this in the above expression, it simplifies to

$$D \le \frac{5}{9}u - \frac{7}{3b^2}u - \frac{1}{27}\left(u^3 + 6\left(1 + \frac{1}{b^2}\right)u^2 + 8\left(1 + \frac{1}{b^2}\right)^3\right)$$

$$= -\frac{7}{3b^2}u - \frac{A(u)}{27}, \qquad (3.C.2)$$

where, by definition,

$$A(u) = u^3 + 6\left(1 + \frac{1}{b^2}\right)u^2 - 15u + 8\left(1 + \frac{1}{b^2}\right)^3. \qquad (3.C.3)$$

Clearly, to establish $D < 0$, it is enough to show that $A(u) > 0$ for all $u > 0$ and $|b| > 1$. Let $\beta = 2(1 + 1/b^2)$, then $|b| > 1$ implies $2 < \beta < 4$ and (3.C.3) simplifies to

$$A(u) = u^3 + 3\beta u^2 - 15u + \beta^3. \qquad (3.C.4)$$

The desired result follows if $A_{min} > 0$ for $u > 0$ and $2 < \beta < 4$. To find its minimum, note that

$$\frac{\partial A(u)}{\partial u} = 3u^2 + 6\beta u - 15 = 0$$

yields the only positive solution

$$u' = -\beta + \sqrt{\beta^2 + 5}$$

and the minimum value for (3.C.4) is given by

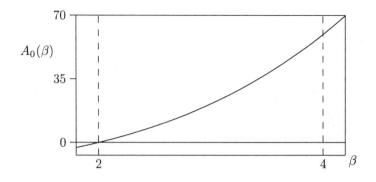

FIGURE 3.8. Positivity of $A_0(\beta)$ for $2 < \beta < 4$ (see (3.C.5)).

$$A(u') = 3\beta^3 - 2(\beta^2 + 5)\sqrt{\beta^2 + 5} + 15\beta \overset{\triangle}{=} A_0(\beta). \qquad (3.C.5)$$

As Fig. 3.8 shows, $A_0(\beta) > 0$ for $2 < \beta < 4$, and this completes the proof.

Appendix 3.D

Existence of the Second Bounded-Real Solution

To investigate the solution due to x_2 in (3.157), notice that under the assumptions $x_1 > 0$, $b > 1$, the second root x_2 and hence $a^{(2)} \overset{\triangle}{=} x_2 b$ are negative and as a result, the condition $|a^{(2)} - b| > 1$ is automatically satisfied. For bounded reality of the $\rho(z)$ so obtained from $a^{(2)}$ and b, it remains to show that $|a^{(2)} + b| \geq 1$. To prove this, let $y = a + b = b(a/b+1) = b(x+1)$ or $x = (y/b) - 1$. Substituting this in (3.151) yields the polynomial

$$V(y) = y^3 - 3by^2 + (3+p)b^2 y + (-1 - p + q)b^3. \qquad (3.D.1)$$

Let y_1, y_2 and y_3 denote the three real roots of this cubic equation. We will show that all these roots have magnitudes greater than unity. In fact, from (3.165) and the assumptions, we have $y_1 = a^{(1)} + b > 1$ ($a^{(1)} = x_1 b$) and further using (3.160), $V(0) = (-1 - p + q)b^3 > 0$. Since one of its roots y_1 is greater than unity and the ordinate-intercept $V(0)$ is positive, to establish the above claim, it is enough to show that $V(1) > 0$ and $V(-1) > 0$. In that case, because of its cubic nature, $V(y)$ must have the shape shown in Fig. 3.9 which corresponds to a strict Hurwitz polynomial. An easy manipulation shows that

$$V(1) = (b - 1)\left(\frac{a_1}{a_n} - 1 - b\right)^2 > 0 \qquad (3.D.2)$$

and

$$V(-1) = (b + 1)\left(\frac{a_1}{a_n} + 1 - b\right)^2 > 0, \qquad (3.D.3)$$

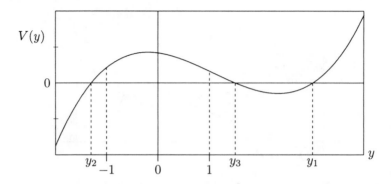

FIGURE 3.9. Structure of $V(y)$ for $b > 1$ and $x_1 > 0$. Here $V(0)$, $V(1)$ and $V(-1)$ are all positive. (See (3.D.1)–(3.D.3).)

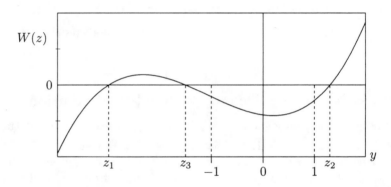

FIGURE 3.10. Structure of $W(z)$ for $b > 1$ and $x_1 < 0$. Here $W(0)$, $W(1)$ and $W(-1)$ are all negative. (See (3.D.4)–(3.D.6).)

which, in particular, establishes that $|y_2| = |a^{(2)} + b| > 1$. Thus, $\rho_2(z) = 1/(a^{(2)} + bz)$ also leads to a bounded-real function and consequently to an admissible solution to the original problem.

To complete the proof, it remains to exhaust the remaining choices for b and x_1, viz., $b > 1$, $x_1 < 0$; $b < -1$, $x_1 > 0$; $b < -1$, $x_1 < 0$. Of these, for example, when $b > 1$ and $x_1 < 0$, we have $a^{(1)} < 0$ and $a^{(2)} > 0$ and $a^{(3)} = x_3 b > 0$, and hence $|a^{(2)} + b| > 1$ is automatic. In that case, we need $|a^{(2)} - b| > 1$ and letting $z = a - b = b(x - 1)$ or $x = (z/b) + 1$, (3.150) leads to the new cubic equation

$$W(z) = z^3 + 3bz^2 + (3 + p)b^2 z + (1 + p + q)b^3. \qquad (3.D.4)$$

Once again since $z_1 = b(x_1 - 1) < -1$ and $W(0) < 0$, to prove $|z_2| > 1$, it is enough to show that $W(1) < 0$ and $W(-1) < 0$. In fact,

$$W(1) = -(b+1)\left(\frac{a_1}{a_n} - 1 + b\right)^2 < 0 \qquad (3.D.5)$$

and

$$W(-1) = -(b-1)\left(\frac{a_1}{a_n} - 1 + b\right)^2 < 0 \qquad (3.D.6)$$

and this situation corresponds to that in Fig. 3.10. Thus, $|z_2| = |a^{(2)} - b| > 1$, etc. Interestingly, the above arguments also establish that $|a^{(3)} \pm b| > 1$. However, this is irrelevant since $|x_3| < 1$. The remaining cases can be dealt with in a similar manner. Notice that, to determine the bounded reality for $\rho_2(z)$, we have made use of the previously established fact that $|a^{(1)} \pm b| > 1$.

Problems

1. Let $A_n(z)$ represent the Levinson polynomial of the first kind.

   (a) If one of the roots of $A_n(z)$ lie on the unit circle, then show that all other roots of $A_n(z)$ are simple and lie on the unit circle.

   (b) If the reflection coefficient $s_k \neq 0$, then show that $A_k(z)$ and $A_{k+1}(z)$ have no roots in common.

2. (a) Show that if every reflection coefficient generated from a power spectral density function $S(\theta)$ is strictly bounded by unity and they are square summable, then this implies the Paley-Wiener criterion (see also (3.80)). Notice that even though the Paley-Wiener criterion implies $\Delta_k > 0$, $k = 0 \to \infty$ (see (2.48)), the converse need not be true since $\Delta_k > 0$, $k = 0 \to \infty$, only implies $|s_k| < 1$, $k = 1 \to \infty$, and not their square summability. For example, a discrete process with an infinite number of components satisfies $\Delta_k > 0$, $k = 0 \to \infty$, but not the Paley-Wiener criterion.

   (b) Use an induction argument on (3.75) to show that the Paley-Wiener criterion implies

   $$Re\,(R_k) > 0\,, \quad k \geq 0\,.$$

   Here R_k represents the characteristic impedance of the $(k+1)^{th}$ line.

3. Let $A_m(z)$, $m \leq n$ represent the Levinson polynomials of the first kind generated from the first $(n+1)$ Fourier coefficients of $S(\theta)$ that satisfy (3.2) and (3.3). Show that the reciprocal polynomials $\tilde{A}_m(z) = z^m A_m^*(1/z^*)$ satisfy

   $$\frac{1}{2\pi} \int_{-\pi}^{\pi} \tilde{A}_n(e^{j\theta})\tilde{A}_m^*(e^{j\theta})S(\theta)d\theta = \delta_{nm}\,, \qquad (3.P.1)$$

and with γ_k as defined in (3.17)

$$\frac{1}{2\pi}\int_{-\pi}^{\pi} A_n(e^{j\theta})A_m^*(e^{j\theta})S(\theta)d\theta = \frac{A_m(0)}{A_n(0)} = \sqrt{\frac{\gamma_n}{\gamma_m}} \le 1, \quad m \le n,$$
(3.P.2)

i.e., the reciprocal polynomials are orthonormal on the unit circle with respect to $S(\theta)$. (Note that $A_n(z)$ are not orthonormal.)
Hint: Use (3.67) to show that

$$\frac{1}{2\pi}\int_{-\pi}^{\pi} \tilde{A}_n(e^{j\theta})e^{jk\theta}S(\theta)d\theta = \begin{cases} 0, & 0 \le k < n \\ \sqrt{\dfrac{\Delta_n}{\Delta_{n-1}}}, & k = n. \end{cases}$$
(3.P.3)

4. Use (3.83) to show that (3.44) can be simplified to give

$$\left| Z(z) - \frac{2B_n(z)}{A_n(z)} \right| \le K \frac{|z|^{n+1}}{1-|z|}, \quad |z| < 1,$$
(3.P.4)

where K is a finite constant. Thus

$$\frac{2B_n(z)}{A_n(z)} \longrightarrow Z(z)$$

uniformly in every circle $|z| \le R < 1$ and similarly, show that the bounded function

$$\frac{h_n(z)}{g_n(z)} \longrightarrow \rho(z) = \frac{Z(z) - R_0}{Z(z) + R_0}$$
(3.P.5)

uniformly in $|z| < 1$.

5. If the reflection coefficients satisfy $s_k = \rho^k$, $k = 1 \to \infty$, where $|\rho| < 1$, then show that all zeros of the Levinson polynomial $A_n(z)$, $n = 1 \to \infty$ lie on the circle of radius $1/\rho$ [50].

6. Let $A_n(z)$, $n = 0 \to \infty$, represent the Levinson polynomials associated with the reflection coefficients $\{s_k\}_{k=1}^{\infty}$. Define

$$s_k' = \lambda^k s_k, \quad |\lambda| = 1, \quad k = 1 \to \infty.$$

Show that the new set of Levinson polynomials are given by $A_n(\lambda z)$, $n = 0 \to \infty$. Thus if $\{s_k\}_{k=1}^{\infty}$ is replaced by $\{(-1)^k s_k\}_{k=1}^{\infty}$, the new set of Levinson polynomials are given by $A_n(-z)$, $n = 0 \to \infty$.

7. Let $A_n(z)$ and $B_n(z)$ represent the Levinson polynomials of the first and second kind associated with the reflection coefficients $\{s_k\}_{k=1}^{n}$. Consider a new set of reflection coefficients

$$s_k' = \lambda s_k, \quad |\lambda| = 1, \quad k = 1 \to n,$$

and let $A_k(\lambda, z)$ and $B_k(\lambda, z)$ represent the corresponding new set of Levinson polynomials. Then show that [35]

(a) $A_k(\lambda, z) = \left(\dfrac{1+\lambda}{2}\right) A_k(z) + \left(\dfrac{1-\lambda}{2}\right) B_k(z) , \quad k = 1 \to n ,$

$B_k(\lambda, z) = \left(\dfrac{1+\lambda}{2}\right) B_k(z) + \left(\dfrac{1-\lambda}{2}\right) A_k(z) , \quad k = 1 \to n ,$

with $A_0(\lambda, z) = 1/\sqrt{r_0}$ and $B_0(\lambda, z) = \sqrt{r_0}/2$.

(b) The new autocorrelation sequence $\{r'_k\}_{k=0}^n$ is given by

$$r'_0 = r_0, \quad r'_1 = \lambda r_1, \quad r'_k = \lambda r_k - \frac{1-\lambda}{r_0} \sum_{i=1}^{k-1} r_i r'_{k-i}, \qquad (3.P.6)$$

where $\{r_k\}_{k=0}^n$ represent the original set of autocorrelations. Thus in particular, when $\{s_k\}_{k=1}^n \to \{-s_k\}_{k=1}^n$, the role of $A_k(z)$ and $B_k(z)$, $k = 1 \to n$, gets interchanged.

8. Consider an MA(1) process with transfer function

$$H(z) = 1 - z .$$

(a) Show that $\Delta_k = k + 2, \ k \geq 0$, and

$$s_k = -\frac{1}{k+1}, \quad k = 1 \to \infty . \qquad (3.P.7)$$

(b) Consider the new process with reflection coefficients

$$s'_k = -s_k = \frac{1}{k+1}, \quad k = 1 \to \infty , \qquad (3.P.8)$$

and $r'_0 = r_0 = 2$. Clearly $\sum_{k=1}^\infty |s'_k|^2 < \infty$. Show that the new autocorrelation sequence r'_k is given by

$$r'_0 = 2, \quad r'_k = 1, \quad k \geq 1 ,$$

and hence $S(\theta)$ is not integrable since $r'_k \nrightarrow 0$ as $k \to \infty$. Show that the new positive function $Z'(z)$ has a pole at $z = 1$. Thus although the Paley-Wiener criterion is satisfied here, the reflection coefficients do not form a summable sequence and the spectral density has an isolated pole at $\theta = 0$.

9. **Toeplitz Inverses** (Gohberg and Semuncul) [42, 43] Consider the Hermitian Toeplitz matrix $\mathbf{T}_n > 0$ generated from $r_0 \to r_n$.

(a) Show that its inverse can be expressed as

$$\mathbf{T}_n^{-1} = \mathbf{M}_0 \mathbf{M}_0^* - \mathbf{M}_n \mathbf{M}_n^* = \mathbf{M}_0^* \mathbf{M}_0 - \mathbf{M}_n^* \mathbf{M}_n , \qquad (3.P.9)$$

where $\mathbf{M}_0$ and $\mathbf{M}_n$ represent upper triangular Toeplitz matrices given by

$$
\mathbf{M}_0 \overset{\triangle}{=}
\begin{bmatrix}
a_0 & a_1 & \cdots & a_n \\
0 & a_0 & \cdots & a_{n-1} \\
\vdots & \vdots & \ddots & \vdots \\
0 & 0 & \cdots & a_0
\end{bmatrix}
, \quad
\mathbf{M}_n \overset{\triangle}{=}
\begin{bmatrix}
0 & a_n^* & \cdots & a_1^* \\
0 & 0 & \cdots & a_2^* \\
\vdots & \vdots & \ddots & \vdots \\
0 & 0 & \cdots & 0
\end{bmatrix}.
$$

Here, a_k, $k = 0 \to n$, represent the coefficients of the nth-degree Levinson polynomial $A_n(z)$ in (3.61) generated from $r_0 \to r_n$.

(b) Alternatively, show that

$$
\mathbf{T}_{n-1}^{-1} = \mathbf{X}_1 \mathbf{X}_1^* - \mathbf{X}_n \mathbf{X}_n^* = \mathbf{X}_1^* \mathbf{X}_1 - \mathbf{X}_n^* \mathbf{X}_n , \tag{3.P.10}
$$

where $\mathbf{X}_1$ and $\mathbf{X}_n$ are upper triangular Toeplitz matrices given by

$$
\mathbf{X}_1 \overset{\triangle}{=}
\begin{bmatrix}
a_0 & a_1 & \cdots & a_n \\
0 & a_0 & \cdots & a_{n-1} \\
\vdots & \vdots & \ddots & \vdots \\
0 & 0 & \cdots & a_0
\end{bmatrix}
, \quad
\mathbf{X}_n \overset{\triangle}{=}
\begin{bmatrix}
a_n^* & a_{n-1}^* & \cdots & a_1^* \\
0 & a_n^* & \cdots & a_2^* \\
\vdots & \vdots & \ddots & \vdots \\
0 & 0 & \cdots & a_n^*
\end{bmatrix}.
$$

(c) Let $A_k(z) = \sum_{i=0}^n a_i^{(k)} z^i$ represent the kth-degree Levinson polynomial and define the lower triangular matrix

$$
\mathbf{A} \overset{\triangle}{=}
\begin{bmatrix}
a_0^{(n)*} & 0 & \cdots & 0 \\
a_1^{(n)*} & a_0^{(n-1)*} & \cdots & 0 \\
\vdots & \vdots & \ddots & \vdots \\
a_n^{(n)*} & a_{n-1}^{(n-1)*} & \cdots & a_0^{(0)*}
\end{bmatrix}.
$$

Show that

$$
\mathbf{A}^* \mathbf{T}_n \mathbf{A} = \mathbf{I}_{n+1} .
$$

(Although $\mathbf{T}_n^{-1} = \mathbf{A}\mathbf{A}^*$, notice that unlike (3.P.9)–(3.P.10), $\mathbf{A}$ is only lower triangular, and not Toeplitz!)

10. Show that there are no real values for a_0, a_1 and b with $|b| > 1$ that satisfy the equation $\cos\phi = 1$, where $\cos\phi$ is as given in (3.159).

11. Complete the analysis for the parametric formulation of the two-step predictor discussed in section 3.4 in the case of complex autocorrelations.

4

ARMA-System Identification and Rational Approximation

4.1 Introduction

The problem of identifying a linear time-invariant (LTI) system from measurements of its output response to a known input excitation, such as a white noise source, is of fundamental importance in many areas in engineering. Such an identification naturally allows one to predict the system outputs, and as a result this problem has considerable impact in areas where forecasting is of extreme importance. Such applications exist in diverse fields such as ocean modeling, economics system theory, seismology, pattern recognition and signal processing, to name a few.

It may be remarked that physical systems such as above may not be linear or time-invariant. Nevertheless, over a reasonable interval, the outputs can be assumed to be generated by an LTI system, and for an accurate description over a long period it may be enough to update the system parameters adaptively. Moreover, the LTI system under consideration may not possess a rational transfer function. In that case it becomes necessary to model such systems with reasonable accuracy through a finite-order recursive model that is optimal in some fashion. Hence, within this approach, physical systems can be parametrized accurately through a finite-order recursive model, provided the system outputs can be used to describe the system under consideration in a satisfactory manner.

Our approach in this chapter is to make use of the theory developed in the previous chapter to obtain a new technique for identifying rational LTI systems from a finite set of their output observations under white noise excitation. This information can be either in the form of their output autocorrelations or output data itself. When the actual system under consideration is not rational, the theory developed in the rational case is made use of in obtaining faithful rational approximations of the original nonrational system.

Let $S(\theta)$ represent the power spectrum associated with the output process $x(n)$ as in (3.1) and r_k, $k = 0 \rightarrow \infty$, its autocorrelations. Then under the integrability condition (3.2) and the physical-realizability condition (3.3), $S(\theta)$ can be uniquely factorized as in (3.5) in terms of its minimum-phase factor $B(z)$ that is analytic together with its inverse in the interior of the unit circle. Such a situation can be represented as in Fig. 2.2, where the input is a stationary white noise process of unit spectral density. Since

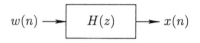

FIGURE 4.1. Minimum-phase representation of a stationary process.

the spectrum can be used to identify the system uniquely only up to its minimum-phase factor, we will assume the unknown system under consideration to be minimum-phase, and the output $x(n)$ is generated by the action of a stationary white noise process $w(n)$ of unit spectral density on this minimum-phase system $H(z)$ (see Fig. 4.1).

In the rational case, $H(z)$ is given by

$$H(z) = \frac{\beta_0 + \beta_1 z + \cdots + \beta_q z^q}{1 + \alpha_1 z + \cdots + \alpha_p z^p} \triangleq \frac{\beta(z)}{\alpha(z)}. \tag{4.1}$$

Here $\alpha(z)$ is a strict Hurwitz polynomial of degree p, and $\beta(z)$ is in general a Hurwitz polynomial of degree q, thus allowing the output spectrum to have zeros on the unit circle. Here, p and q are known as the model order of the system and α_k, $k = 1 \rightarrow p$, and β_k, $k = 0 \rightarrow q$, represent the system parameters. Since there are $p + q + 1$ unknowns, the system identification problem assumes that the first $p+q+1$ autocorrelations $r_0, r_1, \ldots, r_{p+q}$ of the output process are given. Extensive work has been reported on this case, which we will refer to as the exact or deterministic case. A more realistic problem is to identify these parameters from the system output data itself. Since the interrelationship that exists between the system parameters and the output autocorrelations is crucial to our understanding, we postpone the discussion of the known data case till section 4.1.2.

4.1.1 EXACT AUTOCORRELATIONS CASE

In this case, it is not difficult to derive the relationship between the ARMA parameters and the output autocorrelations. In fact, from Fig. 4.1 and the above rational model, we have

$$x(n) = -\sum_{k=1}^{p} \alpha_k x(n-k) + \sum_{k=0}^{q} \beta_k w(n-k) \tag{4.2}$$

and since the present input $w(n)$ is uncorrelated with its own past values as well as the past values of the system output, this gives

$$E[w(n) w^*(n-k)] = 0, \quad k \geq 1, \tag{4.3}$$

and

$$E[w(n) x^*(n-k)] = 0, \quad k \geq 1. \tag{4.4}$$

By making use of (4.2)–(4.4), we obtain

$$
\begin{aligned}
r_k &= E[x(n)\, x^*(n-k)] \\
&= -\sum_{i=1}^{p} \alpha_i\, E[x(n-i)\, x^*(n-k)] + \sum_{i=0}^{q} \beta_i\, E[w(n-i)\, x^*(n-k)] \\
&= -\sum_{i=1}^{p} \alpha_i\, r_{k-i} + \sum_{i=0}^{q} \beta_i\, E[w(n-i)\, x^*(n-k)]
\end{aligned}
$$

and hence in general

$$
\sum_{i=1}^{p} \alpha_i\, r_{k-i} + r_k \neq 0, \quad k \leq q, \tag{4.5}
$$

and

$$
\sum_{i=1}^{p} \alpha_i\, r_{k-i} + r_k = 0, \quad k \geq q+1. \tag{4.6}
$$

The infinite set of linear equations in (4.6) can be used to determine the AR parameters as well as the model order p and q. In particular, notice that the first p equations in (4.6) can be written in matrix form as

$$
\begin{bmatrix}
r_q & r_{q-1} & \cdots & r_{q-p+1} \\
r_{q+1} & r_q & \cdots & r_{q-p+2} \\
\vdots & \vdots & \ddots & \vdots \\
r_{q+p-1} & r_{q+p-2} & \cdots & r_q
\end{bmatrix}
\begin{bmatrix}
\alpha_1 \\ \alpha_2 \\ \vdots \\ \alpha_p
\end{bmatrix}
= -
\begin{bmatrix}
r_{q+1} \\ r_{q+2} \\ \vdots \\ r_{q+p}
\end{bmatrix}, \tag{4.7}
$$

or more compactly as

$$
\mathbf{R}(p,q) \cdot
\begin{bmatrix}
\alpha_1 \\ \alpha_2 \\ \vdots \\ \alpha_p
\end{bmatrix}
= -
\begin{bmatrix}
r_{q+1} \\ r_{q+2} \\ \vdots \\ r_{q+p}
\end{bmatrix}, \tag{4.8}
$$

where

$$
\mathbf{R}(p,q) \triangleq
\begin{bmatrix}
r_q & r_{q-1} & \cdots & r_{q-p+1} \\
r_{q+1} & r_q & \cdots & r_{q-p+2} \\
\vdots & \vdots & \ddots & \vdots \\
r_{q+p-1} & r_{q+p-2} & \cdots & r_q
\end{bmatrix} \tag{4.9}
$$

is a $p \times p$ matrix and the p equations in (4.7) are known as the extended Yule-Walker equations[1] [51–60]. Since α_k, $k = 1 \to p$, are unique, the $p \times p$

---

[1]The p equations in (4.7) with $q = 0$ reduce to the standard Yule-Walker equations for an AR(p) system [56].

matrix $\mathbf{R}(p, q)$ in the above equation must be nonsingular, and in principle the AR parameters, α_k, $k = 0 \rightarrow p$, with $\alpha_0 = 1$, can be obtained from (4.7). Similarly, the first $p + 1$ equations in (4.6) has the matrix representation

$$\begin{bmatrix} r_{q+1} & r_q & \cdots & r_{q-p+1} \\ r_{q+2} & r_{q+1} & \cdots & r_{q-p+2} \\ \vdots & \vdots & \ddots & \vdots \\ r_{q+p+1} & r_{q+p} & \cdots & r_{q+1} \end{bmatrix} \begin{bmatrix} 1 \\ \alpha_1 \\ \vdots \\ \alpha_p \end{bmatrix} = \begin{bmatrix} 0 \\ 0 \\ \vdots \\ 0 \end{bmatrix}. \tag{4.10}$$

Letting

$$\boldsymbol{\alpha}_1 = [\, 1, \, \alpha_1, \, \ldots, \, \alpha_p \,]^T \tag{4.11}$$

and making use of (4.9), the above matrix equation reduces to

$$\mathbf{R}(p+1, q+1)\boldsymbol{\alpha}_1 = \mathbf{0}, \tag{4.12}$$

where $\mathbf{R}(p+1, q+1)$ is a $(p+1) \times (p+1)$ matrix. Clearly since $\boldsymbol{\alpha}_1 \neq \mathbf{0}$, the above coefficient matrix $\mathbf{R}(p+1, q+1)$ must be singular. Thus, combining the above two determinantal conditions, for an ARMA(p, q) system, we have

$$det\,\mathbf{R}(p, q) \neq 0 \tag{4.13}$$

and

$$det\,\mathbf{R}(p+1, q+1) = 0. \tag{4.14}$$

Equation (4.14) also follows directly from (4.6), since the first column of $\mathbf{R}(p+1, q+1)$ is linearly dependent on its remaining p columns. More precisely, if $\mathbf{X}$ represents the last p columns of $\mathbf{R}(p+1, q+1)$, then

$$\mathbf{R}(p+1, q+1) = [\; \mathbf{X}\boldsymbol{\alpha}_1 \mid \mathbf{X} \;] = \mathbf{X}[\; \boldsymbol{\alpha}_1 \mid \mathbf{I}_p \;] \overset{\triangle}{=} \mathbf{XD}, \tag{4.15}$$

where $\boldsymbol{\alpha}_1$ is defined as (4.11). Since $rank\,\mathbf{X} = rank\,\mathbf{D} = p$, from Sylvester's inequality [3], $rank\,\mathbf{R}(p+1, q+1) = p$ or $det\,\mathbf{R}(p+1, q+1) = 0$.

The above observation can be used to establish the more general result [53–60] that for any ARMA(p, q) system,

$$det\,\mathbf{R}(p+k, q+i) = 0, \quad k > 0, \; i > 0. \tag{4.16}$$

In fact, by making use of (4.6), it follows that if $k = i$, the first i columns of $\mathbf{R}(p+k, q+i)$ are linearly dependent on its last p columns. Thus if $\mathbf{Y}$ denotes the last p columns of $\mathbf{R}(p+k, q+i)$, then

$$\mathbf{R}(p+k, q+i) = [\; \mathbf{YB}_1 \mid \mathbf{Y} \;] = \mathbf{Y}[\; \mathbf{B}_1 \mid \mathbf{I}_p \;] \overset{\triangle}{=} \mathbf{YE}, \tag{4.17}$$

where $\mathbf{B}_1$ is a $p \times k$ matrix. Since $rank\,\mathbf{Y} = rank\,\mathbf{E} = p$, once again from Sylvester's inequality, we have

$$rank\,\mathbf{R}(p+k, q+i) = p, \quad k > 0, \; i > 0, \tag{4.18}$$

from which (4.18) follows for $k = i$. If $k < i$, again it is easy to verify that the first i columns of $\mathbf{R}(p + k, q + i)$ linearly depend on the columns of the above $\mathbf{Y}$ matrix, i.e.,

$$\mathbf{R}(p + k, q + i) = \mathbf{Y} \left[\, \mathbf{C}_1 \mid \mathbf{L} \, \right] , \tag{4.19}$$

where $\mathbf{C}_1$ is a $p \times i$ matrix and $\mathbf{L}$ is a $p \times (p + k - i)$ matrix given by

$$\mathbf{L} = \left[\begin{array}{c} \mathbf{I}_{p+k-i} \\ \hline \mathbf{0}_{i-k, p+k-i} \end{array} \right] ,$$

and since $rank \, \mathbf{Y} = p$, (4.16) holds again. Finally, the case $k > i$ can be established by similar considerations on its rows. Notice that if $k < 0$ or $i < 0$, no such linear dependency can be proved for any sets of columns or rows of $\mathbf{R}(p + k, q + i)$ and hence in general, $\mathbf{R}(p + k, q + i)$ will be nonsingular if $k < 0$ or $i < 0$. To summarize the above discussion, for an ARMA(p, q) system [57–60],

$$det \, \mathbf{R}(p + k, q + i) = \left\{ \begin{array}{ll} = 0 , & \text{if } k > 0 \text{ and } i > 0 \\ \neq 0 , & \text{if } k = 0 \text{ or } i = 0 \\ \text{generally } \neq 0 , & \text{if } k < 0 \text{ or } i < 0. \end{array} \right. \tag{4.20}$$

Several techniques have been proposed for model order selection based on the determinantal condition in (4.20) [53–60]. In [57], Chow suggests a hypothesis testing procedure for model order determination based on (4.20). The test proceeds by computing $det \, \mathbf{R}(\hat{p}, \hat{q})$ for some $\hat{p} > 0$ and $\hat{q} = m, \, m + 1, \, \ldots$, where $m > max \, (\hat{p}, q)$. If $det \, \mathbf{R}(\hat{p}, \hat{q})$ is close to zero or the arithmetic average of the above determinants is nearly zero, then $\hat{p} > p$; otherwise, $\hat{p} \leq p$ [57]. Through this procedure, once the AR order is chosen, the corresponding AR parameters can be computed by making use of (4.7). The motivation for MA order selection in [57] comes from (4.5). The AR coefficients obtained above are used in (4.5) and the largest integer $\hat{q}$ for which the inequality in (4.5) holds is taken as the MA order [57]. However, caution must be exercised in using this procedure for MA order selection since the right-hand side of (4.5) can be very small for $k \leq q$ and in particular it can be zero for $k \leq q$. For example, consider the ARMA$(1, 1)$ system given by

$$x(n) = -\alpha_1 x(n - 1) + \beta_0 w(n) + \beta_1 w(n - 1)$$

with $\alpha_1 = -r_1/r_0$. Since $|\alpha_1| < 1$, $\alpha(z)$ is stable and for $k = 0$, (4.5) gives

$$r_0 + \alpha_1 r_1 = \frac{r_0^2 - r_1^2}{r_0} > 0$$

and for $k = 1$, we get

$$r_1 + \alpha_1 r_0 = 0 ,$$

establishing the above remark. Naturally, application of the above procedure in this case will erroneously select the MA order to be zero and the system to be AR.

To avoid the complexity and inefficiency/inaccuracy of computing the determinant for checking the rank of the matrix, Cadzow [53] has proposed a procedure for determining the ARMA model order (p, q) by making use of the singular value decomposition (SVD) of $\mathbf{R}(p + k, q + i)$ for $k > 0$ and $i > 0$. Chan and Wood [59] have also suggested the Gram-Schmidt orthonormalization procedure to check the linear dependency of columns of $\mathbf{R}(\tilde{p} + \tilde{q}, \tilde{q})$ where $\tilde{p} > p$ and $\tilde{q} > q$. Fuchs [60] has proposed a scheme based on the symmetric (Hankel) matrix $\mathbf{H}(p, q)$ given by

$$\mathbf{H}(p, q) = \mathbf{R}(p, q)\mathbf{J}, \tag{4.21}$$

where $\mathbf{R}(p, q)$ as in (4.9) and $\mathbf{J}$ is the $p \times p$ 'identity mirror image' matrix given by

$$\mathbf{J} = \begin{bmatrix} 0 & \cdots & 0 & 1 \\ 0 & \cdots & 1 & 0 \\ \vdots & \cdots & \vdots & \vdots \\ 1 & \cdots & 0 & 0 \end{bmatrix}. \tag{4.22}$$

In that case [60], the order (p, q) of the ARMA model is determined by incrementing p and q systematically from 1 and 0, respectively, and the first value of (p, q), for which $\mathbf{H}(p, q)$ has a zero eigenvalue, is taken as the model order.

Interestingly, if $p > q$, then (4.20) can be used to derive a determinantal test that does not involve q. In that case, p represents the degree of the rational system. If we let $k = 0$ and $i = p - q > 0$, then (4.20) reads

$$det \, \mathbf{R}(p, p) \neq 0 \tag{4.23}$$

and if $k > 0$ and $i = p + k - q > 0$, it gives

$$det \, \mathbf{R}(p + k, p + k) = 0. \tag{4.24}$$

Letting

$$\mathbf{H}_p = \mathbf{R}(p, p)\mathbf{J} = \begin{bmatrix} r_1 & r_2 & \cdots & r_p \\ r_2 & r_3 & \cdots & r_{p+1} \\ \vdots & \vdots & \cdots & \vdots \\ r_p & r_{p+1} & \cdots & r_{2p-1} \end{bmatrix} \tag{4.25}$$

denote the above $p \times p$ Hankel matrix generated from $r_1, r_2, \ldots, r_{2p-1}$, the determinantal conditions in (4.23)–(4.24) take the familiar form [61, 62],

$$rank \, \mathbf{H}_p = rank \, \mathbf{H}_{p+k} = p, \quad k \geq 1, \tag{4.26}$$

i.e., in the case of rational systems, every such Hankel matrix of size greater than or equal to p has rank p, the degree of the rational system. Of course,

(4.26) also follows directly from (4.6) with $k \geq p$ and considering p or more equations from that stage onwards. Alternatively, those p equations can be rewritten as

$$
\begin{bmatrix}
r_0 & r_1 & \cdots & r_{p-1} \\
r_1 & r_2 & \cdots & r_p \\
\vdots & \vdots & \cdots & \vdots \\
r_{p-1} & r_p & \cdots & r_{2(p-1)}
\end{bmatrix}
\begin{bmatrix}
\alpha_p \\
\alpha_{p-1} \\
\vdots \\
\alpha_1
\end{bmatrix}
= -
\begin{bmatrix}
r_p \\
r_{p+1} \\
\vdots \\
r_{2p-1}
\end{bmatrix}
\tag{4.27}
$$

and letting

$$
\mathbf{D}_k =
\begin{bmatrix}
r_0 & r_1 & \cdots & r_k \\
r_1 & r_2 & \cdots & r_{k+1} \\
\vdots & \vdots & \cdots & \vdots \\
r_k & r_{k+1} & \cdots & r_{2k}
\end{bmatrix},
\tag{4.28}
$$

it immediately follows that

$$
rank\,\mathbf{D}_{p-1} = rank\,\mathbf{D}_{p+k} = p, \quad k \geq 0.
\tag{4.29}
$$

Moreover, using (4.27)–(4.28), the denominator polynomial in (4.1) simplifies to

$$
\alpha(z) = 1 + \sum_{k=1}^{p} \alpha_k z^k = 1 + [\, z^p \ z^{p-1} \ \cdots \ z \,]
\begin{bmatrix}
\alpha_p \\
\alpha_{p-1} \\
\vdots \\
\alpha_1
\end{bmatrix}
$$

$$
= 1 - [\, z^p \ z^{p-1} \ \cdots \ z \,] \mathbf{D}_{p-1}^{-1}
\begin{bmatrix}
r_p \\
r_{p+1} \\
\vdots \\
r_{2p-1}
\end{bmatrix}
$$

$$
= \frac{1}{|\mathbf{D}_{p-1}|}
\begin{vmatrix}
r_0 & r_1 & \cdots & r_{p-1} & r_p \\
r_1 & r_2 & \cdots & r_p & r_{p+1} \\
\vdots & \vdots & \cdots & \vdots & \vdots \\
r_{p-1} & r_p & \cdots & r_{2p-2} & r_{2p-1} \\
z^p & z^{p-1} & \cdots & z & 1
\end{vmatrix}.
\tag{4.30}
$$

Interestingly, the Hankel type determinant expression in (4.30) is valid for *any* ARMA(p, q) system irrespective of the exact value of q, provided $p \geq q$. Thus in the case of an AR(p) system, this Hankel determinantal expression involving $r_0 \to r_{2p-1}$ must correspond to the normalized Levinson polynomial $a_p(z)$ of the first kind given in (3.70).

Referring back to the discussion on the system parameter evaluation, the AR parameters obtained in (4.7) can be used to evaluate the MA parameters β_k, $k = 0 \to q$. Since

$$S(\theta) = H(z)H_*(z)|_{z=e^{j\theta}} = \left(r_0 + \sum_{k=1}^{\infty} r_k z^k + \sum_{k=1}^{\infty} r_k^* z^{-k} \right)\Bigg|_{z=e^{j\theta}}$$

and $H(z)$ is rational, the above equality must be true for all z. Thus

$$H(z)H_*(z) = r_0 + \sum_{k=1}^{\infty} r_k z^k + \sum_{k=1}^{\infty} r_{-k} z^{-k} \tag{4.31}$$

and making use of (4.1), we have

$$\beta(z)\beta_*(z) = \alpha(z)\alpha_*(z) \left(r_0 + \sum_{k=1}^{\infty} r_k z^k + \sum_{k=1}^{\infty} r_{-k} z^{-k} \right) . \tag{4.32}$$

Let

$$\beta(z)\beta_*(z) = \sum_{k=-q}^{q} b_k z^k , \quad b_k = \sum_{i=k}^{q} \beta_i \beta_{i-k}^* = b_{-k}^* \tag{4.33}$$

and

$$\alpha(z)\alpha_*(z) = \sum_{k=-p}^{p} c_k z^k , \quad c_k = \sum_{i=k}^{p} \alpha_i \alpha_{i-k}^* = c_{-k}^* , \tag{4.34}$$

where c_k's are known quantities, since α_k's already have been obtained from (4.7). Substituting (4.33)–(4.34) into (4.32) gives

$$b_k = \sum_{i=-p}^{p} c_i r_{k-i} , \quad k = -q \to q , \tag{4.35}$$

and since (4.33) a priori represents a nonnegative polynomial on the unit circle, any standard polynomial factorization technique [7, 65–106] can be used to obtain the MA parameters β_k, $k = 0 \to q$.

In [66] and [67], Parzen and Graupe et al. have made use of the series expansion for any minimum-phase rational function $H(z)$ in $|z| < 1$, to compute the system parameters. Referring back to (4.1), since $H(z)$ is minimum-phase, $1/H(z) = \alpha(z)/\beta(z)$ also has a power series expansion and let

$$\frac{1}{H(z)} = \sum_{k=0}^{\infty} d_k z^k \triangleq D(z) , \quad |z| < 1 . \tag{4.36}$$

Then

$$H(z) = \frac{\beta(z)}{\alpha(z)} = \frac{1}{\displaystyle\sum_{k=0}^{\infty} d_k z^k} , \quad |z| < 1 . \tag{4.37}$$

If the exact d_k's are known, then from (4.37), cross-multiplication and comparison of equal powers of z yield two matrix equations (assuming $p \geq q$), i.e., for MA parameters (with $\beta_0 = 1$ [67]),

$$
\begin{bmatrix}
d_p & d_{p-1} & \cdots & d_{p+1-q} \\
d_{p+1} & d_p & \cdots & d_{p+2-q} \\
\vdots & \vdots & \ddots & \vdots \\
d_{p+q-1} & d_{p+q-2} & \cdots & d_p
\end{bmatrix}
\begin{bmatrix}
\beta_1 \\
\beta_2 \\
\vdots \\
\beta_q
\end{bmatrix}
= -
\begin{bmatrix}
d_{p+1} \\
d_{p+2} \\
\vdots \\
d_{p+q}
\end{bmatrix}
\tag{4.38}
$$

and for AR parameters (with $\alpha_0 = 1$ [67]),

$$
\begin{bmatrix}
\alpha_1 \\
\alpha_2 \\
\vdots \\
\alpha_p
\end{bmatrix}
=
\begin{bmatrix}
d_1 & d_0 & 0 & \cdots & 0 \\
d_2 & d_1 & d_0 & \cdots & 0 \\
\vdots & \vdots & \vdots & \cdots & \vdots \\
d_p & d_{p-1} & d_{p-2} & \cdots & d_{p-q}
\end{bmatrix}
\begin{bmatrix}
1 \\
\beta_1 \\
\vdots \\
\beta_q
\end{bmatrix}.
\tag{4.39}
$$

However, since d_k, $k = 1 \to \infty$, are unknown, they suggest approximating $D(z)$ in (4.36) to $D_T(z)$ using a new set of coefficients, f_k, $k = 0 \to M$ [66, 67], where

$$
D_T(z) = \sum_{k=0}^{M} f_k z^k
$$

with $M \geq p + q$. Any available AR parameter estimation technique, such as the Yule-Walker equations mentioned earlier, can be used to evaluate f_k's from r_k, $k = 0 \to M$. In that case, the $D_T(z)$ so obtained is also free of zeros in $|z| \leq 1$, and its coefficients can be used in (4.38) and (4.39) to compute the AR and MA parameters [66, 67]. However, it may be remarked that f_k's so obtained are solutions to the AR(M) problem and *not* to the original ARMA(p, q) problem, and this approximation should be taken in that context.

Next, we discuss techniques that make use of the output data to solve the system identification problem in the rational case.

4.1.2 Known Data Case

A related and more practical problem is to identify the system from available output data $x(n)$ itself. In this case, the system output is assumed to be known for $x(n)$, $n = 1 \to N$, and the problem becomes statistical. Numerous approaches have been suggested in this case [52–60, 66–88]. An obvious procedure is to repeat the above techniques with the exact (unknown) autocorrelations replaced by their estimated counterparts. Given the system output $x(n)$, assuming it is wide-sense stationary, the estimates of autocorrelations can be computed by [55]

$$
\hat{r}_k = \frac{1}{N-k} \sum_{n=1}^{N-k} x(n+k)x^*(n)
\tag{4.40}
$$

in the case of unbiased estimation, or

$$\hat{r}_k = \frac{1}{N} \sum_{n=1}^{N-k} x(n+k)x^*(n) \tag{4.41}$$

in the case of biased estimation. Use of these estimated autocorrelations in any technique will render the results so obtained random, and the standard statistical procedures such as asymptotic consistency can be imposed on the results to justify this procedure [68]. Naturally, in this approach, the autocorrelation estimation becomes very critical, especially for short data samples.

Parameter Estimation

In this situation, since the problem is statistical, well known parameter estimation techniques can be used to estimate the model order and the system parameters directly. Since the parameters are unknown, but non-random, the maximum likelihood (ML) technique is applicable here, and extensive work has been reported on this subject [55, 70, 72–77, 85–88]. In principle, this procedure computes the (log-)likelihood function

$$L(\mathbf{x}; \boldsymbol{\alpha}, \boldsymbol{\beta}, \sigma^2) = \ln f_x(\mathbf{x}; \boldsymbol{\alpha}, \boldsymbol{\beta}, \sigma^2), \tag{4.42}$$

where

$$\mathbf{x} \overset{\triangle}{=} [x(1), \ \ldots, \ x(N)]^T$$

$$\boldsymbol{\alpha} \overset{\triangle}{=} [\alpha_1, \ \ldots, \ \alpha_p]^T$$

$$\boldsymbol{\beta} \overset{\triangle}{=} [\beta_1, \ \ldots, \ \beta_q]^T,$$

and chooses the most likely values for the unknown parameters, α_i, $i = 1 \to p$, β_j, $j = 1 \to q$ and[2] $\sigma^2 = |\beta_0|^2$ (input noise variance) suggested by the data $\mathbf{x}$, i.e., the ML estimates $\hat{\alpha}_i$, $i = 1 \to p$, $\hat{\beta}_j$, $j = 1 \to q$ and $\hat{\sigma}^2$ are given by maximizing (4.42) with respect to the unknown parameters. In (4.42), $f_x(\mathbf{x}; \boldsymbol{\alpha}, \boldsymbol{\beta}, \sigma^2)$ represents the joint probability density function (p.d.f.) of the observations $\mathbf{x}$ for a given set of parameters $\boldsymbol{\alpha}$, $\boldsymbol{\beta}$ and σ^2. If the input process is assumed to be zero-mean Gaussian, then the output $x(n)$ is also zero-mean Gaussian, and in that case (4.42) becomes [68]

$$f_x(\mathbf{x}; \boldsymbol{\alpha}, \boldsymbol{\beta}, \sigma^2) = \frac{1}{(\pi\sigma^2)^N |\mathbf{R}|} \, exp\left(-\frac{\mathbf{x}^*\mathbf{R}^{-1}\mathbf{x}}{\sigma^2}\right), \tag{4.43}$$

where

$$\mathbf{R} \overset{\triangle}{=} E[\mathbf{x}\mathbf{x}^*] > 0. \tag{4.44}$$

---

[2]Following the convention in the statistical literature, if we let $|\beta_0|^2$ represent the input noise variance σ^2, then the rest of the MA parameters should be normalized with respect to β_0.

Under the above Gaussian assumption, the exact expression for the likelihood function can be simplified [73–75]. Newbold [73] has derived the exact likelihood function using a generalization of the approach used by Box and Jenkins [75] for pure MA processes. Considering the nonsingular transformation from the set of $N + p + q$ input variables $w(n)$ for $1 - q \leq n \leq N$ and $x(n)$ for $1 - p \leq n \leq 0$ to the set of $N + p + q$ variables $w(n)$ for $1 - q \leq n \leq 0$ and $x(n)$ for $1 - p \leq n \leq N$, the equation (4.2) can be rewritten as [73]

$$\begin{bmatrix} \mathbf{w}_2 \\ \mathbf{w}_1 \end{bmatrix} = \begin{bmatrix} \mathbf{0}_{p+1,N} \\ \mathbf{F}_1 \end{bmatrix} \mathbf{x} + \begin{bmatrix} \mathbf{I}_{p+1} \\ \mathbf{F}_2 \end{bmatrix} \mathbf{w}_2, \tag{4.45}$$

where $\mathbf{F}_1$ and $\mathbf{F}_2$ are $N \times N$ and $N \times (p+q)$ matrices, respectively, involving only α_k's and β_k's, and

$$\mathbf{w}_1 \overset{\triangle}{=} [w(1), \ \ldots, \ w(N)]^T$$

$$\mathbf{w}_2 \overset{\triangle}{=} [w(1-q), \ w(2-q), \ \ldots, \ w(0), \ x(1-p), \ x(2-p), \ \ldots, \ x(0)]^T.$$

Suppose a $(p + q) \times (p + q)$ nonsingular matrix $\mathbf{M}$ can be found such that

$$\mathbf{M}\boldsymbol{\Omega}\mathbf{M}^* = \mathbf{I}_{p+q}, \tag{4.46}$$

where $\boldsymbol{\Omega} = E[\mathbf{w}_2\mathbf{w}_2^*]$. In that case, Newbold [73] has shown that the exact likelihood function becomes

$$f_x(\mathbf{x}; \alpha, \beta, \sigma^2) = \frac{1}{(\pi\sigma^2)^N |\mathbf{R}|} exp\left(-\frac{S(\alpha, \beta)}{\sigma^2}\right), \tag{4.47}$$

where

$$S(\alpha, \beta) = \mathbf{x}^*\mathbf{F}^* \left(\mathbf{I}_{N+p+q} - \mathbf{F}_2(\boldsymbol{\Omega}^{-1} + \mathbf{F}_2^*\mathbf{F}_2)^{-1}\mathbf{F}_2^*\right) \mathbf{F}_1\mathbf{x} \tag{4.48}$$

and

$$\mathbf{R} = \mathbf{I}_N - \mathbf{F}_2(\boldsymbol{\Omega}^{-1} + \mathbf{F}_2^*\mathbf{F}_2)^{-1}\mathbf{F}_2^*. \tag{4.49}$$

Newbold also gives an explicit formula for the case of ARMA$(1,1)$ which is highly nonlinear in α_1, β_1 and σ^2 [73]. From (4.49), it is not difficult to see that $\mathbf{R}$ is a highly nonlinear function of α and β, and for a small number of observations $\mathbf{R}$ usually is found to be of critical importance. However, under the assumptions that (i) the data records are sufficiently large, (ii) the data is real and Gaussian, and (iii) the poles and zeros of the rational model are not too close to the unit circle, (4.47) is dominated by the exponent term $S(\alpha, \beta)/\sigma^2$. Box and Jenkins [75] have suggested that the determinant term $|\mathbf{R}|$ in (4.47) may be disregarded in this case, and with the remaining expression in the exponent, a least squares estimation problem can be formulated. To examine this, rewrite (4.43) as

$$f(\mathbf{x}; \alpha, \beta, \sigma^2) \overset{\triangle}{=} f(x(N), x(N-1), \ldots, x(p), \ldots, x(1))$$

$$= f(x(N), \ldots, x(p+1) | x(p), \ldots, x(1)) \, f(x(p), \ldots, x(1)), \qquad (4.50)$$

where the dependency on $\boldsymbol{\alpha}$, $\boldsymbol{\beta}$ and σ^2 on the right-hand side of (4.50) is understood. Letting $\mathbf{x}_p \overset{\triangle}{=} [x(1), \ldots, x(p)]^T$, this allows us to write the unconditional p.d.f. on the right-hand side of (4.50) as

$$f(\mathbf{x}_p) = \frac{1}{(\pi\sigma^2)^p |\mathbf{R}_{\alpha,\beta}|} \, exp \left(-\frac{\mathbf{x}_p^* \mathbf{R}_{\alpha,\beta}^{-1} \mathbf{x}_p}{\sigma^2} \right), \qquad (4.51)$$

where $E[\mathbf{x}_p \mathbf{x}_p^*] \overset{\triangle}{=} \sigma^2 \mathbf{R}_{\alpha,\beta}$ represents the autocorrelation matrix of the Gaussian vector $\mathbf{x}_p$. To evaluate the conditional p.d.f. in (4.50), we can again rewrite it as

$$f(x(N), \ldots, x(p+1) | \mathbf{x}_p) = f(x(N) | x(N-1), \ldots, x(p+1), \mathbf{x}_p)$$
$$\times f(x(N-1) | x(N-2), \ldots, x(p+1), \mathbf{x}_p)$$
$$\times \cdots \times f(x(p+2) | x(p+1), \mathbf{x}_p)$$
$$\times f(x(p+1) | \mathbf{x}_p). \qquad (4.52)$$

From (4.2), a typical term of the right-hand side of (4.52), say,

$$f(x(j) | x(j-1), \ldots, x(p+1), \mathbf{x}_p), \quad p+1 \le j \le N,$$

is Gaussian with mean value $-\sum_{k=1}^p \alpha_k x(j-k)$ and variance

$$E \left(\left| \sum_{k=0}^q \beta_k w(j-k) \right|^2 \right) = \sigma^2 \sum_{k=0}^q \beta_k^2 \overset{\triangle}{=} \sigma_\beta^2.$$

Thus, (4.52) simplifies to

$$f(x(N), \ldots, x(p+1) | \mathbf{x}_p)$$
$$= \prod_{j=p+1}^N \frac{1}{\pi\sigma_\beta^2} \, exp \left(-\frac{1}{\sigma_\beta^2} \left| x(j) + \sum_{k=1}^p \alpha_k x(j-k) \right|^2 \right)$$
$$= \frac{1}{(\pi\sigma_\beta^2)^{(N-p)}} \, exp \left(-\frac{1}{\sigma_\beta^2} \sum_{j=p+1}^N \left| x(j) + \sum_{k=1}^p \alpha_k x(j-k) \right|^2 \right). \quad (4.53)$$

Substituting (4.51)–(4.53) into the log-likelihood function gives

$$L(\mathbf{x}; \boldsymbol{\alpha}, \boldsymbol{\beta}, \sigma^2) = \ln f(\mathbf{x}; \boldsymbol{\alpha}, \boldsymbol{\beta}, \sigma^2)$$

$$= -(N-p) \ln (\pi\sigma_\beta^2) - p \ln (\pi\sigma^2)$$

$$-\ln |\mathbf{R}_{\alpha,\beta}| - \frac{\mathbf{x}_p^* \mathbf{R}_{\alpha,\beta}^{-1} \mathbf{x}_p}{\sigma^2} - \frac{S(\boldsymbol{\alpha})}{\sigma_\beta^2}, \qquad (4.54)$$

where

$$S(\boldsymbol{\alpha}) = \sum_{j=p+1}^{N} \left| x(j) + \sum_{k=1}^{p} \alpha_k x(j-k) \right|^2 . \tag{4.55}$$

Maximization of L with respect to $\boldsymbol{\alpha}$, $\boldsymbol{\beta}$ and σ^2, in general, will give rise to a coupled set of highly nonlinear equations and seldom can they be solved directly for the unknown parameters. However, for large N, if we make use of Box and Jenkins' suggestion [75], then we can disregard the unconditional p.d.f. in (4.50) to eliminate terms involving $\mathbf{R}_{\alpha,\beta}$ in (4.54). In that case, from (4.54) and (4.55), maximization with respect to $\boldsymbol{\alpha}$ and $\boldsymbol{\beta}$ reduces to minimization of $S(\boldsymbol{\alpha})$, and this yields

$$\frac{\partial S}{\partial \alpha_m} = 2 \sum_{j=p+1}^{N} \left(x(j) + \sum_{k=1}^{p} \alpha_k x(j-k) \right) x^*(j-m) = 0, \quad m = 1 \to p,$$

which for large N simplifies to

$$\begin{bmatrix} \hat{r}_{11} & \hat{r}_{12} & \cdots & \hat{r}_{1p} \\ \hat{r}_{21} & \hat{r}_{22} & \cdots & \hat{r}_{2p} \\ \vdots & \vdots & \cdots & \vdots \\ \hat{r}_{p1} & \hat{r}_{p2} & \cdots & \hat{r}_{pp} \end{bmatrix} \begin{bmatrix} \hat{\alpha}_1 \\ \hat{\alpha}_2 \\ \vdots \\ \hat{\alpha}_p \end{bmatrix} = \begin{bmatrix} \hat{r}_{01} \\ \hat{r}_{02} \\ \vdots \\ \hat{r}_{0p} \end{bmatrix}, \tag{4.56}$$

where

$$\hat{r}_{ik} = \frac{1}{N-p} \sum_{j=p+1}^{N} x(j-i)\, x^*(j-k) \tag{4.57}$$

represents unbiased estimates of r_{k-i} since $E[\hat{r}_{ik}] = r_{k-i}$. Thus for large data records, these equations are approximately the same as the Yule-Walker equations for $q = 0$, with the exact autocorrelations replaced by their estimated counterparts (set $q = 0$ in (4.7)). Since these approximate maximum likelihood estimates in (4.56) always (irrespective of the actual value of q) coincide with the solutions obtained by using the estimated autocorrelations in the Yule-Walker equations for an AR(p) model, discarding the term involving $\mathbf{R}_{\alpha,\beta}$ in (4.54) can be justified only for an AR(p) model. In the AR(p) case, the noise variance estimate is given by

$$\hat{\sigma}^2 = \hat{r}_0 + \sum_{j=1}^{p} \hat{\alpha}_j \hat{r}_j . \tag{4.58}$$

To illustrate the highly nonlinear nature of the equations for the ML estimates of $\boldsymbol{\alpha}$ and $\boldsymbol{\beta}$, we consider the simplest AR model with $p = 1$. With the assumption that $x(n)$ and $w(n)$ in (4.2) are zero-mean Gaussian, we have the likelihood function given by (4.54) with $q = 0$. To derive the exact likelihood function for this case, notice that $\mathbf{R}_{\alpha,\beta} = (1 - \alpha_1^2)^{-1}$ and

$$S(\alpha_1) = \sum_{m=2}^{N} |x(m) + \alpha_1 x(m-1)|^2 .$$

With these in (4.54), maximization of the log-likelihood function with respect to α_1 yields a cubic equation in α_1. The solution for $\hat{\alpha}_1$ need not be unique in general. Moreover, none of the solutions of this cubic equation may give rise to a stable system. In general, we can expect the degree of nonlinearity to increase as the order p increases, and eventually the ML estimation of ARMA parameters using the exact likelihood function will involve a set of highly nonlinear equations in α and β and will lend little insight into a practical estimation procedure [55, 75].

A closely related method in this context is the iterated least squares approach [80–82]. The method assumes the model order (p, q) with $p > q$, and it is derived from the lattice filter structure which is an orthogonalizing filter that transforms the original observations into two orthogonal sequences of forward and backward prediction errors, both spanning the same space as that of the original data. It is shown in [80–82] that the parameters of an ARMA(p, q) model can be estimated by first obtaining an AR model of order $p + q$ and then solving a set of linear equations. This approach is similar to that of Graupe et al. [67]. To state the final results [80–82], at first solve for the auxiliary variables γ_j, $j = 1 \to q$, from

$$
\begin{bmatrix}
\alpha_p^{(p+q-1)} & \alpha_{p-1}^{(p+q-2)} & \cdots & \alpha_{p-q+1}^{(p)} \\
\alpha_{p+1}^{(p+q-1)} & \alpha_p^{(p+q-2)} & \cdots & \alpha_{p-q+2}^{(p)} \\
\vdots & \vdots & \cdots & \vdots \\
\alpha_{p+q-1}^{(p+q-1)} & \alpha_{p+q-2}^{(p+q-2)} & \cdots & \alpha_p^{(p)}
\end{bmatrix}
\begin{bmatrix}
\gamma_1 \\
\gamma_2 \\
\vdots \\
\gamma_q
\end{bmatrix}
= -
\begin{bmatrix}
\alpha_{p+1}^{(p+q)} \\
\alpha_{p+2}^{(p+q)} \\
\vdots \\
\alpha_{p+q}^{(p+q)}
\end{bmatrix} . \tag{4.59}
$$

Then, the AR parameters, α_i, $i = 1 \to p$ with $\alpha_0 = 1$, are obtained by

$$
\begin{bmatrix}
\alpha_1 \\
\alpha_2 \\
\vdots \\
\alpha_q \\
\vdots \\
\alpha_p
\end{bmatrix}
=
\begin{bmatrix}
\alpha_1^{(p+q)} \\
\alpha_2^{(p+q)} \\
\vdots \\
\alpha_q^{(p+q)} \\
\vdots \\
\alpha_p^{(p+q)}
\end{bmatrix}
+
\begin{bmatrix}
1 & 0 & \cdots & 0 \\
\alpha_1^{(p+q-1)} & 1 & \cdots & 0 \\
\vdots & \vdots & \ddots & \vdots \\
\alpha_{q-1}^{(p+q-1)} & \alpha_{q-2}^{(p+q-2)} & \cdots & 1 \\
\vdots & \vdots & \cdots & \vdots \\
\alpha_{p-1}^{(p+q-1)} & \alpha_{p-2}^{(p+q-2)} & \cdots & \alpha_{p-q}^{(p)}
\end{bmatrix}
\begin{bmatrix}
\gamma_1 \\
\gamma_2 \\
\vdots \\
\gamma_q
\end{bmatrix}
\tag{4.60}
$$

and the MA parameters β_j, $j = 1 \to q$ with $\beta_0 = 1$, are approximated by

$$
\beta_j = \gamma_j \left[\prod_{i=1}^{j} (1 - s_{p+q+1-i}^2), \right]^{-1/2} . \tag{4.61}
$$

Here, s_m, $m = 1 \to p + q$, are the first $(p + q)$ reflection coefficients of the ARMA(p, q) or AR(∞) model that can be derived from Burg's algorithm [39], and $\alpha_k^{(m)}$ is the kth coefficient of the AR(m) model that is the best m^{th}-order approximation of the AR(∞) model. Notice that parameters from different orders are used, and this is to be distinguished from the method of Graupe et al. [67]. The parameters $\alpha_k^{(m)}$, $k = q \rightarrow m$, $m = q \rightarrow p + q$, can be obtained by using the Levinson recursion

$$\alpha_k^{(m+1)} = \begin{cases} 1, & k = 0 \\ \alpha_k^{(m)} + s_{m+1}\alpha_{m-k+1}^{(m)}, & 1 \leq k \leq m \\ s_{m+1}, & k = m + 1. \end{cases} \tag{4.62}$$

Model Order Selection

So far, we have been discussing parameter estimation under the assumption that the model order (p, q) is known. This is invariably not the case, and several criteria based on information measures have been proposed for model order determination [72, 83–86]. Of these, Akaike's technique [72] for determining the model order by minimizing the Kullback-Leibler's information theoretic distance [83] (or minimization of cross-entropy [84]) between an assumed probability density function (p.d.f.) and the true p.d.f. of the data seems to be the most frequently used one. This procedure, known as the AIC (Akaike Information Criterion), assumes the process to be Gaussian, and is given by [55, 72]

$$\text{AIC}(\hat{p}, \hat{q}) = N \cdot \ln \hat{\sigma}^2 + 2(\hat{p} + \hat{q}), \tag{4.63}$$

where $(\hat{p}, \hat{q})$ is the assumed order-pair, N is the number of data points and $\hat{\sigma}^2$ is the ML estimate of the input noise variance. It is easy to show that using the forward and backward predictors,

$$\hat{\sigma}^2 = \frac{\displaystyle\sum_{i=\hat{p}+1}^{N} \left\{ \left(\sum_{k=0}^{\hat{p}} \hat{\alpha}_k x(i-k) \right)^2 + \left(\sum_{k=0}^{\hat{p}} \hat{\alpha}_{\hat{p}-k} x(i-k) \right)^2 \right\}}{2(N-\hat{p})\displaystyle\sum_{k=0}^{\hat{q}} |\hat{\beta}_k|^2}, \tag{4.64}$$

where $\hat{\alpha}_k$ and $\hat{\beta}_k$ are the ML estimates ($\hat{\beta}_0 = 1$) of the ARMA parameters with assumed model order $\hat{p}$ and $\hat{q}$, respectively. The model order $(\hat{p}, \hat{q})$ is selected as the one that minimizes the AIC. Schwartz [85] and Rissanen [86] have also proposed techniques for model order determination. Based on the asymptotic behavior of Bayes estimation, Schwartz has proposed the minimum of [85]

$$N \cdot \ln \hat{\sigma}^2 + (\hat{p} + \hat{q}) \ln N \tag{4.65}$$

to be the "best" model order. Rissanen's technique, known as the minimum description length (MDL), computes [86]

$$\text{MDL}(\hat{p}, \hat{q}) = N \cdot \ln \hat{\sigma}^2 + (\hat{p} + \hat{q} + 1) \ln (N + 2)$$

$$+ 2\ln(\hat{p} + 1)(\hat{q} + 1) + \sum_{i=1}^{\hat{p}+\hat{q}} \ln\left(\hat{\theta}_i^2 \frac{\partial^2 \ln \hat{\sigma}^2}{\partial \hat{\theta}_i^2} \right), \tag{4.66}$$

where $\theta_0 = \sigma^2$, $\theta_i = \alpha_i$, $i = 1 \to p$, and $\theta_{p+j} = \beta_j$, $j = 1 \to q$. Note that all of the above techniques for model order determination involve a two-dimensional search. Another possibility is to filter the output sequence with the estimated inverse filter to generate an estimate of the white noise process. If the correct model order has been reached, the corresponding estimated autocorrelations will be approximately zero for all lags except the zeroth one. Methods for testing the whiteness of the filtered process and determining the model order have been proposed in this context [87, 88].

It may be remarked that none of the techniques mentioned above for model order selection (AIC, MDL, etc.), exploit the key feature of a rational system — the concept of its degree. This is because statistical procedures in themselves have no mechanism to bring out the rationality of the system explicitly; it always works with the probability density function both in the rational case as well as the nonrational case. Next, we examine a new approach to the problem of system identification based on the theory developed in chapter 3. This new approach evaluates the model order as well as the system parameters in a fundamentally different fashion, and it yields stable rational minimum-phase system transfer functions without the a priori knowledge of the model order (p, q) [31]. This is achieved by exploiting the key feature of a rational system — its degree — and bringing out certain associated invariant characteristics of the system.

4.2   ARMA-System Identification – A New Approach

Given any integrable power spectral density

$$S(\theta) = \sum_{k=-\infty}^{\infty} r_k e^{jk\theta} \geq 0 \tag{4.67}$$

that satisfies the causality requirement, from (3.1)–(3.5), there exists a unique function

$$H(z) = \sum_{k=0}^{\infty} b_k z^k \tag{4.68}$$

that is analytic together with its inverse in $|z| < 1$ and free of poles on $|z| = 1$ such that

$$S(\theta) = |H(e^{j\theta})|^2, \quad a.e. \tag{4.69}$$

Moreover, from Schur's theorem ((2.118)–(2.119)), associated with every such spectral density $S(\theta)$, there exists a unique positive function

$$Z(z) = r_0 + 2\sum_{k=1}^{\infty} r_k z^k \tag{4.70}$$

such that

$$S(\theta) = Re\, Z(e^{j\theta}) = |H(e^{j\theta})|^2\,, \quad a.e. \tag{4.71}$$

Since

$$\left.\frac{Z(z) + Z_*(z)}{2}\right|_{z=e^{j\theta}} = Re\, Z(e^{j\theta})\,,$$

the all z-extension of (4.71) reads

$$\frac{Z(z) + Z_*(z)}{2} = H(z)H_*(z)\,. \tag{4.72}$$

The representation in (4.72) shows the one-to-one correspondence between positive functions that are free of poles on $|z| = 1$ and minimum-phase transfer functions. In particular, the transfer function $H(z)$ is rational iff $Z(z)$ is a rational positive function.

To prove this, first assume that $Z(z)$ is a rational positive function of degree p. Then

$$Z(z) = \frac{b(z)}{a(z)}\,, \tag{4.73}$$

where $max\,(\delta(a), \delta(b)) = p$ and

$$\frac{Z(z) + Z_*(z)}{2} = \frac{a(z)b_*(z) + b(z)a_*(z)}{2a(z)a_*(z)}\,.$$

Since $Re\, Z(e^{j\theta}) \geq 0$, $a(z)b_*(z) + b(z)a_*(z)$ is nonnegative on the unit circle, and it can be factorized into the form $2\beta(z)\beta_*(z)$ with $\delta(\beta) \leq p$. Thus

$$a(z)b_*(z) + b(z)a_*(z) = 2\beta(z)\beta_*(z)\,. \tag{4.74}$$

Here, $\beta(z)$ can be made unique by choosing it to be free of zeros in $|z| < 1$. Thus

$$\frac{Z(z) + Z_*(z)}{2} = \frac{\beta(z)\beta_*(z)}{a(z)a_*(z)} = H(z)H_*(z)\,, \tag{4.75}$$

where

$$H(z) = \frac{\beta(z)}{a(z)}\,.$$

Notice that from (4.73), any pole of $Z(z)$ on the unit circle must be a simple zero of $a(z)$, and from (4.74) $\beta(z)$ also will have that zero at least up to multiplicity one. As a result, the above $H(z)$ is free of poles[3] in $|z| \leq 1$ and can be made unique by choosing it to be minimum-phase. Conversely, let $H(z)$ represent a rational minimum-phase transfer function of degree p that is free of poles on $|z| = 1$. Then

$$H(z) = \frac{\beta(z)}{\alpha(z)} = \frac{\beta_0 + \beta_1 z + \cdots + \beta_q z^q}{1 + \alpha_1 z + \cdots + \alpha_p z^p}\,, \quad p \geq q\,, \tag{4.76}$$

---

[3]If such pole-zero cancellations occur in $H(z)$, naturally the degree of the resulting minimum-phase $H(z)$ will be less than that of the $Z(z)$ in (4.73).

where $\alpha(z)$ is a strict Hurwitz polynomial, and with

$$|H(e^{j\theta})|^2 = \sum_{k=-\infty}^{\infty} r_k \, e^{jk\theta}, \tag{4.77}$$

we have

$$Z(z) = r_0 + 2 \sum_{k=1}^{\infty} r_k \, z^k \tag{4.78}$$

is a positive function. However, from (4.76)–(4.77) the autocorrelations satisfy (see, for example, (4.6))

$$r_k = -\sum_{i=1}^{p} \alpha_i \, r_{k-i}, \quad k \geq q+1,$$

and since $p \geq q$, substituting this into (4.78), we get

$$Z(z) = r_0 + 2 \sum_{k=1}^{p} r_k \, z^k + 2 \sum_{k=p+1}^{\infty} \left(-\sum_{i=1}^{p} \alpha_i r_{k-i}\right) z^k$$

$$= r_0 + 2 \sum_{k=1}^{p} r_k z^k - 2 \sum_{i=1}^{p} \alpha_i z^i \sum_{k=p+1}^{\infty} r_{k-i} z^{k-i}. \tag{4.79}$$

But

$$2 \sum_{k=p+1}^{\infty} r_{k-i} \, z^{k-i} = 2 \sum_{m=p-i+1}^{\infty} r_m \, z^m = Z(z) - r_0 - 2 \sum_{k=1}^{p-i} r_k \, z^k \tag{4.80}$$

and with (4.80) in (4.79), we obtain

$$Z(z) \cdot \sum_{i=0}^{p} \alpha_i \, z^i = r_0 + \sum_{i=1}^{p} (r_0 \alpha_i + 2r_i) z^i + 2 \sum_{i=1}^{p} \alpha_i z^i \sum_{k=1}^{p-i} r_k z^k$$

$$= r_0 + (r_0\alpha_1 + 2r_1)z + \sum_{i=2}^{p} \left(r_0\alpha_i + 2r_i + 2\sum_{k=1}^{i-1} \alpha_k r_{i-k}\right) z^i \triangleq b(z)$$

or

$$Z(z) = \frac{b(z)}{\alpha(z)}, \tag{4.81}$$

a rational positive function of degree p that is free of poles on $|z| = 1$. From (4.74)–(4.75), clearly $H(z)$ will contain all zeros of $Z(z)$, if any, on the unit circle. However, the converse is not true in general, since $H(e^{j\theta_0}) = 0$ only implies $Re \, Z(e^{j\theta_0}) = 0$ and the imaginary part of $Z(e^{j\theta_0})$ need not be zero.

Moreover, $Z(z)$ so obtained in (4.81) is unique up to an additive Foster function. Otherwise, let $Z_1(z)$ be any other positive function solution of (4.72). Then

$$Z_1(z) - Z(z) = -(Z_1(z) - Z(z))_* . \tag{4.82}$$

Hence $Z_1(z) - Z(z)$ is analytic in $|z| < 1$ as well as $|z| > 1$, and its poles are all restricted to lie on $|z| = 1$. These poles are simple, and since $Z(z)$ is free of poles on the unit circle, the residue of $Z_1(z) - Z(z)$ at these poles equal precisely those of $Z_1(z)$. Thus the residues are positive, and from (2.165)–(2.169) and (4.82), $Z_1(z) - Z(z)$ is positive and Foster, i.e.,

$$Z_1(z) = Z(z) + Z_F(z) ,$$

or, $Z(z)$ in (4.81) is unique up to an arbitrary additive Foster function. Thus, in particular, there exists a one-to-one correspondence between minimum phase rational transfer functions and rational positive functions that are free of poles on $|z| = 1$. Let $Z(z)$ represent such a rational positive function associated with the $H(z)$ in (4.76).

The Schur algorithm described in section 2.2.2 can be used to generate new positive functions $Z_1(z)$, $Z_2(z)$, ... that are of no greater complexity (in terms of their degree) than the $Z(z)$ so obtained. In fact, let

$$d_0(z) = \frac{Z(z) - R_0}{Z(z) + R_0^*} \tag{4.83}$$

as in (2.178) represent the reflection coefficient of $Z(z)$ in (4.81) normalized to $R_0 = Z(0)$. Since $d_0(0) = 0$, the function

$$d_1(z) = \frac{1}{z} d_0(z) \tag{4.84}$$

is analytic in $|z| \leq 1$ and further from the maximum modulus theorem [26], it is also bounded by unity inside the closed unit circle. Thus $d_1(z)$ is a bounded function and letting $Z_1(z)$ represent the positive function associated with $d_1(z)$ that is normalized with respect to R_0, we have

$$d_1(z) = \frac{Z_1(z) - R_0}{Z_1(z) + R_0^*} = \frac{1}{z} \cdot \frac{Z(z) - R_0}{Z(z) + R_0^*} . \tag{4.85}$$

Since $Z(z)$ is rational, so is $d_1(z)$ and $Z_1(z)$ and in general $\delta(Z_1(z)) \leq \delta(Z(z))$. However, from Richards' theorem, $\delta(Z_1(z)) < \delta(Z(z))$ iff

$$Z(z) + Z_*(z)|_{z=0} = 0 \tag{4.86}$$

as in (2.187) and in that case $\delta(Z_1(z)) = \delta(Z(z)) - 1$. Thus the degree of the new positive function $Z_1(z)$ is strictly less than the degree of $Z(z)$ by unity, provided $Z(z)$ satisfies the 'even part' condition (4.86).

4.2.1   ARMA-System Characterization

To investigate the above degree reduction condition in the case of rational systems, we can make use of (4.72), (4.76) and (4.81). In that case, $\delta(Z(z)) = p$ and

$$\frac{Z(z) + Z_*(z)}{2} = H(z)H_*(z) = \frac{\beta(z)\beta_*(z)}{a(z)a_*(z)} = z^{p-q}\frac{\beta(z)\tilde{\beta}(z)}{a(z)\tilde{a}(z)}, \qquad (4.87)$$

where $\tilde{a}(z) \triangleq z^p a^*(1/z^*)$, $\tilde{\beta}(z) \triangleq z^q \beta^*(1/z^*)$ represent the respective reciprocal polynomials. Since

$$\frac{\beta(0)\tilde{\beta}(0)}{a(0)\tilde{a}(0)} = \frac{\beta_0 \beta_q^*}{a_p^*} \neq 0,$$

from (4.87), $z = 0$ is a zero of $(Z(z) + Z_*(z))/2$ of exact order $p - q$. Thus

$$\delta(Z_1(z)) = p - 1 \quad \text{iff} \quad p - q > 0 \qquad (4.88)$$

and hence if $p > q$, the new rational positive function $Z_1(z)$ is exactly of one degree less than $Z(z)$. In that case, under the identification $z = e^{-2p\tau}$, (4.85) yields the interesting configuration in Fig. 2.7, where $Z(z)$ is realized as the input impedance of an ideal line of characteristic impedance R_0 and one-way delay $\tau > 0$, closed on the positive function $Z_1(z)$. Equation (4.85) that relates the reflection coefficient $d_1(z)$ of the load $Z_1(z)$ (normalized to $Z(0)$) to the reflection coefficient $d_0(z)$ of the input positive function $Z(z)$ in (4.81) (normalized to $Z(0)$), also provides the transforming properties of the line and together with (4.88) forms the basis for the new approach to the system identification problem [89]. Thus from (4.87), if $p - q \geq 1$, the new positive function $Z_1(z)$ is of one degree less than the original $Z(z)$, and this process of degree reduction can be repeated if $p - q \geq 2$. In fact, from (4.85)

$$Z(z) = \frac{(R_0 + zR_0^*)Z_1(z) + (1 - z)|R_0|^2}{(1 - z)Z_1(z) + (R_0^* + zR_0)}$$

and as a result

$$H(z)H_*(z) = \frac{Z(z) + Z_*(z)}{2} = \frac{z(R_0 + R_0^*)^2}{M(z)N(z)} \cdot \frac{Z_1(z) + Z_{1*}(z)}{2}, \qquad (4.89)$$

where

$$M(z) = (R_0^* + zR_0) + (1 - z)Z_1(z)$$

and

$$N(z) = (R_0^* + zR_0) - (1 - z)Z_{1*}(z).$$

Note first that $M(0) = Z^*(0) + Z_1(0) \neq 0$, since $Re\, M(0) = Re\,(Z^*(0) + Z_1(0)) > 0$. Secondly, any pole of $N(z)$ in $|z| < 1$ is a pole of $Z_{1*}(z)$ and

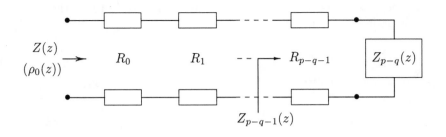

FIGURE 4.2. ARMA (p, q) system and its irreducible representation. Here $\delta(Z(z)) = p$, $\delta((Z_r(z)) = p - r$, $r < p - q$ and $\delta(Z_r(z)) = q$, $r \geq p - q$.

therefore not a zero of $(Z_1 + Z_{1*})/N(z)$. Thus, if $p - q \geq 2$, then from (4.87) and (4.89), $z = 0$ must be a zero of $Z_1(z) + Z_{1*}(z)$. As a consequence,

$$\lim_{z \to 0} N(z) = \lim_{z \to 0} [Z^*(0) - Z_{1*}(z)] = Z^*(0) + Z_1(0) = M(0)$$

because $Z_{1*}(z) \to -Z_1(0)$ as $z \to 0$. Thus $N(0)$ is also nonzero. It now follows that $z = 0$ is a zero of $(Z_1 + Z_{1*})/2$ of order $p - q - 1$. Repeating the procedure that led to (4.85), we obtain a new positive function $Z_2(z)$ with $\delta(Z_2) = \delta(Z_1) - 1$, and after $p - q$ such cycles with degree reduction we have the configuration shown in Fig. 4.2. Clearly, at every such cycle, the new positive function $Z_r(z)$ satisfies $\delta(Z_r) = p - r$ and

$$\frac{\left(Z_r(z) + Z_{r*}(z)\right)\big|_{z=0}}{2} = 0, \quad r = 0 \to p - q - 1, \tag{4.90}$$

where $Z_0(z) \equiv Z(z)$ in (4.81). The positive termination $Z_{p-q}(z)$ in Fig. 4.2 has degree q, and the characteristic impedances $R_0 = Z(0) = r_0 > 0$ and

$$R_r = Z_r(0), \quad r = 1 \to p - q - 1, \tag{4.91}$$

have positive real parts (see Problem 3.2(b)). Moreover, and this is a *key* observation which ultimately will enable us to find p and q, because $z = 0$ is *not* a zero of the 'even part'

$$\frac{Z_{p-q}(z) + Z_{(p-q)*}(z)}{2},$$

additional line extractions do *not* produce any more degree reduction and hence

$$\delta(Z_r(z)) = q, \quad \text{for every } r \geq p - q. \tag{4.92}$$

From the basic transformation formula (4.85), more generally at the rth stage, as in (2.191) we have

$$d_{r+1}(z) \triangleq \frac{Z_{r+1}(z) - R_r}{Z_{r+1}(z) + R_r^*} = \frac{1}{z} \cdot \frac{Z_r(z) - R_r}{Z_r(z) + R_r^*}, \quad r \geq 0. \tag{4.93}$$

For renormalization convenience, define, as before (see (2.197)), the modified bounded functions

$$\rho_r(z) = d_r(z)e^{-j2\phi_{r-1}}, \quad r \geq 0, \tag{4.94}$$

where ϕ_r represents the sum of the phases of $(R_i + R^*_{i-1})$, $i = 1 \to r$, i.e.,

$$\lambda_i = R_i + R^*_{i-1} = |\lambda_i|e^{j\theta_i}$$

and

$$\phi_r = \theta_1 + \theta_2 + \cdots + \theta_r.$$

Then with

$$s_r = \rho_r(0), \tag{4.95}$$

these bounded functions are related as in (2.200) through the single-step Schur update algorithm

$$\rho_{r+1}(z) = \frac{s_{r+1} + z\rho_{r+2}(z)}{1 + zs^*_{r+1}\rho_{r+2}(z)}, \quad r \geq 0. \tag{4.96}$$

Since the above argument shows that there is no more degree reduction in $Z_r(z)$ or $\rho_r(z)$ for $r \geq p - q$, from (4.93)–(4.94) we have the degree constraint

$$\delta\left(\rho_{r+1}(z)\right) = \delta\left(Z_{r+1}(z)\right) = \delta\left(Z_r(z)\right) = q, \quad \text{for } r \geq p - q, \tag{4.97}$$

where the last two equalities follow from (4.92). Further, equating the first and last part of (4.93) and making use of (4.94), we also have

$$\delta\left(z\rho_{r+1}(z)\right) = \delta\left(Z_r(z)\right) = q, \quad \text{for } r \geq p - q, \tag{4.98}$$

and hence from (4.97)–(4.98),

$$\delta\left(\rho_{r+1}(z)\right) = \delta\left(z\rho_{r+1}(z)\right) = q, \quad r \geq p - q. \tag{4.99}$$

Equation (4.90) together with (4.99) represent the key feature of a rational ARMA(p, q) system $H(z)$ in (4.76). The associated rational positive function $Z(z)$ in (4.81) upon line extraction undergoes degree reduction for the first $p - q$ stages (one degree at every stage) and thereafter remains *invariant* with respect to its degree. Thus, from (4.99) the bounded functions $\rho_{r+1}(z)$ and $z\rho_{r+1}(z)$ have degrees equal to q, for any $r \geq p - q$. As a result, every such $\rho_{r+1}(z)$ is expressible in irreducible form as the ratio of a numerator polynomial of degree $\leq q - 1$ and a denominator polynomial of degree equal to q, i.e.,

$$\rho_{r+1}(z) = \frac{h_0 + h_1z + \cdots + h_{q-1}z^{q-1}}{1 + g_1z + \cdots + g_{q-1}z^{q-1} + g_qz^q}, \quad r \geq p - q. \tag{4.100}$$

This invariant character of the bounded functions with respect to their degree beyond a certain stage is characteristic of $\text{ARMA}(p, q)$ systems, and is exploited later on for model order identification and system parameter evaluation.

To make use of the above synthesis procedure for generating a new positive/bounded function together with their degree constraints, in the ARMA system identification problem, it is necessary to reformulate these observations analytically with the help of the Fourier coefficients in the expansion (4.77). This is easily accomplished by making use of the basic interpolation formula derived in section 3.3, for the class of all spectral extensions that match the given set of Fourier coefficients. As we show below, together with the degree reduction arguments given above, this procedure turns out to be a powerful tool for system identification and model order determination. To begin with, we summarize the interpolation formula derived in section 3.3, in the following theorem.

Theorem 4.1: Let $\mathbf{T}_n$ represent the Hermitian Toeplitz matrix

$$\mathbf{T}_n = \begin{bmatrix} r_0 & r_1 & \cdots & r_n \\ r_1^* & r_0 & \cdots & r_{n-1} \\ \vdots & \vdots & \ddots & \vdots \\ r_n^* & r_{n-1}^* & \cdots & r_0 \end{bmatrix}$$

generated from the given autocorrelations $r_0, r_1, \ldots, r_n$, and let Δ_n represent the determinant of $\mathbf{T}_n$. Then, every Δ_k, $k = 0 \to n$, is positive [4];

1) The Levinson polynomial[4]

$$A_n(z) = \frac{1}{\sqrt{\Delta_n \Delta_{n-1}}} \begin{vmatrix} & & & r_n \\ & \mathbf{T}_{n-1} & & \vdots \\ & & & r_1 \\ z^n & \cdots & z & 1 \end{vmatrix}$$

$$\stackrel{\triangle}{=} a_0 + a_1 z + \cdots + a_n z^n, \quad n \geq 0, \tag{4.101}$$

given by (3.67) is free of zeros in the closed unit circle $|z| \leq 1$. Further these polynomials can be recursively generated from the single-step update rule

$$\sqrt{1 - |s_k|^2} \, A_k(z) = A_{k-1}(z) - z s_k \tilde{A}_{k-1}(z), \quad k \geq 1, \tag{4.102}$$

starting with $A_0(z) = 1/\sqrt{r_0}$, where the junction reflection coefficients s_k

---

[4] $\Delta_{-1} \stackrel{\triangle}{=} 1$ and $A_0(z) \stackrel{\triangle}{=} 1/\sqrt{r_0}$.

satisfy (see (3.69))

$$s_k = \left\{ A_{k-1}(z) \sum_{i=1}^{k} r_i z^i \right\}_k A_{k-1}(0), \quad k \geq 1. \tag{4.103}$$

(Here, $\{\ \}_k$ represents the coefficient of z^k in the expansion inside $\{\ \}$.)

2) For every $n \geq 0$, the class of all spectral extensions that interpolate the given autocorrelations $r_0, r_1, \ldots, r_n$ is given by [12]

$$S(\theta) = \frac{1 - |\rho_{n+1}(e^{j\theta})|^2}{|D_n(e^{j\theta})|^2}, \tag{4.104}$$

where

$$D_n(z) = A_n(z) - z\rho_{n+1}(z)\tilde{A}_n(z), \tag{4.105}$$

$\tilde{A}_n(z)$ represents the reciprocal polynomial of $A_n(z)$, and $\rho_{n+1}(z)$ is an *arbitrary* bounded function that represents the termination after $n + 1$ line extractions (see Fig. 3.2).

Discussion: In the case of rational spectral extensions, $\rho_{n+1}(z)$ is *rational*, and let $\Gamma_{n+1}(z)$ denote the unique rational bounded solution of the equation

$$1 - \rho_{n+1}(z)\rho_{n+1*}(z) = \Gamma_{n+1}(z)\Gamma_{n+1*}(z) \tag{4.106}$$

that is analytic and free of zeros in $|z| < 1$. From the discussion that follows (3.55), $D_n(z)$ is also analytic and nonzero in $|z| < 1$ for *any* $\rho_{n+1}(z)$ that is bounded. Hence, from (4.104)

$$S(\theta) = |H(e^{j\theta})|^2 = \sum_{k=-n}^{n} r_k e^{jk\theta} + o(e^{j(n+1)\theta}) \tag{4.107}$$

where up to an arbitrary phase term,

$$H(z) = \frac{\Gamma_{n+1}(z)}{A_n(z) - z\rho_{n+1}(z)\tilde{A}_n(z)} = \frac{\Gamma_{n+1}(z)}{D_n(z)} \tag{4.108}$$

is the unique minimum-phase rational factor that satisfies (4.107), and it represents the basic interpolatory formula valid for all $n \geq 0$, since the first $2n + 1$ Fourier coefficients of $S(\theta)$ equal r_k, $|k| = 0, 1, \ldots, n$, *irrespective* of the choice of the rational bounded function $\rho_{n+1}(z)$.

4.2.2   ARMA-PARAMETER ESTIMATION

In particular, since the unknown transfer function in (4.76) also *must* follow from (4.107)–(4.108) for a specific choice of a rational bounded function $\rho_{n+1}(z)$, to make use of the above theorem in the reconstruction of $H(z)$, it is necessary to know $\rho_{n+1}(z)$ for at least one value of n. To start with, although $\rho_{n+1}(z)$ is not known for any value of n, its functional form in (4.100), that is valid for every $n \geq p-q$, is extremely useful in this respect. From (4.100) for any $n \geq p - q$, since $\rho_{n+1}(z)$ has always $2q$ unknowns, as shown below, the smallest n suitable for its evaluation turns out to be $n = p + q$. Indeed, let

$$\rho_{n+1}(z) \overset{\triangle}{=} \rho_{p+q+1}(z)$$

represent the bounded function after extracting $p + q + 1$ lines in Fig. 4.2 and assume, *temporarily*, that p and q are known. In that case, from (4.100)

$$\rho_{p+q+1}(z) = \frac{h(z)}{g(z)} \overset{\triangle}{=} \frac{h_0 + h_1 z + \cdots + h_{q-1} z^{q-1}}{1 + g_1 z + \cdots + g_{q-1} z^{q-1} + g_q z^q}, \tag{4.109}$$

where $g(z)$ is a strict Hurwitz polynomial of degree q with $g(0) = 1$ and $\delta(h) \leq q - 1$. The coefficients of $h(z)$ and $g(z)$ constitute $2q$ unknowns.[5] From (4.105)–(4.109) and (4.76),

$$\frac{\displaystyle\sum_{k=0}^{q} \beta_k z^k}{\displaystyle\sum_{k=0}^{p} \alpha_k z^k} = \frac{\beta(z)}{\alpha(z)} = H(z) = \frac{\psi(z)}{A_{p+q}(z)g(z) - zh(z)\tilde{A}_{p+q}(z)}, \tag{4.110}$$

where $\psi(z)$ is the unique Hurwitz polynomial solution of the equation[6]

$$\psi(z)\psi_*(z) = g(z)g_*(z) - h(z)h_*(z) \tag{4.111}$$

and $A_{p+q}(z)$ represents the Levinson polynomial of degree $p + q$ obtained by substituting the data $r_0, r_1, \ldots, r_{p+q}$ into (4.101) or (4.102)–(4.103). Obviously, $\delta(\psi) = q$ and the denominator polynomial in (4.110) has *formal* degree $p + 2q$ since $\delta(\tilde{A}_n) = n$. But $\delta(\alpha(z)) = p$ in (4.110), and hence the coefficients of the higher-order terms beyond z^p must be zeros. Thus by equating the coefficients of $z^{p+1}, z^{p+2}, \ldots z^{p+2q}$ to zero there, we derive $2q$ linear equations that determine the $2q$ unknowns g_k, $k = 1 \to q$, and h_k, $k = 0 \to q-1$, uniquely. Since $\rho_{p+q+1}(z)$ exists as a bounded function, the above linear equations do guarantee a unique solution at the correct stage. In matrix form, these equations become

$$\mathbf{A}\mathbf{x} = \mathbf{b}, \tag{4.112}$$

---

[5] The coefficients h_{k-1}, g_k, $k = 1 \to q$, in (4.100) and (4.109) are generic, and of course they depend on the present value of n.

[6] If $H(z)$ is free of zeros on $|z| = 1$, then $\psi(z)$ will be a strict Hurwitz polynomial.

where $\mathbf{A}$ is a $2q \times 2q$ matrix given by

$$
\begin{bmatrix}
a_p & a_{p-1} & \cdots & a_{p-q+1} & -a_q^* & -a_{q+1}^* & \cdots & -a_{2q-1}^* \\
a_{p+1} & a_p & \cdots & a_{p-q+2} & -a_{q-1}^* & -a_q^* & \cdots & -a_{2q-2}^* \\
\vdots & \vdots & \cdots & \vdots & \vdots & \vdots & \cdots & \vdots \\
a_{p+q-1} & a_{p+q-2} & \cdots & a_p & -a_1^* & -a_2^* & \cdots & -a_q^* \\
a_{p+q} & a_{p+q-1} & \cdots & a_{p+1} & -a_0^* & -a_1^* & \cdots & -a_{q-1}^* \\
0 & a_{p+q} & \cdots & a_{p+2} & 0 & -a_0^* & \cdots & -a_{q-2}^* \\
0 & 0 & \cdots & a_{p+3} & 0 & 0 & \cdots & -a_{q-3}^* \\
\vdots & \vdots & \cdots & \vdots & \vdots & \vdots & \cdots & \vdots \\
0 & 0 & \cdots & a_{p+q-1} & 0 & 0 & \cdots & -a_1^* \\
0 & 0 & \cdots & a_{p+q} & 0 & 0 & \cdots & -a_0^*
\end{bmatrix}
$$

$$
\tag{4.113}
$$

$$
\mathbf{x} \triangleq [g_1 \; g_2 \; \cdots \; g_q \; h_0 \; h_1 \; \cdots \; h_{q-1}]^T \tag{4.114}
$$

and

$$
\mathbf{b} \triangleq -[a_{p+1} \; a_{p+2} \; \cdots \; a_{p+q} \; 0 \; 0 \; \cdots \; 0]^T . \tag{4.115}
$$

Here, a_k's denote the coefficients of the Levinson polynomial $A_{p+q}(z)$, i.e.,

$$
A_{p+q}(z) = a_0 + a_1 z + a_2 z^2 + \cdots + a_{p+q} z^{p+q} . \tag{4.116}
$$

With the h_k's and g_k's so determined, [7] the denominator polynomial $D(z) = g(z)A_{p+q}(z) - zh(z)\tilde{A}_{p+q}(z)$ in (4.110) takes the explicit form

$$
D(z) = \tilde{a}_0 + \tilde{a}_1 z + \tilde{a}_2 z^2 + \cdots + \tilde{a}_p z^p , \tag{4.117}
$$

where

$$
\tilde{a}_0 = a_0 \tag{4.118}
$$

and (with $g_0 = 1$)

$$
\tilde{a}_i = \begin{cases} \displaystyle\sum_{k=0}^{i} g_k a_{i-k} - \sum_{k=0}^{i-1} h_k a_{p+q+k-i+1}^* , & 1 \le i \le q \\ \displaystyle\sum_{k=0}^{q} g_k a_{i-k} - \sum_{k=0}^{q-1} h_k a_{p+q+k-i+1}^* , & q+1 \le i \le p. \end{cases} \tag{4.119}
$$

The factorization in (4.111) can now be carried out in any number of ways [7]. However, if $H(z)$ is known to be free of zeros on $|z| = 1$, then

---

[7]At the correct stage (p, q), since $\rho_{p+q+1}(z)$ exists as a unique bounded-real function, (4.112) *always* has a unique solution. However, at a former/later stage, such need not be the case, and if it so happens, that particular stage is skipped. (See the discussion after (4.129).)

$\psi(z)$ is strict Hurwitz, and to exploit the polynomial nature of $\psi(z)$, it is best to consider the function

$$\frac{1}{|\psi(e^{j\theta})|^2} = \frac{1}{|g(e^{j\theta})|^2 - |h(e^{j\theta})|^2} \tag{4.120}$$

and let c_k, $k = 0 \to q$ represent its first q autocorrelations. Thus

$$c_k = \frac{1}{2\pi} \int_{-\pi}^{\pi} \frac{1}{|g(e^{j\theta})|^2 - |h(e^{j\theta})|^2} \, e^{-jk\theta} d\theta, \quad k \geq 0. \tag{4.121}$$

Clearly, from (3.60) and (4.120) the degree q Levinson polynomial associated with these autocorrelations agrees with $\psi(z)$ up to multiplication by a constant of unit magnitude, and hence it is identified as $\psi(z)$.

Finally, on comparing (4.110), (4.117)–(4.119) together with (4.76), we see that the normalized denominator polynomial $\alpha(z) = D(z)/\tilde{a}_0$ and the numerator polynomial $\beta(z) = \psi(z)/\tilde{a}_0$. From (4.107)–(4.108) and (4.110), the power spectrum associated with this ARMA(p,q) system $H(z)$ matches the autocorrelations r_0, r_1, ..., r_{p+q}.

The above analysis assumes that p and q are known. However, these quantites are usually unknown, and to identify them, as shown below, the invariant characteristics of the bounded function at the next stage can be exploited.

4.2.3   MODEL ORDER SELECTION

In the final and crucial step that follows, a new method for determining the model order p and q is proposed. Having determined $\rho_{p+q+1}(z)$ using (4.112), the bounded function $\rho_{p+q+2}(z)$ at the next stage also can be independently determined in a similar manner with $n = p + q + 1$. Notice that $\rho_{p+q+2}(z)$ has the same generic form as in (4.100) and it makes use of additional information r_{p+q+1} (or s_{p+q+1}) through $A_{p+q+1}(z)$. Moreover, since $\rho_{p+q+1}(z)$ and $\rho_{p+q+2}(z)$ are related through (4.96), this immediately gives q conditions in terms of their coefficients that are true *only* from stage $p-q+1$ onwards. It may be remarked that, although (4.96) is valid for any two $\rho_{r+1}(z)$ and $\rho_{r+2}(z)$ that appear in the line extraction process, in the case of ARMA(p,q) systems the particular form in (4.100) is true only for $r \geq p - q$. However, since the earliest stage where these quantities can be computed from the autocorrelations turns out to be for $r = p+q$, the $q+1$ conditions so obtained are the first direct consequence of the ARMA(p,q) nature of the problem.

To formalize the above observation into a sequential procedure, starting with $n = 1$, $m = 0$, and for any $n \geq 1$ and $m \leq n$, write

$$\rho_{n+m+1}(z) = \frac{h_0 + h_1 z + \cdots + h_{m-1} z^{m-1}}{1 + g_1 z + \cdots + g_{m-1} z^{m-1} + g_m z^m} \overset{\triangle}{=} \frac{h_{m-1}(z)}{g_m(z)}, \tag{4.122}$$

$$\rho_{n+m+2}(z) = \frac{f_0 + f_1 z + \cdots + f_{m-1} z^{m-1}}{1 + e_1 z + \cdots + e_{m-1} z^{m-1} + e_m z^m} \triangleq \frac{f_{m-1}(z)}{e_m(z)}, \quad (4.123)$$

and proceed as follows. From (4.109)–(4.110), clearly, the bounded functions $\rho_{n+m+1}(z)$ and $\rho_{n+m+2}(z)$ have the above form whenever $n \geq p$ and $m = q$.

1) Use the coefficients r_k, $k = 0 \to n + m + 1$, to compute $A_{n+m}(z)$ and $A_{n+m+1}(z)$. Then, find the h_i's, g_i's and f_i's, e_i's by forcing the respective polynomials

$$g_m(z) A_{n+m}(z) - z h_{m-1}(z) \tilde{A}_{n+m}(z) \quad (4.124)$$

and

$$e_m(z) A_{n+m+1}(z) - z f_{m-1}(z) \tilde{A}_{n+m+1}(z) \quad (4.125)$$

to have degree n. Notice that in the case of (4.124), the above degree restriction gives rise to $2m$ linear equations in $2m$ unknowns and they have the same representation as in (4.112)–(4.116) with p and q there replaced by n and m, respectively. However, imposing the above degree constraint on the polynomial (4.125) leads to an overdetermined system of $2m + 1$ equations for $2m$ unknowns. Explicitly,

$$\mathbf{B}\mathbf{y} = \mathbf{c}, \quad (4.126)$$

where $\mathbf{B}$ is $(2m + 1) \times 2m$ and is given by

$$
\left[
\begin{array}{cccc|cccc}
b_n & b_{n-1} & \cdot\cdot & b_{n-m+1} & -b^*_{m+1} & -b^*_{m+2} & \cdot\cdot & -b^*_{2m} \\
b_{n+1} & b_n & \cdot\cdot & b_{n-m+2} & -b^*_m & -b^*_{m+1} & \cdot\cdot & -b^*_{2m-1} \\
\vdots & \vdots & \cdot\cdot & \vdots & \vdots & \vdots & \cdot\cdot & \vdots \\
b_{n+m-1} & b_{n+m-2} & \cdot\cdot & b_n & -b^*_2 & -b^*_3 & \cdot\cdot & -b^*_{m+1} \\
b_{n+m} & b_{n+m-1} & \cdot\cdot & b_{n+1} & -b^*_1 & -b^*_2 & \cdot\cdot & -b^*_m \\
b_{n+m+1} & b_{n+m} & \cdot\cdot & b_{n+2} & -b^*_0 & -b^*_1 & \cdot\cdot & -b^*_{m-1} \\
0 & b_{n+m+1} & \cdot\cdot & b_{n+3} & 0 & -b^*_0 & \cdot\cdot & -b^*_{m-2} \\
\vdots & \vdots & \cdot\cdot & \vdots & \vdots & \vdots & \cdot\cdot & \vdots \\
0 & 0 & \cdot\cdot & b_{n+m} & 0 & 0 & \cdot\cdot & -b^*_1 \\
0 & 0 & \cdot\cdot & b_{n+m+1} & 0 & 0 & \cdot\cdot & -b^*_0
\end{array}
\right]
$$

$$(4.127)$$

$$\mathbf{y} \triangleq [e_1\ e_2\ \cdots\ e_m\ f_0\ f_1\ \cdots\ f_{m-1}]^T \quad (4.128)$$

$$\mathbf{c} \triangleq -[b_{n+1}\ b_{n+2}\ \cdots\ b_{n+m}\ b_{n+m+1}\ 0\ \cdots\ 0]^T \quad (4.129)$$

and the b_k's denote the coefficients of the Levinson polynomial

$$A_{n+m+1}(z) = b_0 + b_1 z + \cdots + b_{n+m+1} z^{n+m+1}.$$

Once the e's and f's are so determined,[8] the bounded functions in (4.122) and (4.123) are known completely. If $\rho_{n+m+1}(z)$ and/or $\rho_{n+m+2}(z)$ do not exist as bounded-real functions, obviously there is no ARMA(n, m) system that matches the given autocorrelations, and this particular stage is skipped and the indices are updated. However, they are guaranteed to exist at the correct stage $n = p$ and $m = q$.

2) Since $s_{n+m+1} = \rho_{n+m+1}(0) = h_0$, according to (4.106),

$$\rho_{n+m+1}(z) = \frac{z\rho_{n+m+2}(z) + h_0}{1 + h_0^* z\rho_{n+m+2}(z)} . \tag{4.130}$$

Or, expressed in terms of the coefficients of $h_{m-1}(z)$, $g_m(z)$ and $f_{m-1}(z)$, $e_m(z)$, and equating the ratios of like powers on both sides of (4.130), we obtain the m conditions

$$\varepsilon_0(n, m) \stackrel{\triangle}{=} f_{m-1} + h_0 e_m = 0 \tag{4.131}$$

and

$$\varepsilon_k \stackrel{\triangle}{=} \frac{f_{k-1} + h_0 e_k}{e_k + h_0^* f_{k-1}} - \frac{h_k}{g_k} = 0 , \quad k = 1 \to m , \tag{4.132}$$

whenever $n \geq p$ and $m = q$. (Here $f_{-1} \stackrel{\triangle}{=} 0$. Since the ratio of the constant terms on both sides of (4.130) equals h_0, (4.132) is trivially true for $k = 0$.) Notice that these conditions are a direct consequence of (4.90) and (4.100) and reflect the ARMA(n, m) nature of the problem. Since the first stage where (4.131) and (4.132) are satisfied occurs at $n = p$ and $m = q$, by updating sequentially, p and q are found here as the smallest integers $n \geq 1$ and $m \leq n$, respectively, for which

$$\varepsilon_0(n, m) \approx 0 \tag{4.133}$$

and, more generally,

$$\varepsilon(n, m) \stackrel{\triangle}{=} \left(\sum_{k=0}^{m-1} |\varepsilon_k|^2 \right)^{1/2} \approx 0 , \tag{4.134}$$

where $\varepsilon_m \stackrel{\triangle}{=} \varepsilon_0(n, m)$ given in (4.131).

To summarize, an ARMA(p, q) system has $p + q + 1$ unknowns, and in the present approach these parameters are reconstructed using a two-step procedure. In the first step, the central unknown quantity behind

---

[8]The unknowns e's and f's in (4.125), (4.126) can be determined either by using a pseudo-inverse solution for $\mathbf{B}$ (if it exists) or by making use of only the first $2m$ most significant equations in (4.126). (We have implemented the latter in our computations in section 4.2.3, since the last equation in (4.126) simplifies into (4.131).)

an ARMA(p, q) system is shown to be a rational bounded function as in (4.100), with numerator degree at most equal to $q - 1$ and denominator degree equal to q, that exhibits invariant characteristics with respect to its degree. The degree restrictions of the original ARMA system allow the $2q$ unknowns associated with this bounded function to be computed in terms of the coefficients of the Levinson polynomial $A_{p+q}(z)$. In the final step, the parameters of the bounded function so obtained are used to generate simultaneously the AR and MA parameters as in (4.110)–(4.119). The degree restrictions are further utilized to derive a new model order selection criterion that fully takes rationality into account.

The key feature of a rational system — its degree — is fully exploited by this new algorithm in a consistent manner by making use of the invariant properties contained in (4.90) and (4.100), and the numerical examples presented in the next section offer convincing proof of its remarkable effectiveness and versatility.

4.2.4 SIMULATION RESULTS

The illustrative examples chosen in this section deal with the reconstruction of rational transfer functions and have been designed to highlight all important facets of the algorithm described in the previous section. In the first part, the autocorrelations are assumed to be known exactly, and in the second part, they are estimated from samples of the output response of the system to a white noise input.

Known Autocorrelations (Figs. 4.3–4.9)

In the first step, the system is assumed to be ARMA(n, m), $m \leq n$, and initialization begins with $n = 1$ and $m = 0$. The formula

$$r_k = \frac{1}{2\pi} \int_{-\pi}^{\pi} |H(e^{j\theta})|^2 e^{-jk\theta}\, d\theta = \frac{1}{\pi} \int_0^{\pi} |H(e^{j\theta})|^2 \cos k\theta\, d\theta, \quad k \geq 0,$$

$$(4.135)$$

is then used in the case of real transfer functions $H(z) = \beta(z)/\alpha(z)$ to generate the autocorrelations. Computation of the strict Hurwitz polynomials $A_{n+m}(z)$ and $A_{n+m+1}(z)$, followed by that of the functions $\rho_{n+m+1}(z)$ and $\rho_{n+m+2}(z)$, using (4.112) and (4.126) allows $\varepsilon_0(n, m)$ and $\varepsilon(n, m)$ to be computed from (4.131)–(4.132) and (4.134), provided both $\rho_{n+m+1}(z)$ and $\rho_{n+m+2}(z)$ exist as b.r. functions. The heavy dots on all curves in Figs. 4.3b–4.9b indicate the presence of such a stage (n, m) (if such is not the case, that particular stage is skipped) and as mentioned earlier, they are *guaranteed* to exist at the correct stage $n = p$, $m = q$. Since the first stage where $\varepsilon_0(n, m)$ and $\varepsilon(n, m)$ equal zero occurs only at $n = p$ and $m = q$, sequential updating of n and m continues until *substantial* relative minima in the values of $\varepsilon_0(n, m)$ and $\varepsilon(n, m)$ are observed to occur for the first time.[9] The corresponding pair (n, m) is then identified with (p, q).

The polynomials $g_m(z)$ and $h_{m-1}(z)$ are defined by (4.122), and the procedure in (4.120)–(4.121) can be used to determine the Hurwitz/strict Hurwitz polynomial $\psi(z) = \psi_0 + \psi_1 z + \cdots + \psi_m z^m$. Finally using (4.110), (4.117)–(4.119)

$$H_e(z) \triangleq \frac{\psi(z)}{D(z)} \triangleq \frac{\tilde{\beta}_0 + \tilde{\beta}_1 z + \cdots + \tilde{\beta}_{m-1} z^{m-1} + \tilde{\beta}_m z^m}{1 + \tilde{\alpha}_1 z + \cdots + \tilde{\alpha}_{n-1} z^{n-1} + \tilde{\alpha}_n z^n} \tag{4.136}$$

is the reconstructed version of $H(z)$, where $\tilde{\beta}_k = \psi_k / \tilde{a}_0$, $k = 0 \to m$, $\tilde{\alpha}_k = \tilde{a}_k / \tilde{a}_0$, $k = 1 \to n$. To facilitate comparison, the reconstructed spectrum

$$\tilde{S}(\theta) = \left| H_e(e^{j\theta}) \right|^2$$

and the actual spectrum (whenever available)

$$S(\theta) = \left| H(e^{j\theta}) \right|^2$$

are plotted on the same diagram (Figs. 4.3a–4.9a).

Notice that, even if Fig. 4.5b is used to interpret the true model order to be ARMA$(6, 5)$, computation of the ARMA parameters in that case reveals $\tilde{\beta}_5 = 0$, and the rest of the parameters coincide with the computed values in Fig. 4.5 indicating the true model order to be ARMA$(6, 4)$.

The theoretical development in section 4.2 presupposes that $q \leq p$, a totally unnecessary restriction, introduced solely to promote quick understanding of the main ideas. For example, in Fig. 4.7, $p = 2$ and $q = 4$. In that case, when the stage $n = 4$, $m = 4$ is reached (refer to Fig. 4.7b), $\varepsilon_0(4, 4)$ and $\varepsilon(4, 4)$ dip together, a sure indication that the correct model order has been found. Note that coefficients of z^3 and z^4 in $D(z)$ are of order 10^{-15}. Since the coefficient of z^3 in $\psi(z)$ is of order 10^{-15}, $H(z)$ is identified here up to sign. Additional examples, such as ARMA$(4, 6)$ and MA(2) have been worked out in Figs. 4.8–4.9, and the fidelity of these reconstructions is truly impressive [89].

Every case $q > p$ (e.g., an MA(q) process or an ARMA(p, q) with $p < q$) is detected as an ARMA(n, n) situation in which $n = q$. This means, of course, that some coefficients of $D(z)$ are filled in *automatically* to raise its degree to q. Nevertheless, all superfluous coefficients are computed as *totally negligible quantities!* This unique property is valid proof of the fundamental character of the present algorithm.

---

[9]Since $x(n) \sim \text{AR}(p) \iff s_k = 0$, $k \geq p + 1$, this possibly can be established/eliminated directly.

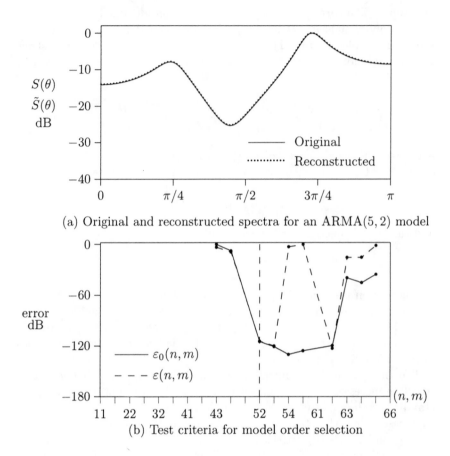

(a) Original and reconstructed spectra for an ARMA$(5,2)$ model

(b) Test criteria for model order selection

FIGURE 4.3. Reconstruction of an ARMA$(5,2)$ model from exact auto-correlations. The original model corresponds to

$$H(z) = \frac{1.440 - 0.416z + z^2}{1 + 0.1899z + 0.1183z^2 - 0.0821z^3 + 0.4957z^4 + 0.1315z^5}.$$

Poles of $H(z)$ are at $1.2\angle \pm 45°$, $1.15\angle \pm 130°$ and $4.0\angle 180°$.
Zeros of $H(z)$ are at $1.2\angle \pm 80°$.
    The reconstructed model is given by

$$H_e(z) = \frac{1.4400 - 0.4160z + z^2}{1.0000 + 0.1899z + 0.1183z^2 - 0.0821z^3 + 0.4957z^4 + 0.1315z^5}.$$

Poles of $H_e(z)$ are at $1.2005\angle\pm44.993°$, $1.1500\angle\pm129.979°$ and $3.9898\angle180°$.
Zeros of $H_e(z)$ are at $1.200\angle \pm 80.0183°$.

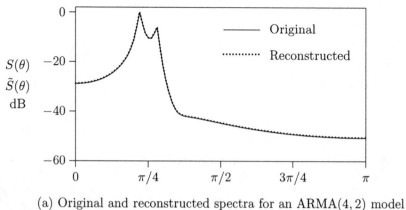

(a) Original and reconstructed spectra for an ARMA(4, 2) model

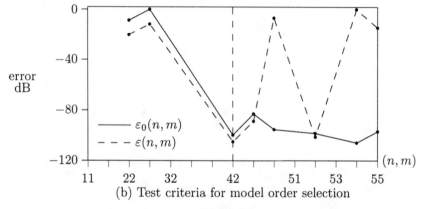

(b) Test criteria for model order selection

FIGURE 4.4. Reconstruction of an ARMA(4, 2) model from exact auto-correlations. The original model corresponds to

$$H(z) = \frac{1.2345 - 1.0706z + z^2}{1 - 2.7596z + 3.8078z^2 - 2.6504z^3 + 0.9224z^4}.$$

Poles of $H(z)$ are at $1.0204\angle \pm 39.6°$ and $1.0204\angle \pm 50.4°$.
Zeros of $H(z)$ are at $1.1111\angle \pm 61.2°$.
   The reconstructed model is given by

$$H_e(z) = \frac{1.2345 - 1.0706z + 0.9999z^2}{1 - 2.7596z + 3.8078z^2 - 2.6504z^3 + 0.9224z^4}.$$

Poles of $H_e(z)$ are at $1.0205\angle \pm 39.62°$ and $1.0203\angle \pm 50.38°$.
Zeros of $H_e(z)$ are at $1.1111\angle \pm 61.20°$.

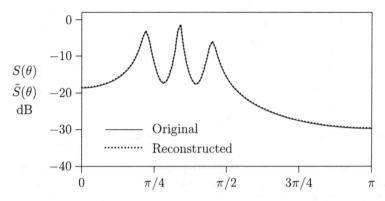

(a) Original and reconstructed spectra for an ARMA(6, 4) model

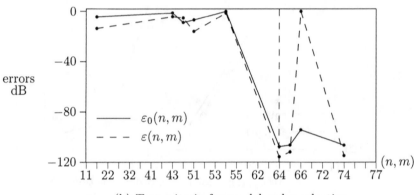

(b) Test criteria for model order selection

FIGURE 4.5. Reconstruction of an ARMA(6, 4) model from exact auto-correlations. The original model corresponds to

$$H(z) = \frac{1.5516 - 2.6678z + 3.5097z^2 - 2.1354z^3 + z^4}{1 - 2.7483z + 4.9765z^2 - 5.564z^3 + 4.6265z^4 - 2.3855z^5 + 0.8099z^6}.$$

Poles of $H(z)$ are at $1.051\angle \pm 39.6°$, $1.021\angle \pm 61.2°$ and $1.04\angle \pm 80.5°$.
Zeros of $H(z)$ are at $1.111\angle \pm 50.4°$ and $1.121\angle \pm 71.3°$.
The reconstructed model is given by

$$H_e(z) = \frac{1.5516 - 2.6678z + 3.5097z^2 - 2.1354z^3 + 0.9999z^4}{1 - 2.7483z + 4.9765z^2 - 5.5640z^3 + 4.6265z^4 - 2.3855z^5 + 0.8099z^6}.$$

Poles of $H_e(z)$ are at $1.0433\angle \pm 39.86°$, $1.0198\angle \pm 60.51°$ and $1.0443\angle \pm 80.64°$. Zeros of $H_e(z)$ are at $1.1111\angle \pm 50.40°$ and $1.1211\angle \pm 71.30°$.

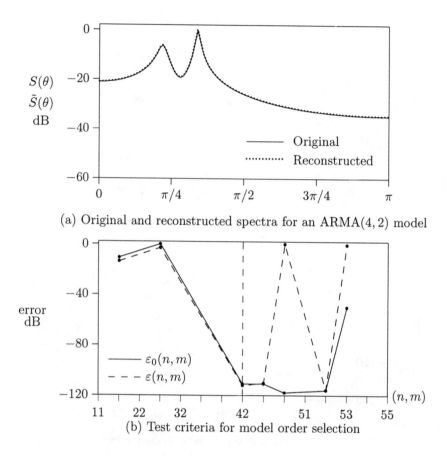

(a) Original and reconstructed spectra for an ARMA$(4, 2)$ model

(b) Test criteria for model order selection

FIGURE 4.6. Reconstruction of an ARMA$(4, 2)$ model from exact auto-correlations. The original model corresponds to

$$H(z) = \frac{1.2345 - 1.4164z + z^2}{1 - 2.4119z + 3.2531z^2 - 2.2658z^3 + 0.8710z^4}.$$

Poles of $H(z)$ are at $1.05\angle \pm 39.6°$ and $1.02\angle \pm 61.2°$.
Zeros of $H(z)$ are at $1.111\angle \pm 50.4°$.
    The reconstructed model is given by

$$H_e(z) = \frac{1.2345 - 1.4164z + 0.9999z^2}{1 - 2.4119z + 3.2531z^2 - 2.2658z^3 + 0.8710z^4}.$$

Poles of $H_e(z)$ are at $1.05\angle \pm 39.60°$ and $1.0205\angle \pm 61.20°$.
Zeros of $H_e(z)$ are at $1.1111\angle \pm 50.40°$.

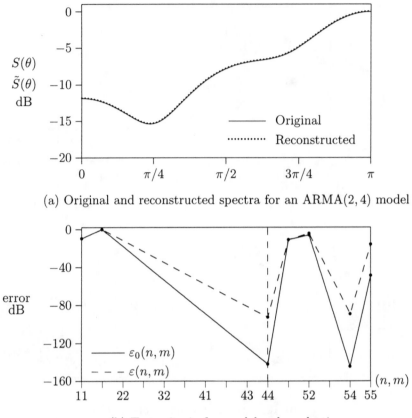

(a) Original and reconstructed spectra for an ARMA$(2,4)$ model

(b) Test criteria for model order selection

FIGURE 4.7. Reconstruction of an ARMA$(2,4)$ model from exact auto-correlations. The original model corresponds to

$$H(z) = \frac{8 - 2z + z^2 + 1.5z^4}{1 + 0.5z + 0.1z^2}.$$

Poles of $H(z)$ are at $3.16\angle \pm 142.2°$.
Zeros of $H(z)$ are at $1.414\angle \pm 45°$ and $1.633\angle \pm 127.8°$.
    The reconstructed model is given by

$$H_e(z) = \frac{8.0 - 2z + 1.0z^2 + 1.06 \times 10^{-13}z^3 + 1.5z^4}{1 + 0.5z - 0.01z^2 + 3.4 \times 10^{-15}z^3 + 1.73 \times 10^{-15}z^4}.$$

Poles of $H_e(z)$ are at $3.162\angle \pm 142.24°$ and $3.6 \times 10^6 \angle \pm 89.99°$.
Zeros of $H_e(z)$ are at $1.414\angle \pm 45°$ and $1.633\angle \pm 127.76°$.

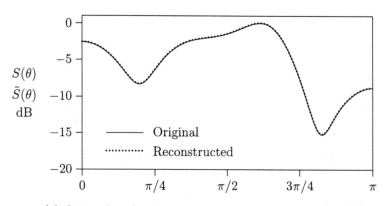

(a) Original and reconstructed spectra for an ARMA(4, 6) model

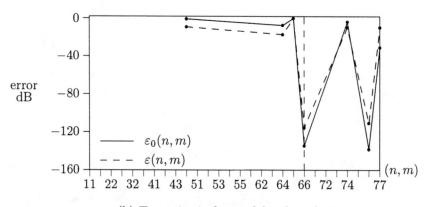

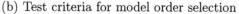

(b) Test criteria for model order selection

FIGURE 4.8. Reconstruction of an ARMA(4, 6) model from exact auto-correlations. The original model corresponds to

$$H(z) = \frac{6 + z - 1.2z^2 + 2z^4 + z^6}{1 - 0.125z + 0.25z^2 - 0.1875z^3 + 0.125z^4}.$$

Poles of $H(z)$ are at $1.94\angle \pm 42.4°$ and $1.46\angle \pm 117.7°$.
Zeros of $H(z)$ are at $1.23\angle \pm 35.4°$, $1.75\angle \pm 91.1°$ and $1.14\angle \pm 147.5°$.
    The reconstructed model $H_e(z)$ is given by

$$\frac{6 + z - 1.20z^2 + 5.4 \times 10^{-14}z^3 + 2.0z^4 - 4.3 \times 10^{-14}z^5 + z^6}{1 + 0.125z - 0.25z^2 - 0.1875z^3 + 0.125z^4 - 4.5 \times 10^{-15}z^5 - 1.4 \times 10^{-15}z^6}.$$

Poles of $H_e(z)$ are at $1.937\angle \pm 42.44°$, $1.46\angle \pm 117.72°$ and $9.5 \times 10^6 \angle 0°$
and $9.5 \times 10^6 \angle 180°$.
Zeros of $H_e(z)$ are at $1.226\angle \pm 35.39°$, $1.75\angle \pm 91.13°$ and $1.144\angle \pm 147.49°$.

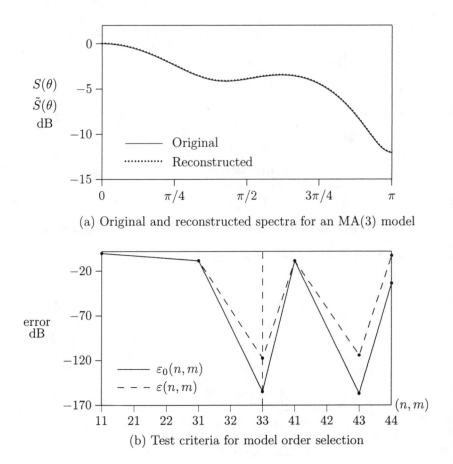

(a) Original and reconstructed spectra for an MA(3) model

(b) Test criteria for model order selection

FIGURE 4.9. Reconstruction of a MA(3) model from exact autocorrelations. The original model corresponds to

$$H(z) = 5 + 2z + z^3.$$

Poles of $H(z)$ are at ∞.
Zeros of $H(z)$ are at $1.94\angle \pm 70°$ and $1.33\angle 180°$.
    The reconstructed model is given by

$$H_e(z) = \frac{5 + 2z - 1.06 \times 10^{-15}z^2 + z^3}{1 - 7.3 \times 10^{-17}z + 5.9 \times 10^{-17}z^2 + 6.0 \times 10^{-17}z^3}.$$

Poles of $H_e(z)$ are at $2.55 \times 10^5 \angle \pm 60.00°$ and $2.55 \times 10^5 \angle 180°$.
Zeros of $H_e(z)$ are at $1.94\angle \pm 69.98°$ and $1.328\angle 180°$.

Known Output Data (Figs. 4.10–4.15)

The original $H(z)$ under test is implemented recursively and then driven at the input by a stationary zero-mean white noise source[10] $w(n)$ with variance σ^2, to generate the required output data stream $x(n)$. Thus,

$$x(n) = -\sum_{k=1}^{p} \alpha_k x(n-k) + \sum_{k=0}^{q} \beta_k w(n-k).$$

In all these examples we have used the statistic

$$\hat{r}_k = \frac{1}{N} \sum_{n=N_0+1}^{N_0+N-k} x(n+k)x^*(n), \quad k \geq 0, \tag{4.137}$$

to estimate the autocorrelations r_k from the data samples $x(N_0+1)$, $x(N_0+2)$, ..., $x(N_0+N)$. The number of data samples N should be large enough so that the Toeplitz matrix formed from the estimated autocorrelations $\hat{r}_0 \to \hat{r}_{n+m+1}$ is positive definite. Such a set of $\hat{r}_k$'s have been used here to compute the estimates of the Levinson polynomials $A_{n+m}(z)$, $A_{n+m+1}(z)$ as well as those of the bounded functions $\rho_{n+m+1}(z)$ and $\rho_{n+m+2}(z)$, whenever they exist. As before, the heavy dots on Figs. 4.11b–4.15b indicate the presence of such a stage, and in the exact case since the first stage where $\varepsilon_0(n,m)$ and $\varepsilon(n,m)$ equal zero occurs at $n = p$, $m = q$, their estimated counterparts are also expected to be small at that stage. Accompanying the curves of $\varepsilon_0(n,m)$ or $\varepsilon(n,m)$ shown in Figs. 4.10b–15b are also plots of the associated Akaike Information Criterion $\mathrm{AIC}(n,m)$.

Recall that in Akaike's approach [55, 72], p and q are determined as the integers n and m that minimize the information criterion (see (4.63)–(4.64))

$$\mathrm{AIC}(n,m) = N \cdot \ln \hat{\sigma}^2 + 2(n+m), \tag{4.138}$$

where $\hat{\sigma}^2$ is the maximum likelihood estimate of the input noise variance using the forward and backward predictors as in (4.64) with $\hat{p} = n$ and $\hat{q} = m$. To maintain positivity (almost always) for the first term in (4.138), it is desirable to choose the input noise variance σ^2 to be greater than unity[11].

The estimated transfer function is given by

$$\hat{H}(z) = \frac{\hat{\beta}(z)}{\hat{\alpha}(z)} = \frac{\hat{\beta}_0 + \hat{\beta}_1 z + \cdots + \hat{\beta}_m z^m}{\hat{\alpha}_0 + \hat{\alpha}_1 z + \cdots + \hat{\alpha}_n z^n} \tag{4.139}$$

and

$$\hat{S}(\theta) = |\hat{H}(e^{j\theta})|^2.$$

---

[10]It has been our experience that standard "white" noise sources are inadequate and must be subjected to considerable "prewhitening" to be really useful.

[11]Notice that $\hat{\sigma}^2 < 1$ could give rise to large negative values for the first term in (4.138), which in turn might shift the true minimum value there.

Here, $\hat{\alpha}_k$'s and $\hat{\beta}_k$'s represent the coefficients of the estimates $\hat{\alpha}(z)$ and $\hat{\beta}(z)$ at stage (n, m) that have been computed using (4.112)–(4.121).

To establish some basis for comparison, Figs. 4.11–4.14 refer to examples taken from the open literature. Figures 4.11–4.12 refer to ARMA$(4, 2)$ examples discussed in [55] and represent our reconstruction from data samples. The autocorrelations are determined from an output stream of data samples and all necessary details are indicated in the figures. Figures 4.13–4.14 correspond to two examples taken from [81, 82]. In all these cases the variable z^{-1} has been changed into z to achieve conformity with our formulation. With exact covariances, $\varepsilon_0(n, m)$ shows a substantial dip at the correct stage, but with r_k's estimated from a finite number of output samples, even though the dip is reduced, the discrimination ratio is still significantly large compared to the behavior exhibited by the AIC in Figs. 4.10b–4.15b. Furthermore, as an examination of Figs. 4.10b and 4.14b reveals, AIC underdetermines the order of $H(z)$ [93] and the indicator function $\varepsilon_0(n, m)$ has been found to be superior to AIC(n, m) in all essential respects.

The efficacy of the criterion $\varepsilon_0(n, m)$ in pinpointing the model order is remarkable. Almost all other comparable techniques [67, 81, 82, 90–92], postulate the order, a priori, and then attempt to fit the model parameters. It may be remarked that techniques such as AIC, MDL [86], etc., that are usually used for model order selection have no mechanism to exploit the key feature of a rational system – its degree. This is because statistical procedures in themselves have no inherent capability to bring out the rationality of the system explicitly; they always work with the likelihood function or its variations both in the rational as well as the nonrational case. The superior performance of $\varepsilon_0(n, m)$ in model order determination in the present case can be attributed to the exploitation of rationality at every stage.

Next, we examine nonrational systems and techniques for rationalizing these systems in some optimal manner.

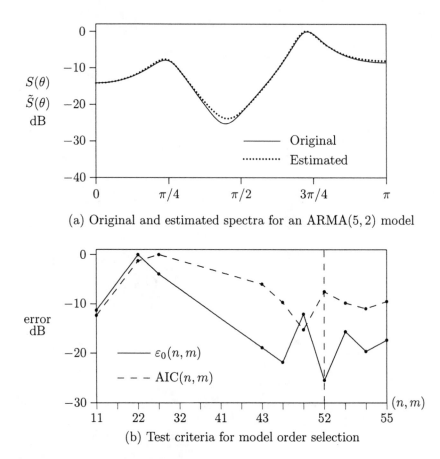

(a) Original and estimated spectra for an ARMA$(5, 2)$ model

(b) Test criteria for model order selection

FIGURE 4.10. Estimation of the ARMA$(5, 2)$ model given in Fig. 4.3 from eight realizations of 500 data samples. The estimated model averaged over these eight realizations is given by

$$\hat{H}(z) = \frac{1.4526 - 0.3341z + 0.9211z^2}{1.00 + 0.2460z + 0.1039z^2 - 0.0951z^3 + 0.4853z^4 + 0.1777z^5}$$

and the associated spectrum is shown superimposed in (a). Poles of $\hat{H}(z)$ are at $1.1975\angle \pm 44.23°$, $1.145\angle \pm 129.40°$ and $2.9935\angle 180°$. Zeros of $\hat{H}(z)$ are at $1.2558\angle \pm 81.69°$.

The estimated model computed from a single realization of 4000 samples is given by

$$\hat{H}(z) = \frac{1.4399 - 0.4168z + 0.999z^2}{1 + 0.1899z + 0.1182z^2 - 0.08189z^3 + 0.4955z^4 + 0.1312z^5}.$$

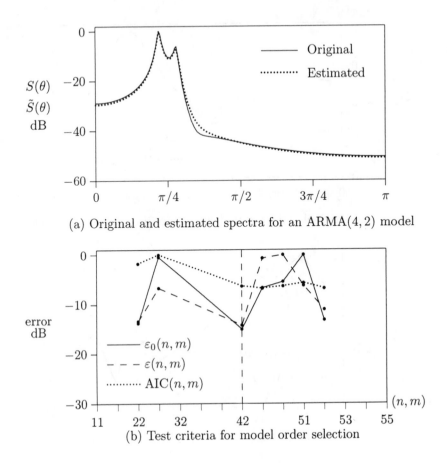

(a) Original and estimated spectra for an ARMA$(4, 2)$ model

(b) Test criteria for model order selection

FIGURE 4.11. Estimation of the ARMA$(4, 2)$ model given in Fig. 4.4 from 900 data samples. The estimated model is given by

$$\hat{H}(z) = \frac{1.3532 - 1.1248z + 0.9703z^2}{1 - 2.7533z + 3.7957z^2 - 2.6406z^3 + 0.9205z^4} .$$

Poles of $\hat{H}(z)$ are at $1.0189\angle \pm 39.62°$ and $1.0229\angle \pm 50.59°$.
Zeros of $\hat{H}(z)$ are at $1.1809\angle \pm 60.61°$.

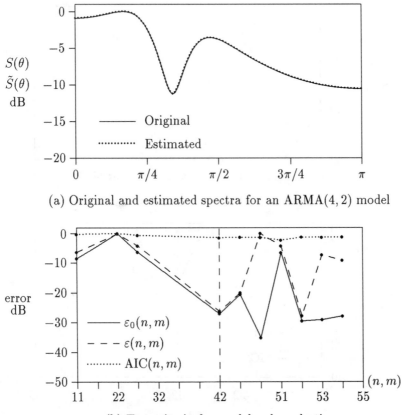

(a) Original and estimated spectra for an ARMA$(4, 2)$ model

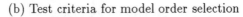

(b) Test criteria for model order selection

FIGURE 4.12. Estimation of the ARMA$(4, 2)$ model from 8000 data samples. The original model corresponds to

$$H(z) = \frac{1.2345 - 1.0706z + z^2}{1 - 1.3678z + 1.3353z^2 - 0.6706z^3 + 0.2401z^4}.$$

Poles of $H(z)$ are at $1.4286\angle \pm 43.2°$ and $1.4286\angle \pm 75.6°$.
Zeros of $H(z)$ are at $1.1111\angle \pm 61.2°$.
   The estimated model is given by

$$\hat{H}(z) = \frac{1.2348 - 1.0706z + 0.9983z^2}{1 - 1.3687z + 1.3342z^2 - 0.67z^3 + 0.2399z^4}.$$

Poles of $\hat{H}(z)$ are at $1.4277\angle \pm 43.16°$ and $1.4298\angle \pm 75.64°$.
Zeros of $\hat{H}(z)$ are at $1.1122\angle \pm 61.18°$.

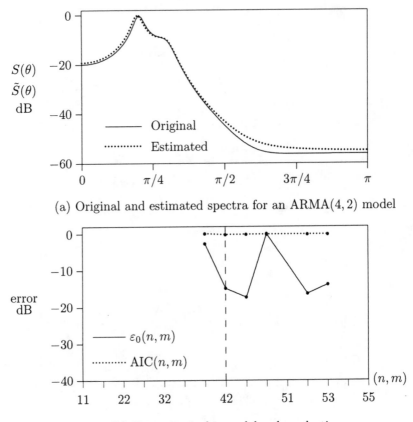

(a) Original and estimated spectra for an ARMA$(4, 2)$ model

(b) Test criteria for model order selection

FIGURE 4.13. Estimation of an ARMA$(4, 2)$ model from 600 data samples. The original model corresponds to

$$H(z) = \frac{2.0409 + 0.8829z + z^2}{1 - 2.5953z + 3.3390z^2 - 2.2001z^3 + 0.7311z^4}.$$

Poles of $H(z)$ are at $1.0526\angle \pm 36°$ and $1.1111\angle \pm 54°$.
Zeros of $H(z)$ are at $1.4286\angle \pm 108°$.
   The estimated model is given by

$$\hat{H}(z) = \frac{2.1265 + 0.8003z + 0.999z^2}{1 - 2.5966z + 3.3426z^2 - 2.2041z^3 + 0.7329z^4}.$$

Poles of $\hat{H}(z)$ are at $1.0522\angle \pm 35.97°$ and $1.1102\angle \pm 54.03°$.
Zeros of $\hat{H}(z)$ are at $1.4582\angle \pm 105.93°$.

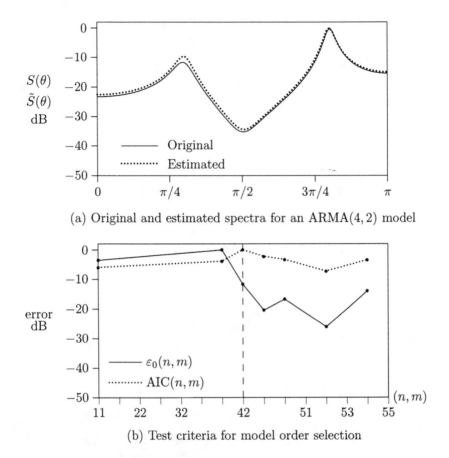

(a) Original and estimated spectra for an ARMA$(4, 2)$ model

(b) Test criteria for model order selection

FIGURE 4.14. Estimation of an ARMA$(4, 2)$ model from 400 data samples. The original model corresponds to

$$H(z) = \frac{1.3842 + z^2}{1 + 0.4791z + 0.0862z^2 + 0.2902z^3 + 0.7311z^4}.$$

Poles of $H(z)$ are at $1.1111 \angle \pm 54°$ and $1.0526 \angle \pm 144°$.
Zeros of $H(z)$ are at $1.1765 \angle \pm 90°$.
 The estimated model is given by

$$\hat{H}(z) = \frac{1.4425 + 0.0254z + 1.0409^2}{1 + 0.4640z + 0.1086z^2 + 0.3178z^3 + 0.7483z^4}.$$

Poles of $\hat{H}(z)$ are at $1.0955 \angle \pm 54.40°$ and $1.0552 \angle \pm 143.67°$.
Zeros of $\hat{H}(z)$ are at $1.1772 \angle \pm 90.59°$.

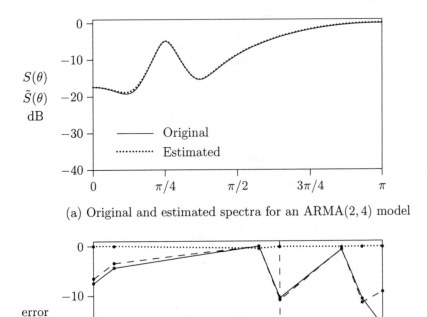

(a) Original and estimated spectra for an ARMA$(2,4)$ model

(b) Test criteria for model order selection

FIGURE 4.15. Estimation of an ARMA$(2,4)$ model from data samples. The original model corresponds to

$$H(z) = \frac{2.0736 - 4.5928z + 5.0862z^2 - 3.1894z^3 + z^4}{1 - 1.2627z + 0.7972z^2}.$$

Poles of $H(z)$ are at $1.12\angle \pm 45°$.
Zeros of $H(z)$ are at $1.2\angle \pm 25°$ and $1.2\angle \pm 65°$.
   The estimated model is an ARMA$(4,4)$ given by

$$\hat{H}(z) = \frac{2.0805 - 4.5654z + 5.0360z^2 - 3.1380z^3 + 0.9707z^4}{1 - 1.2369z + 0.7874z^2 + 0.0069z^3 + 0.0018z^4}.$$

Poles of $\hat{H}(z)$ are at $1.1176\angle \pm 45.28°$ and $20.892\angle \pm 97.35°$.
Zeros of $\hat{H}(z)$ are at $1.2205\angle \pm 25.10°$ and $1.1995\angle \pm 64.78°$.
   In this case, the autocorrelations are estimated by averaging over 40 realizations each containing 2000 data samples.

4.3 Rational Approximation of Nonrational Systems

A nonrational system has a transfer function that, unlike the rational systems, cannot be expressed as the ratio of two polynomials (of finite degree). Such systems occur in practice in a variety of contexts such as in distributed structures, partial differential equations and systems regulated by differential equations with delays. For example, e^{-z}, $(1-z)ln(1-z)$, $z/(e^z-1)$ all represent nonrational systems, and in any rearrangement their numerators and/or denominators will contain an infinite number of terms. Thus, in general

$$H(z) = \frac{\sum_{k=0}^{\infty} \beta_k z^k}{\sum_{k=0}^{\infty} \alpha_k z^k} = \sum_{k=0}^{\infty} b_k z^k \qquad (4.140)$$

where the power series is valid in some contour in the complex z-plane. If the system is known to be stable, for example, then the above power series is analytic in $|z| < 1$. The input-output relation of nonrational systems can be expressed in terms of the infinite recursion

$$x(n) = -\sum_{k=1}^{\infty} \alpha_k x(n-k) + \sum_{k=0}^{\infty} \beta_k w(n-k) = \sum_{k=0}^{\infty} b_k w(n-k). \qquad (4.141)$$

The problem now is how does one simulate such a system?

Obviously every term in (4.141) *cannot* be faithfully simulated in a practical setup and the straightforward choice of truncating the above infinite series is not to be recommended for several reasons. First, the truncated system in general need not be minimum-phase or stable (as the original system) and moreover, if the terms do not decay fast, truncation will lead to severe round-off error. Even though approximating a nonrational function by truncating its power series (polynomial approximation) is not attractive, nevertheless, it may be possible to realize the same approximation by making use of a 'simpler' rational function. This is, of course, the idea behind Padé approximations [46, 49], where the rational function

$$H_r(z) = \frac{\beta_0 + \beta_1 z + \cdots + \beta_q z^q}{\alpha_0 + \alpha_1 z + \cdots + \alpha_p z^p} = \frac{B(z)}{A(z)} \qquad (4.142)$$

is said to be a Padé approximation to $H(z) = \sum_{k=0}^{\infty} h_k z^k$ if at least the first $p + q + 1$ terms in the power series expansion of $H(z)$ and $B(z)/A(z)$ match exactly, i.e.,

$$H(z) - \frac{B(z)}{A(z)} = O(z^{p+q+1}). \qquad (4.143)$$

It is not difficult to show that such an approximation is unique. If not, let $B_1(z)/A_1(z)$ and $B_2(z)/A_2(z)$ be both Padé approximations to $H(z)$ where $\delta(B_1) = \delta(B_2) = q$, $\delta(A_1) = \delta(A_2) = p$. Then, from (4.143), $B_1(z)/A_1(z)$ and $B_2(z)/A_2(z)$ both agree up to $p + q + 1$ terms in their power series expansion, i.e.,

$$\frac{B_1(z)}{A_1(z)} - \frac{B_2(z)}{A_2(z)} = O(z^{p+q+1}) \tag{4.144}$$

which gives

$$B_1(z)A_2(z) - B_2(z)A_1(z) = O(z^{p+q+1}).$$

But the left-hand side is a polynomial of degree at most equal to $p + q$ and hence for (4.144) to hold, it must be identically zero, i.e.,

$$B_1(z)A_2(z) - B_2(z)A_1(z) \equiv 0$$

or

$$\frac{B_1(z)}{A_1(z)} = \frac{B_2(z)}{A_2(z)} = H(z) + O(z^{p+q+1}).$$

Thus, Padé approximations are unique. Unfortunately, since such approximations only involve a finite number of terms $(b_0, b_1, \ldots, b_{p+q})$ of the actual impulse response in determining the rational approximation, the crucial asymptotic characteristics of the original nonrational system can never be faithfully represented by this approach. Furthermore, nothing in the above proof guarantees such approximations to be stable. For example, the ARMA(1, 1) Padé approximation to the minimum-phase function e^{-2z} is not stable, since in that case

$$\frac{B_1(z)}{A_1(z)} = \frac{1-z}{1+z} = e^{-2z} + O(z^3)$$

and $A_1(z)$ is not strict Hurwitz.

Following the approach developed in the previous chapter for the rational case, our solution to this problem tries to impose Padé-like constraints on the magnitude square of the transfer function.

4.3.1   Padé-like Approximation

To be specific, let

$$S(\theta) = |H(e^{j\theta})|^2 = \sum_{k=-\infty}^{\infty} r_k e^{jk\theta} \tag{4.145}$$

represent the power spectral density and the autocorrelations of the given nonrational function. Then a rational system as in (4.142) is said to be a

Padé-like approximation to the above nonrational system if

$$|H_r(e^{j\theta})|^2 = \left| \frac{\beta_0 + \beta_1 e^{j\theta} + \cdots + \beta_q e^{jq\theta}}{1 + \alpha_1 e^{j\theta} + \cdots + \alpha_p e^{jp\theta}} \right|^2$$

$$= \sum_{k=-n}^{n} r_k e^{jk\theta} + O(e^{j(n+1)\theta}), \qquad (4.146)$$

where $n \geq p + q$, i.e., the Fourier coefficients of the ARMA(p,q) rational approximation must match with the Fourier coefficients of the given nonrational system *at least* up to the first $p + q + 1$ terms.

More generally, as remarked earlier, any power spectral density $S(\theta) \geq 0$ that satisfies the integrability condition (3.2) and the physical-realizability condition (3.3) can be factorized in terms of its minimum-phase Wiener factor

$$H(z) = \sum_{k=0}^{\infty} b_k z^k$$

that is analytic in $|z| < 1$ such that

$$S(\theta) = |H(e^{j\theta})|^2 = \sum_{k=-\infty}^{\infty} r_k e^{jk\theta}. \qquad (4.147)$$

Naturally if $S(\theta)$ is nonrational, so is $H(z)$ as well as the corresponding positive function

$$Z(z) = r_0 + 2 \sum_{k=1}^{\infty} r_k z^k. \qquad (4.148)$$

From the class of all spectral extension formula in (4.104), every extension that matches at least the first $p + q + 1$ autocorrelations $r_0, r_1, \ldots, r_{p+q}$ is parametrized by an arbitrary bounded function $\rho(z)$, and, moreover, the associated Wiener factors are given by (up to multiplication by a constant of unit magnitude)

$$H_\rho(z) = \frac{\Gamma(z)}{A_n(z) - z\rho(z)\tilde{A}_n(z)}, \qquad (4.149)$$

where $n \geq p+q$. Here $\Gamma(z)$ represents the minimum-phase factor associated with the factorization

$$|\Gamma(e^{j\theta})|^2 = 1 - |\rho(e^{j\theta})|^2 \qquad (4.150)$$

that is guaranteed to exist under the causality criterion (see (3.51))

$$\frac{1}{2\pi} \int_{-\pi}^{\pi} \ln\left(1 - |\rho(e^{j\theta})|^2\right) d\theta > -\infty, \qquad (4.151)$$

and $A_n(z)$ is the Levinson polynomial at stage n generated from $r_0 \to r_n$. Clearly, if $S(\theta)$ in (4.147) is nonrational, then it must follow from (4.149) for a specific nonrational $\rho(z)$. From Fig. 3.2, physically, $\rho(z)$ represents the unique renormalized bounded function that remains after extracting $n + 1$ lines from the positive function $Z(z)$ in (4.148).

More interestingly, if an ARMA(p, q) rational function approximation as in (4.146) holds for some $H_r(z)$, then that too must follow from (4.149) for a specific *rational* $\rho(z)$. In that case, since $\delta(\Gamma(z)) = q$, from (4.150), $\delta(\rho(z)) = q$. Let

$$\rho(z) = \frac{h(z)}{g(z)}, \tag{4.152}$$

then (4.149) simplifies to

$$H_r(z) = \frac{\psi(z)}{A_n(z)g(z) - zh(z)\tilde{A}_n(z)} = \frac{\psi(z)}{D(z)}, \tag{4.153}$$

where

$$\psi(z)\psi_*(z) = g(z)g_*(z) - h(z)h_*(z) \tag{4.154}$$

and

$$D(z) = A_n(z)g(z) - zh(z)\tilde{A}_n(z), \quad n \geq p + q. \tag{4.155}$$

However, since $\delta(D(z)) = p$ and $\delta(\tilde{A}_n(z)) = n \geq p + q$, from (4.155), we must have $\delta(h(z)) \leq q - 1$. As a result, $\delta(g(z)) = q$ and the bounded function $\rho(z)$ in (4.152) must have the representation

$$\rho(z) = \frac{h_0 + h_1 z + \cdots + h_{q-1}z^{q-1}}{1 + g_1 z + \cdots + g_{q-1}z^{q-1} + g_q z^q}. \tag{4.156}$$

Notice that the form in (4.156) is identical to that in (4.100), but due to quite different considerations.

The formal degree of $D(z)$ in (4.153) is still $n + q$, and to respect the ARMA(p, q) nature of $H_r(z)$, the coefficients of the higher-order terms beyond z^p must be zeros there. This results in $n + q - p$ linear equations in $2q$ unknowns and since $n \geq p+q$, these equations are at least $2q$ in number. Clearly, the minimum number of these equations is obtained for $n = p + q$, and referring back to (4.110)–(4.117), in that case they are exactly the same as (4.112). However, unlike the rational case, in the present situation, these equations need not have a solution for every p, q. Even if there exists a solution in some cases, $g(z)$ so obtained need not be strict Hurwitz and further $\rho(z)$ need not turn out to be a bounded function. If any of the above possibilities occur for some p and q, then there exists no such ARMA(p, q) rational approximation as in (4.146) to the given nonrational function. Remarkably, if for some p and q, $\rho(z)$ so obtained turns out to be a bounded function, then $H_r(z)$ in (4.153) represents a minimum-phase ARMA(p, q) system whose first $p + q + 1$ autocorrelations are guaranteed to match with those of the given nonrational spectral density function $S(\theta)$. In this case, $H_r(z)$ is said to be a Padé-like approximation to the given transfer function $H(z)$. To sum up, if $\delta(D(z)) = p$, $\delta(g(z)) = q$ and $r_0 \to r_n$ represent the autocorrelations that match with those of the rational approximation $H_r(z)$, then the choice $\rho(z) = h(z)/g(z)$ in (4.156) is optimal if

1. $\rho(z)$ is bounded,

2. $n \geq p + q$,

3. $p < n$ (If $p = n$, then $n \geq p + q$ implies $q = 0$ and hence $g(z) = $ constant, $h(z) \equiv 0$. Thus $\rho(z) \equiv 0$ and it results in an AR(n) transfer function that satisfies the above conditions trivially.)

In that case the rational approximation $H_r(z)$ is automatically minimum-phase. To summarize, we have the following theorem [89].

Theorem 4.2 [12]: Given the autocorrelations $r_0 \to r_{p+q}$ that form a positive definite sequence, the necessary and sufficient condition to fit an ARMA(p, q) model as in (4.146) is that the system of linear equations involving the Levinson polynomial $A_{p+q}(z)$ in (4.112) yield a bounded solution for the function $\rho(z)$ defined in (4.156).

We now offer a tentative analysis which suggests strongly that optimal $\rho(z)$'s should always exist for every $S(\theta)$, at least for certain values of n. The numerical evidence presented in the next section in support of this conjecture appears to be overwhelming.

Theorem 4.3: Let $S(\theta)$ denote a nonnegative periodic function of θ of period 2π that satisfies the integrability condition as well as the Paley-Wiener criterion. Then, for all n, except a finite number at most, there exist polynomial solutions $g(z)$, $h(z)$ and $D(z)$ of (4.155) meeting the above last two requirements $n \geq p + q$ and $p < n$ for optimality.

Proof: Suppose first that $\delta(A_n) < n$. Then, $g(z) \equiv 1$, $h(z) \equiv 0$ and $D(z) \equiv A_n(z)$ constitute an optimal choice because $\rho(z) \equiv 0$, $q = 0$, $p = \delta(A_n)$, and $p + q = p < n$. In this case, $H_r(z) = 1/A_n(z)$ is AR(p). Assume therefore, $\delta(A_n) = n \geq 1$. Clearly, due to the presence of the factor z in (4.155), $\delta(g) = q \geq 1$ and $\delta(h) \leq q - 1$ are necessary conditions for the existence of an optimal $\rho(z)$. We may therefore write

$$h(z) = h_0 + h_1 z + \cdots + h_{q-1} z^{q-1} \tag{4.157}$$

and in general

$$g(z) = g_0 + g_1 z + \cdots + g_q z^q, \tag{4.158}$$

subject to the constraints $g_q \neq 0$ and $n \geq p + q$. From (4.155),

$$\frac{A_n(z)}{\tilde{A}_n(z)} - \frac{zh(z)}{g(z)} = \frac{D(z)}{g(z)\tilde{A}_n(z)}. \tag{4.159}$$

Since $\delta(\tilde{A}_n) = n$, $\delta(g\tilde{A}_n) \geq p + 2q$ and

$$\frac{D(z)}{g(z)\tilde{A}_n(z)} \sim O(z^{-2q}) \tag{4.160}$$

for large z. Moreover, since $A_n(z)$ is strict Hurwitz, the function

$$b_n(z) \triangleq \frac{A_n(z)}{\tilde{A}_n(z)} \tag{4.161}$$

is free of zeros in $|z| \leq 1$, analytic in $|z| \geq 1$ and has unit magnitude in $|z| = 1$. By the maximum modulus theorem [26], $|b_n(z)| \leq 1$, $|z| \geq 1$. Thus, $b_n(z)$ admits a Laurent expansion

$$b_n(z) = \sum_{k=0}^{\infty} d_k z^{-k}, \quad d_0 \neq 0, \tag{4.162}$$

convergent for all z in $|z| \geq 1$. It now follows from (4.159), (4.160) and (4.162) that

$$z(h_0 + h_1 z + \cdots + h_{q-1}z^{q-1}) - (g_0 + g_1 z + \cdots + g_q z^q) \sum_{k=0}^{\infty} d_k z^{-k} \sim O(z^{-q}),$$
$$\tag{4.163}$$

and by comparing coefficients we obtain

$$\begin{aligned}
h_{q-1} &= d_0 g_q \\
h_{q-2} &= d_1 g_q + d_0 g_{q-1} \\
&\vdots \\
h_0 &= d_{q-1}g_q + d_{q-2}g_{q-1} + \cdots + d_0 g_1
\end{aligned} \tag{4.164}$$

and

$$0 = g_0 d_k + g_1 d_{k+1} + \cdots + g_q d_{k+q}, \quad k = 0 \to q-1. \tag{4.165}$$

The q equations given by (4.165) can be rewritten in matrix form as

$$\begin{bmatrix} d_0 & d_1 & \cdots & d_q \\ d_1 & d_2 & \cdots & d_{q+1} \\ \vdots & \vdots & \cdots & \vdots \\ d_{q-1} & d_q & \cdots & d_{2q-1} \end{bmatrix} \mathbf{x} = \mathbf{0}_q, \tag{4.166}$$

where $\mathbf{x} \triangleq (g_0, g_1, \ldots, g_q)^T$. This homogeneous system of q equations in $q+1$ unknowns always possesses a nontrivial solution $\mathbf{x}$.

To construct $g(z)$ strict Hurwitz, it is necessary to guarantee at the outset that $g(0) = g_0 \neq 0$. Clearly, solutions $\mathbf{x}$ in which $g_0 \neq 0$ exist if

$$H_q \triangleq \begin{bmatrix} d_1 & d_2 & \cdots & d_q \\ d_2 & d_3 & \cdots & d_{q+1} \\ \vdots & \vdots & \cdots & \vdots \\ d_q & d_{q-1} & \cdots & d_{2q-1} \end{bmatrix} \tag{4.167}$$

is nonsingular. Similarly, to maintain $\delta(g(z)) = q$, it is necessary to have $g_q \neq 0$, and it is guaranteed if

$$
D_{q-1} \overset{\triangle}{=} \begin{bmatrix} d_0 & d_1 & \cdots & d_{q-1} \\ d_1 & d_2 & \cdots & d_q \\ \vdots & \vdots & \cdots & \vdots \\ d_{q-1} & d_q & \cdots & d_{2q-2} \end{bmatrix}
\tag{4.168}
$$

is nonsingular. Consequently, $g_0 g_q \neq 0$ is assured if

$$
det\left(D_{q-1}H_q\right) \neq 0.
\tag{4.169}
$$

Interestingly, such is the case infinitely often, and this can be established by making use of a key lemma derived in Appendix 4.A. From the lemma, if there exists $n_i = \delta(A_{n_i})$, $i \geq 0$, for which (4.169) is unsatisfiable for every q, $1 \leq q < n_i$, then we must have

$$
\left| \frac{Z(z) - Z(0)}{Z(z) + Z^*(0)} \right| \leq |z|^{n_i}, \quad |z| < 1,
\tag{4.170}
$$

where $Z(z)$ is as given in (4.148). To conclude the proof, if such n_i's are infinite in number, then clearly $n_i \to \infty$, $|z|^{n_i} \to 0$ in $|z| < 1$ and $Z(z) = Z(0)$, a constant. By implication, $S(\theta)$ and all polynomials $A_{n_i}(z)$ are also constants. But $\delta(A_{n_i}) = n_i \geq 1$, a contradiction. Consequently, such n_i's are at most finite in number. Q.E.D.

Unfortunately, the above proof says nothing regarding the bounded character of $\rho(z) = h(z)/g(z)$ in the general case. However, for $q = 1$, the question of boundedness can be answered affirmatively [89]. In particular, it can be shown that optimal Padé-like approximations of the form $ARMA(p, 1)$ exist in general, iff the junction reflection coefficients s_p satisfy

$$
\left| \frac{s_p}{s_{p+1}} \right| \geq 1 + |s_p|, \text{ for some } p \geq 1.
\tag{4.171}
$$

To see this, consider a stage $(p, 1)$, with $p \geq 1$. In that case, $\rho(z)$ must be of the form

$$
\rho(z) = \frac{h_0}{(1 + g_1 z)}
\tag{4.172}
$$

and with $A_{p+1}(z) = \sum_{k=0}^{p+1} a_k^{(p+1)} z^k$, the two unknowns h_0 and g_1 can be obtained by solving the $2q(= 2)$ equations in (4.112) given by

$$
\begin{bmatrix} a_p^{(p+1)} & -a_1^{(p+1)*} \\ a_{p+1}^{(p+1)} & -a_0^{(p+1)*} \end{bmatrix} \begin{bmatrix} g_1 \\ h_0 \end{bmatrix} = - \begin{bmatrix} a_{p+1}^{(p+1)} \\ 0 \end{bmatrix}.
$$

This gives (since $a_0^{(k)} > 0$)

$$\frac{1}{g_1} = \frac{a_1^{(p+1)*}}{a_0^{(p+1)}} - \frac{a_p^{(p+1)}}{a_{p+1}^{(p+1)}} \tag{4.173}$$

and

$$h_0 = \frac{(a_{p+1}^{(p+1)})^2}{a_1^{(p+1)*} a_{p+1}^{(p+1)} - a_0^{(p+1)} a_p^{(p+1)}} . \tag{4.174}$$

These equations can be simplified using the Levinson recursion formula in (4.102) given by

$$\sqrt{1 - |s_{p+1}|^2}\, a_k^{(p+1)} = a_k^{(p)} - s_{p+1} a_{p-k+1}^{(p)*}, \quad k = 0 \to p+1$$

(with $a_{p+1}^{(p)} = 0$). On substituting these into (4.173) and (4.174) we obtain

$$\frac{1}{g_1} = \frac{1 - |s_{p+1}|^2}{s_{p+1}} \cdot \frac{a_p^{(p)}}{a_0^{(p)}} = -(1 - |s_{p+1}|^2) \frac{s_p}{s_{p+1}} \tag{4.175}$$

and

$$h_0 = \frac{s_{p+1}^2}{(1 - |s_{p+1}|^2) s_p} . \tag{4.176}$$

Clearly, for $\rho(z)$ in (4.172) to be bounded, it must be free of poles in $|z| \le 1$ and bounded by unity there (see also (3.111)–(3.112)). Of these, the first condition gives

$$\left| \frac{1}{g_1} \right| = (1 - |s_{p+1}|^2) \left| \frac{s_p}{s_{p+1}} \right| > 1,$$

which simplifies to

$$\left| \frac{s_p}{s_{p+1}} \right| > \frac{1}{1 + |s_{p+1}|} + |s_p|. \tag{4.177}$$

Similarly, the second condition is satisfied iff $\rho(z)$ is bounded by unity where the denominator takes its minimum most value inside the closed unit circle. Thus, since

$$|1 + g_1 z| > 1 - |g_1|, \quad |z| \le 1,$$

we must have

$$\frac{|h_0|}{1 - |g_1|} \le 1.$$

Using (4.175) and (4.176), the above inequality simplifies to

$$|s_{p+1}| \le (1 - |s_{p+1}|^2) \left| \frac{s_p}{s_{p+1}} \right| - 1.$$

Since $|s_p| < 1$, the above expression reduces to

$$1 \leq (1 - |s_{p+1}|) \left| \frac{s_p}{s_{p+1}} \right|,$$

or equivalently

$$|s_{p+1}| \leq \frac{|s_p|}{1 + |s_p|}, \tag{4.178}$$

which is the same as[12]

$$\left| \frac{s_p}{s_{p+1}} \right| \geq 1 + |s_p|. \tag{4.179}$$

Notice that (4.177) is subsumed in (4.179) and hence (4.179) forms the necessary and sufficient condition for an ARMA$(p, 1)$ Padé-like approximation to exist.

Interestingly, from (3.80) as a consequence of the Paley-Wiener criterion, it follows that $s_k \to 0$ as $k \to \infty$. As a result, s_p must decrease as $p \to \infty$, or the ratio $|s_p/s_{p+1}|$ must exceed unity infinitely often! Although the necessary condition $|s_p/s_{p+1}| > 1$ is satisfied by *every* integrable power spectral density that satisfies the Paley-Wiener criterion, this is not true in the case of (4.179). For example, consider the spectral density, whose junction reflection coefficients satisfy

$$s_k = -\frac{\ln k}{k}, \quad k \geq 1. \tag{4.180}$$

Clearly, $0 < |s_k| < 1$, $|s_k/s_{k+1}| > 1$, for $k \geq 3$ and $\sum_{k=1}^{\infty} |s_k|^2 < \infty$, implying the Paley-Wiener criterion. However,

$$\left| \frac{s_p}{s_{p+1}} \right| - (1 + |s_p|) = \frac{(p+1)\ln p}{p \ln (p+1)} - \frac{\ln p}{p} - 1$$

$$= \frac{p(\ln p - \ln (p+1)) + \ln p(1 - \ln (p+1))}{p \ln (p+1)} < 0, \quad p \geq 1,$$

which violates the above sufficiency condition for every p. Thus, no Padé-like approximation of ARMA$(p, 1)$-type exists for this example. However, higher-order Padé-like approximations can exist in this case and the question of the bounded character of the rational $\rho(z)$ in the general case is still open.

---

[12]Incidentally, every ARMA$(p, 1)$ system satisfies (4.177)–(4.179) for any k beyond p. Thus the reflection coefficients s_k of an ARMA$(p, 1)$ system form a monotone decreasing sequence for $k \geq p$ and $s_k > 0$, $k \geq p$. (See also Problem 4.3.)

4.3.2 NUMERICAL RESULTS

As pointed out earlier, if an ARMA(p,q) approximation $H_r(z)$ exists, equation (4.112) can be used to determine the desired bounded function $\rho_{p+q+1}(z)$ and the system parameters. Once again, the bounded function $\rho_{p+q+2}(z)$ at the next stage, if it exists, can be determined by making use of (4.126) and in a sequential manner $\varepsilon_0(n,m)$ as well as $\varepsilon(n,m)$ can be evaluated as in (4.131)–(4.134). The heavy dots on the curves in Figs. 4.16b–4.25b indicate the existence of optimal rational approximations to $H(z)$, where $\rho_{p+q+1}(z)$ exists as a bounded function. (Notice that for an ARMA(p,q) Padé-like approximation the existence of $\rho_{p+q+1}(z)$ as a bounded function is necessary and sufficient, and, hence, the presence or absence of $\rho_{p+q+2}(z)$ as a bounded function is irrelevant except for computing $\varepsilon_0(n,m)$ and $\varepsilon(n,m)$.)

When $S(\theta) = |H(e^{j\theta})|^2$ is known in advance, as in the examples under discussion, the percent-error

$$\eta(n,m) \overset{\triangle}{=} \sup_\theta \frac{|S(\theta) - K_r(\theta)|}{S(\theta)} \qquad (4.181)$$

also functions as a trustworthy measure of fidelity. Here $r = n + m$, and

$$K_r(\theta) = \left|H_r(e^{j\theta})\right|^2 \qquad (4.182)$$

represents the approximated ARMA(n,m) spectrum.

The transfer function $H(z) = e^{-z}$ in Fig. 4.16 is clearly nonrational and defines the nonrational power spectral density

$$S(\theta) = |H(e^{j\theta})|^2 = e^{-2\cos\theta}. \qquad (4.183)$$

As shown in Fig. 4.16, the present approach produces an optimal ARMA$(3,2)$ approximation $H_r(z)$ given by

$$H_r(z) = \frac{1 - 0.410z + 0.0535z^2}{1 + 0.5898z + 0.1434z^2 + 0.0151z^3},$$

which is (necessarily) both stable and minimum-phase.[13] Moreover, the first $3 + 2 + 1 = 6$ Fourier coefficients of $|H_r(e^{j\theta})|^2$ also coincide with the corresponding Fourier coefficients of $S(\theta)$. Hence, $K_r(\theta)$ is truly a Padé approximation to $S(\theta)$. Observe that plots of $\eta(n,m)$ and $\varepsilon_0(n,m)$ are displayed in the later half of each figure, whereas $K_r(\theta)$ and $S(\theta)$ appear superimposed in their former parts. Each heavy dot on the $\eta(n,m)$ and the accompanying solid curve indicates the existence of a corresponding optimal Padé-like approximation to $H(z)$ such as ARMA$(3,3)$, $(4,1)$, $(4,2)$, $(5,3)$, $(5,4)$, etc. This abundance of optimal approximations is also evident in the remaining figures.

---

[13]See Appendix 4.B for Padé approximation of the delay function e^{-p}.

All of these involve enormously complicated transcendental transfer functions $H(z)$ that satisfy the integrability condition and the Paley-Wiener criterion. However, the latter do not preclude either logarithmic or essential singularities on $|z| = 1$. Of these, Figs. 4.19, 4.20 and 4.23 exhibit logarithmic singularities at $\theta = \pi$, $\theta = 0$ and $\theta = 0$, respectively. Notice that the transfer functions shown in Figs. 4.19 and 4.20 are not Wiener factors since they both have zeros in $|z| < 1$ (at $z = 0$). For example, Fig. 4.19 corresponds to

$$H(z) = a(z)ln\,(1 + z)\,, \tag{4.184}$$

where

$$a(z) = \frac{1.2345 - 1.4164z + z^2}{1 - 2.4119z + 3.2531z^2 - 2.2658z^3 + 0.8710z^4}$$

is minimum-phase. Hence the associated Wiener factor can be easily determined by noting that

$$ln\,(1 + z) = z - \frac{z^2}{2} + \frac{z^3}{3} - \frac{z^4}{4} + \cdots\,, \quad |z| < 1\,,$$

so that

$$\frac{ln\,(1 + z)}{z} = 1 - \frac{z}{2} + \frac{z^2}{3} - \frac{z^3}{4} + \cdots\,, \quad |z| < 1\,, \tag{4.185}$$

and clearly (4.185) is free of zeros in $|z| < 1$. Further, on $|z| = 1$,

$$|ln\,(1 + z)| = \left|\frac{ln\,(1 + z)}{z}\right|$$

and hence the Wiener factor for (4.184) is given by

$$a(z)\left(1 - \frac{z}{2} + \frac{z^2}{3} - \frac{z^3}{4} + \cdots\right)\,, \quad |z| < 1\,. \tag{4.186}$$

Figure 4.19 gives a Padé-like approximation to the above Wiener factor to be an ARMA$(7, 4)$ transfer function given by

$$H_r(z) = \frac{1.2407 + 0.1848z - 0.3899z^2 + 0.7921z^3 + 0.3599z^4}{1 - 0.61z - 0.17z^2 + 1.48z^3 - 0.47z^4 - 0.16z^5 + 0.55z^6 + 0.1z^7}\,. \tag{4.187}$$

Similarly, Fig. 4.23 shows a classic example where

$$S(\theta) = -ln\,(sin\,(\theta/2))\,. \tag{4.188}$$

This nonrational integrable spectral density has a logarithmic singularity at $\theta = 0$ and clearly it satisfies the Paley-Wiener criterion. In this case, the Fourier coefficients are given by [94]

$$r_0 = ln\,2\,, \quad r_k = \frac{1}{2k}\,, \quad |k| \geq 1\,.$$

As Fig. 4.23a shows, the Padé-like approximation on the power spectrum gives an ARMA(5, 5) minimum-phase transfer function to be an excellent choice. From Fig. 4.23b, there are several other comparable minimum-phase approximations to this problem as well.

Figure 4.24 shows an example with an essential singularity at $z = 1$. In that case (see also (2.53)),

$$H(z) = 1 + e^{-1/(1-z)}. \tag{4.189}$$

Notice that $H(z)$ is analytic in $|z| < 1$ since $1/(1 - z)$ has a positive real part in $|z| < 1$. On $|z| = 1$, the factor $e^{1/(1-z)}$ is continuous everywhere except at $\theta = 0$. To see this, consider a function

$$H_1(z) = e^{-1/(1-z)}. \tag{4.190}$$

With $z = re^{j\theta}$, we have

$$H_1(re^{j\theta}) = exp\left(\frac{-(1 - r\cos\theta)}{1 + r^2 - 2r\cos\theta}\right) \cdot exp\left(\frac{jr\sin\theta}{1 + r^2 - 2r\cos\theta}\right), \tag{4.191}$$

and by setting $\theta = 0$, this reduces to

$$H_1(r) = e^{-1/(1-r)}.$$

Thus, the interior radial limit at $z = 1$ of $H_1(z)$ is given by

$$H_1(1) = \lim_{r \to 1-0} H_1(r) = e^{-\infty} = 0,$$

i.e., the interior radial limit of $H_1(z)$ at $z = 1$ is zero. However, from (4.191), its tangential limit (with $r = 1$),

$$\lim_{\theta \to 0} H_1(e^{j\theta}) = \lim_{\theta \to 0} e^{-j\cot(\theta/2)},$$

does not exist and hence $H_1(z)$ as well as $H(z)$ are not continuous at $\theta = 0$. Thus $H(z)$ is analytic in $|z| < 1$ and continuous everywhere on $|z| = 1$ except at $z = 1$. Moreover, at $z = 1$, it also exhibits an essential singularity and naturally the behavior of the function in the neighborhood of $z = 1$ *outside* the unit circle will reflect this fact. Remarkably, as Fig. 4.24 shows, the ARMA(12, 7) Padé-like approximation captures all essential features of this transcendental transfer function everywhere except possibly in the close neighborhood of $\theta = 0$.

Figure 4.25 shows another example with an essential singularity. In this case

$$H(z) = 1 + (1 + z)e^{-(1-z)/(1+z)}, \tag{4.192}$$

and it is analytic in $|z| < 1$, continuous everywhere on the unit circle with an essential singularity at $z = -1$. Nevertheless, an ARMA(14, 11) model serves as an excellent Padé-like approximation in this case, except possibly in the close neighborhood of $\theta = \pi$.

The present approach also can be used for the design of IIR stable filters that match the given gain specifications. To illustrate this, Fig. 4.26 deals with the design of a bandpass filter with gain specifications as shown below:

$$G(\theta) = \begin{cases} \delta, & 0 \leq \theta \leq k_1\pi, \ k_4\pi \leq \theta \leq \pi \\ 1 + (1 - \delta)\dfrac{\theta/\pi - k_2}{k_2 - k_1}, & k_1\pi \leq \theta \leq k_2\pi \\ 1.0, & k_2\pi \leq \theta \leq k_3\pi \\ 1 + (1 - \delta)\dfrac{\theta/\pi - k_3}{k_4 - k_3}, & k_3\pi \leq \theta \leq k_4\pi, \end{cases}$$

$$(4.193)$$

where $k_1 = 0.35$, $k_2 = 0.4$, $k_3 = 0.7$, $k_4 = 0.75$ and $\delta = 0.001$. Once again, application of the Padé-like approximation procedure on this gain function, for example, gives rise to an ARMA$(24, 24)$ minimum-phase filter. Figure 4.26a shows the given bandpass gain function as well as its realization using the above recursive filter.

As a final example, Fig. 4.27 shows the power spectrum associated with the reflection coefficients

$$s_k = -\frac{1}{1 + k^{2/3}}, \quad k \geq 1.$$

In this case, since

$$\left| \frac{s_k}{s_{k+1}} \right| - (1 + |s_k|) = \frac{(k + 1)^{2/3} - (1 + k^{2/3})}{1 + k^{2/3}} < 0, \quad k \geq 1,$$

from (4.179), the ARMA$(p, 1)$ Padé-like approximation does not exist for any $p \geq 1$. However, higher-order Padé-like approximations do exist in this situation, and this is illustrated in Fig. 4.27 for an ARMA$(9, 6)$ approximation with all details as indicated there.

An interesting application of this rational approximation procedure will be to explore ways in which this algorithm can be adapted to produce "low-order" rational transfer function models for distributed structures, such as large flexible beams, plates, etc. In fact, it is not even clear how to generate effective computer simulations that faithfully preserve the transcendental character of the problem. To make matters worse, damping is practically negligible and the transfer functions can also possess poles on $|z| = 1$, the boundary of the unit circle. Thus, the algorithm must be extended to encompass the case of discrete spectra. This entire subject is wide open and deserves further study.

The numerical evidence presented here seems to reinforce the opinion that optimality is the rule, rather than the exception. Nevertheless, a rigorous proof is still lacking and the issue remains unresolved.

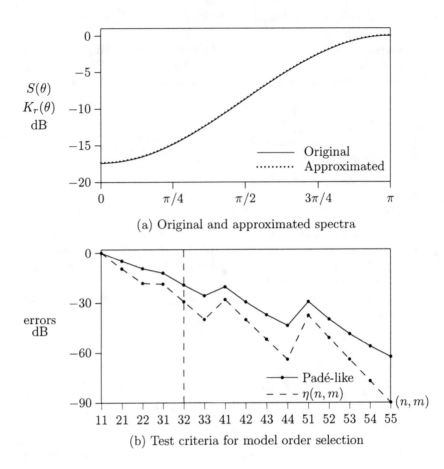

(a) Original and approximated spectra

(b) Test criteria for model order selection

FIGURE 4.16. Rational approximation of a nonrational system. The original nonrational system transfer function is given by

$$H(z) = e^{-z}.$$

The approximated $\mathrm{ARMA}(3,2)$ model is given by

$$H_r(z) = \frac{1 - 0.410z + 0.0535z^2}{1 + 0.5898z + 0.1434z^2 + 0.0151z^3}.$$

Poles of $H_r(z)$ are at $3.8012\angle 180°$ and $4.1735\angle \pm 132.99°$.
Zeros of $H_r(z)$ are at $4.3215\angle \pm 27.59°$.

$$\rho(z) = \frac{-1.07 \times 10^{-3} + 3.77 \times 10^{-4}z}{1 - 0.4102z + 0.0535z^2}.$$

Poles of $\rho(z)$ are at $4.3215\angle \pm 27.59°$.

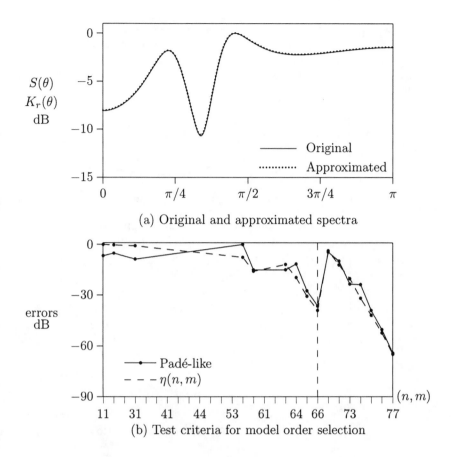

(a) Original and approximated spectra

(b) Test criteria for model order selection

FIGURE 4.17. Rational approximation of a nonrational system. The original nonrational system is given by

$$H(z) = \frac{1.2349 - 1.0700z + z^2}{1 - 1.592z + 1.8056z^2 - 1.0544z^3 + 0.4386z^4} e^{-(z-1)}.$$

The approximated ARMA$(6,6)$ model is given by

$$H_r(z) = \frac{3.357 - 5.224z + 5.456z^2 - 2.642z^3 + 0.72z^4 - 0.12z^5 + 0.011z^6}{1 - 1.282z + 1.34z^2 - 0.539z^3 + 0.162z^4 + 0.107z^5 + 0.012z^6}.$$

Poles of $H_r(z)$ are at $1.229\angle\pm43.19°$, $1.229\angle\pm75.58°$ and $5.982\angle\pm158.03°$.
Zeros of $H_r(z)$ are at $3.549\angle\pm22.19°$, $1.111\angle\pm61.22°$ and $4.452\angle\pm68.04°$.

$$\rho(z) = \frac{-0.062 + 0.101z - 0.054z^2 + 0.015z^3 - 0.026z^4 + 0.0002z^5}{1 - 1.552z + 1.611z^2 - 0.778z^3 - 0.212z^4 - 0.035z^5 + 0.003z^6}.$$

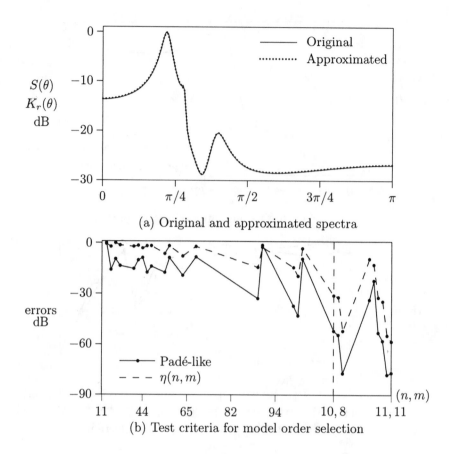

(a) Original and approximated spectra

(b) Test criteria for model order selection

FIGURE 4.18. Rational approximation of a nonrational system. The original nonrational system is given by

$$H(z) = \frac{1.111 + z}{1 - 1.468z + 0.907z^2} e^z$$

$$+ \frac{1.21 - 1.1z + z^2}{1 - 1.863z + 2.569z^2 - 1.641z^3 + 0.809z^4} e^{-(z-1)}.$$

The approximated ARMA(10, 8) model is given by $H_r(z) = B_m(z)/A_n(z)$ where

$$A_n(z) = 1 - 3.40z + 6.41z^2 - 7.39z^3 + 5.76z^4 - 2.72z^5 + 0.65z^6$$
$$+ 0.083z^7 - 0.037z^8 + 0.0009z^9 + 0.0008z^{10}$$

and

$$B_m(z) = 5.53 - 13.01z + 21.54z^2 - 20.10z^3 + 15.02z^4 - 6.53z^5$$
$$+ 2.61z^6 - 0.30z^7 + 0.12z^8.$$

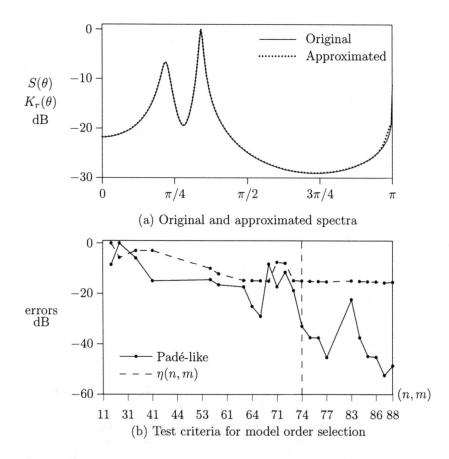

(a) Original and approximated spectra

(b) Test criteria for model order selection

FIGURE 4.19. Rational approximation of a nonrational system with a logarithmic singularity at $z = -1$. The original nonrational system is given by

$$H(z) = \frac{1.2345 - 1.4164z + z^2}{1 - 2.4119z + 3.2531z^2 - 2.2658z^3 + 0.871z^4} \, ln\,(1 + z).$$

Poles of the rational part of $H(z)$ are at $1.05\angle \pm 39.6°$ and $1.02\angle \pm 61.2°$. Zeros of the rational part of $H(z)$ are at $1.111\angle \pm 50.4°$.

The approximated $ARMA(7, 4)$ model is given by

$$H_r(z) = \frac{1.2368 + 0.1437z - 0.3656z^2 + 0.7830z^3 + 0.3402z^4}{1 - 0.65z - 0.11z^2 + 1.44z^3 - 0.49z^4 - 0.13z^5 + 0.53z^6 + 0.09z^7}.$$

Poles of $H_r(z)$ are at $1.050\angle \pm 39.61°$, $1.020\angle \pm 61.20°$, $1.056\angle 180°$ and $1.571\angle 180°$. Zeros of $H_r(z)$ are at $1.111\angle \pm 50.48°$, $1.146\angle 180°$ and $2.569\angle 180°$.

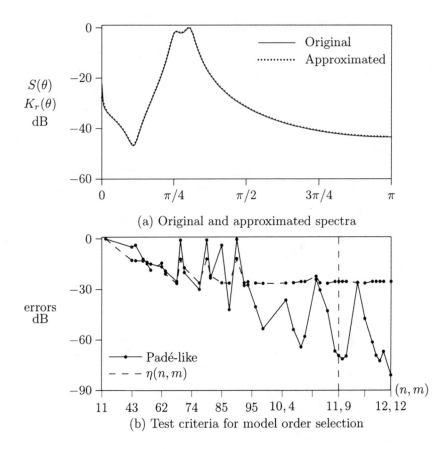

(a) Original and approximated spectra

(b) Test criteria for model order selection

FIGURE 4.20. Rational approximation of a nonrational system with a logarithmic singularity at $z = 1$. The original nonrational system is given by

$$H(z) = \frac{1.1025 - 1.9734z + z^2}{1 - 2.4394z + 3.2856z^2 - 2.2126z^3 + 0.8227z^4} \, ln\,(1 - z).$$

Poles of the rational part of $H(z)$ are at $1.05\angle \pm 45°$ and $1.05\angle \pm 55°$. Zeros of the rational part of $H(z)$ are at $1.05\angle \pm 20°$.

The approximated ARMA(11, 9) model is given by $H_r(z) = B_m(z)/A_n(z)$ where

$$
\begin{aligned}
A_n(z) \;=\; & 1 - 4.56z + 9.07z^2 - 9.06z^3 + 2.07z^4 + 6.44z^5 - 9.43z^6 \\
& + 6.60z^7 - 2.65z^8 + 0.59z^9 + 0.062z^{10} - 0.0021z^{11}
\end{aligned}
$$

and

$$
\begin{aligned}
B_m(z) \;=\; & 1.10 - 3.76z + 4.07z^2 + 0.209z^3 - 3.93z^4 + 3.35z^5 - 1.23z^6 \\
& + 0.203z^7 - 0.013z^8 + 0.0006z^9 \,.
\end{aligned}
$$

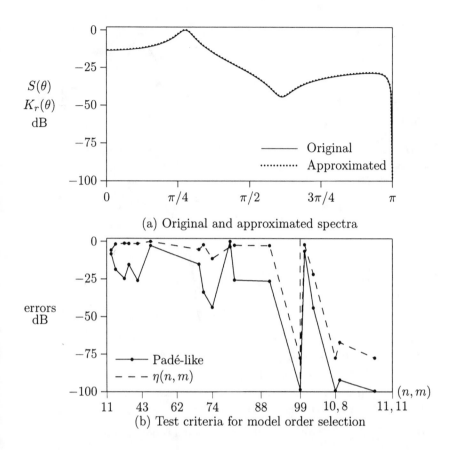

(a) Original and approximated spectra

(b) Test criteria for model order selection

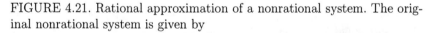

FIGURE 4.21. Rational approximation of a nonrational system. The original nonrational system is given by

$$H(z) = \frac{1.1881 + 0.7456z + z^2}{1 - 0.2644z - 0.2448z^2 + 0.7938z^3} \, ln\,(2+z).$$

Poles of the rational part of $H(z)$ are at $1.08\angle \pm 50°$ and $1.08\angle 180°$. Zeros of the rational part of $H(z)$ are at $1.09\angle \pm 110°$.

The approximated ARMA$(9, 9)$ model is given by $H_r(z) = B_m(z)/A_n(z)$ where

$$\begin{aligned}
A_n(z) \;=\; & 1 - 0.2250z - 1.0404z^2 + 0.2323z^3 + 0.5316z^4 - 0.3504z^5 \\
& -0.4407z^6 - 0.1401z^7 - 0.0167z^8 - 0.0006z^9
\end{aligned}$$

and

$$\begin{aligned}
B_m(z) \;=\; & 0.824 + 1.514z + 0.913z^2 - 0.342z^3 - 1.337z^4 - 1.110z^5 \\
& -0.397z^6 - 0.062z^7 - 0.0040z^8 - 0.00001z^9 \,.
\end{aligned}$$

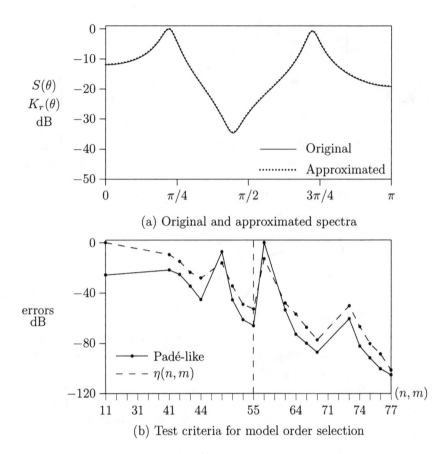

(a) Original and approximated spectra

(b) Test criteria for model order selection

FIGURE 4.22. Rational approximation of a nonrational system. The original nonrational system is given by

$$H(z) = \frac{1.1881 - 0.3786z + z^2}{1 - 0.2171z - 0.0264z^2 - 0.2090z^3 + 0.7488z^4} \, ln\,(3+z).$$

Poles of the rational part of $H(z)$ are at $1.08\angle \pm 40°$ and $1.07\angle \pm 130°$. Zeros of the rational part of $H(z)$ are at $1.09\angle \pm 80°$.

The approximated ARMA$(5,5)$ model is given by

$$H_r(z) = \frac{1.305 + 0.313z + 0.901z^2 + 0.599z^3 + 0.030z^4 - 0.002z^5}{1 + 0.038z - 0.029z^2 - 0.202z^3 + 0.696z^4 + 0.191z^5}.$$

Poles of $H_r(z)$ are at $1.080\angle \pm 40.00°$, $1.070\angle \pm 130°$ and $3.927\angle 180°$. Zeros of $H_r(z)$ are at $1.090\angle \pm 80.00°$, $2.009\angle 180°$, $10.623\angle 180°$ and $29.497\angle 0°$.

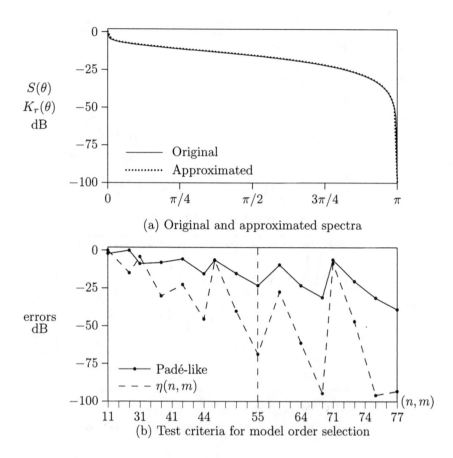

(a) Original and approximated spectra

(b) Test criteria for model order selection

FIGURE 4.23. Rational approximation of a nonrational system with a logarithmic singularity at $\theta = 0$. The original nonrational spectrum is given by

$$S(\theta) = -\ln\left(\sin\left(\theta/2\right)\right).$$

The approximated ARMA(5, 5) model is given by

$$H_r(z) = \frac{0.458 - 0.531z - 0.276z^2 + 0.521z^3 - 0.177z^4 + 0.014z^5}{1 - 2.499z + 2.222z^2 - 0.833z^3 + 0.119z^4 - 0.004z^5}.$$

Poles of $H_r(z)$ are at $1.049\angle 0°$, $1.300\angle 0°$, $2.000\angle 0°$, $4.333\angle 0°$ and $21.317\angle 0°$. Zeros of $H_r(z)$ are at $1.069\angle 0°$, $1.0002\angle 180°$, $1.404\angle 0°$, $2.525\angle 0°$ and $8.636\angle 0°$.

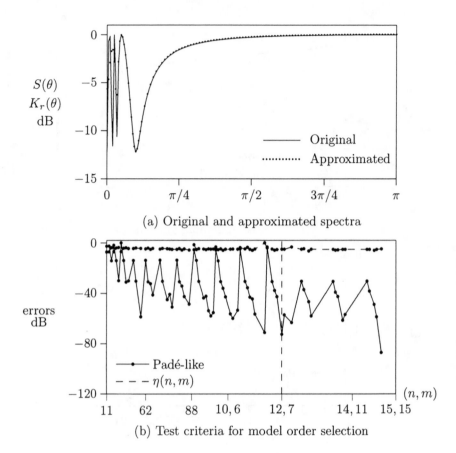

(a) Original and approximated spectra

(b) Test criteria for model order selection

FIGURE 4.24. Rational approximation of a nonrational system with an essential singularity at $z = 1$. The original nonrational system corresponds to

$$H(z) = 1 + e^{-1/(1-z)}.$$

The approximated ARMA(12, 7) model is given by $H_r(z) = B_m(z)/A_n(z)$ where

$$\begin{aligned}
A_n(z) &= 1 - 2.90z + 1.57z^2 + 2.76z^3 - 3.27z^4 + 0.13z^5 + 1.08z^6 \\
&\quad - 0.36z^7 + 1.8 \times 10^{-8}z^8 - 2.9 \times 10^{-9}z^9 - 3.2 \times 10^{-9}z^{10} \\
&\quad - 1.1 \times 10^{-8}z^{11} - 1.2 \times 10^{-8}z^{12}
\end{aligned}$$

and

$$\begin{aligned}
B_m(z) &= 1.37 - 4.34z + 3.02z^2 + 3.69z^3 - 5.57z^4 + 0.78z^5 + 1.79z^6 \\
&\quad - 0.72z^7.
\end{aligned}$$

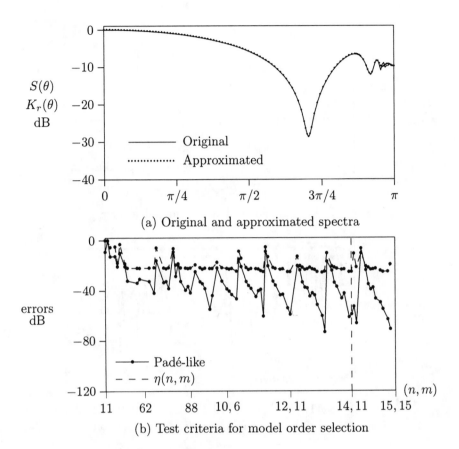

(a) Original and approximated spectra

(b) Test criteria for model order selection

FIGURE 4.25. Rational approximation of a nonrational system with an essential singularity at $z = -1$. The original nonrational system corresponds to

$$H(z) = 1 + (1 + z)e^{-(1-z)/(1+z)}.$$

The approximated ARMA(14, 11) model is given by $H_r(z) = B_m(z)/A_n(z)$ where

$$
\begin{aligned}
A_n(z) = \;& 1 + 2.84z + 1.03z^2 - 4.12z^3 - 4.18z^4 + 0.89z^5 + 2.70z^6 \\
& + 0.72z^7 - 0.42z^8 - 0.22z^9 + 0.02z^{10} - 4.9 \times 10^{-7}z^{11} \\
& + 7.8 \times 10^{-7}z^{12} - 4.6 \times 10^{-7}z^{13} - 6.7 \times 10^{-8}z^{14}
\end{aligned}
$$

and

$$
\begin{aligned}
B_m(z) = \;& 1.37 + 4.99z + 5.28z^2 - 2.65z^3 - 10.20z^4 - 6.58z^5 + 2.78z^6 \\
& + 5.51z^7 + 1.67z^8 - 0.88z^9 - 0.57z^{10} - 0.058z^{11}.
\end{aligned}
$$

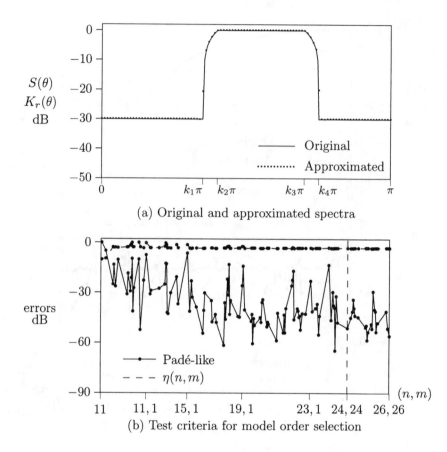

(a) Original and approximated spectra

(b) Test criteria for model order selection

FIGURE 4.26. Rational approximation of a bandpass gain function. The bandpass gain function specifications are given in (4.193) and the approximated ARMA(24, 24) model is given by $H_r(z) = B_m(z)/A_n(z)$ where

$$
\begin{aligned}
A_n(z) = {} & 1 + 2.7z + 8.71z^2 + 16.82z^3 + 34.19z^4 + 52.64z^5 + 82.13z^6 \\
& + 105.7z^7 + 135.8z^8 + 149.4z^9 + 163.1z^{10} + 154.8z^{11} + 145.5z^{12} \\
& + 119.3z^{13} + 96.79z^{14} + 68.0z^{15} + 47.43z^{16} + 28.0z^{17} + 16.54z^{18} \\
& + 7.91z^{19} + 3.84z^{20} + 1.37z^{21} + 0.52z^{22} + 0.11z^{23} + 0.029z^{24}
\end{aligned}
$$

and

$$
\begin{aligned}
B_m(z) = {} & 0.12 + 0.28z + 0.7z^2 + 1.17z^3 + 2.06z^4 + 2.79z^5 + 3.81z^6 \\
& + 4.4z^7 + 5.04z^8 + 4.99z^9 + 4.89z^{10} + 4.22z^{11} + 3.59z^{12} \\
& + 2.67z^{13} + 1.97z^{14} + 1.27z^{15} + 0.8z^{16} + 0.44z^{17} + 0.24z^{18} \\
& + 0.1z^{19} + 0.1z^{20} + 0.02z^{21} + 0.01z^{22} + 0.001z^{23} + 0.0003z^{24} .
\end{aligned}
$$

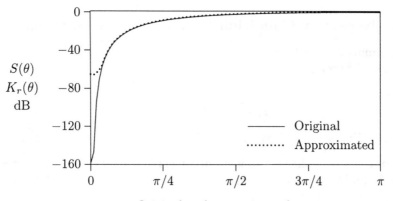

Original and approximated spectra

FIGURE 4.27. Rational approximation of a power spectrum associated with the reflection coefficients

$$s_k = -\frac{1}{1 + k^{2/3}}, \quad k \geq 1.$$

The first twenty autocorrelations $(r_0 \rightarrow r_{19})$ of this process are given by: 1.0000, -0.5000, -0.0399, -0.0138, 0.0133, 7.639×10^{-3}, 3.693×10^{-3}, 1.559×10^{-3}, 5.299×10^{-4}, 8.058×10^{-5}, -8.931×10^{-5}, -1.338×10^{-4}, -1.275×10^{-4}, -1.045×10^{-4}, -7.939×10^{-5}, -5.75×10^{-5}, -4.024×10^{-5}, -2.735×10^{-5}, -1.806×10^{-5}, -1.156×10^{-5}, -7.096×10^{-6}.

The approximated ARMA(9,6) model matches the first sixteen autocorrelations above and is given by

$$H_r(z) = B_m(z)/A_n(z),$$

where

$$
\begin{aligned}
A_n(z) = {} & 1 - 1.717z + 0.517z^2 + 0.566z^3 - 0.418z^4 + 0.0899z^5 \\
& -0.0054z^6 + 5.0 \times 10^{-5}z^7 - 3.6 \times 10^{-7}z^8 + 8.0 \times 10^{-7}z^9
\end{aligned}
$$

and

$$
\begin{aligned}
B_m(z) = {} & 0.505 - 1.706z + 1.828z^2 - 0.226z^3 - 0.788z^4 + 0.444z^5 \\
& -0.056z^6.
\end{aligned}
$$

Poles of $H_r(z)$ are at $1.283\angle 0°$, $1.496\angle 180°$, $1.377\angle 0°$, $2.042\angle 0°$, $3.520\angle 0°$, $11.659\angle \pm 14.44°$ and $21.99\angle \pm 131.01°$.
Zeros of $H_r(z)$ are at $1.033\angle \pm 2.329°$, $1.496\angle 180°$, $1.0336\angle \pm 6.403°$ and $5.261\angle 0°$.

Appendix 4.A

A Necessary Condition for Padé-like Approximation

Lemma: Let $r \geq 1$ be prescribed, and assume that $\delta(A_r) = r$. If (4.169) is false for every q, $1 \leq q < r$, then

$$\left| \frac{Z(z) - Z(0)}{Z(z) + Z^*(0)} \right| \leq |z|^r, \quad |z| < 1, \tag{4.A.1}$$

where $Z(z)$ is the positive function defined in (4.148).

Proof: No generality is lost if the argument is carried through with $r = 4$. We begin by noting that $\delta(A_4) = 4$ implies $b_4(\infty) = d_0 \neq 0$ and that

$$det\,(D_{q-1} H_q) = 0 \tag{4.A.2}$$

for $q = 1 \rightarrow 3$ yields

$$d_0 d_1 = 0$$

$$det \begin{bmatrix} d_0 & d_1 \\ d_1 & d_2 \end{bmatrix} \cdot det \begin{bmatrix} d_1 & d_2 \\ d_2 & d_3 \end{bmatrix} = 0 \tag{4.A.3}$$

and

$$det \begin{bmatrix} d_0 & d_1 & d_2 \\ d_1 & d_2 & d_3 \\ d_2 & d_3 & d_4 \end{bmatrix} \cdot det \begin{bmatrix} d_1 & d_2 & d_3 \\ d_2 & d_3 & d_4 \\ d_3 & d_4 & d_5 \end{bmatrix} = 0. \tag{4.A.4}$$

Since $d_0 \neq 0$, $d_1 = 0$ and (4.A.3) reduces to

$$det \begin{bmatrix} d_0 & 0 \\ 0 & d_2 \end{bmatrix} = 0 \text{ or } det \begin{bmatrix} 0 & d_2 \\ d_2 & d_3 \end{bmatrix} = 0.$$

In both cases $d_2 = 0 = d_1$ and (4.A.4) is possible only if

$$det \begin{bmatrix} d_0 & 0 & 0 \\ 0 & 0 & d_3 \\ 0 & d_3 & d_4 \end{bmatrix} = 0 \text{ or } det \begin{bmatrix} 0 & 0 & d_3 \\ 0 & d_3 & d_4 \\ d_3 & d_4 & d_5 \end{bmatrix} = 0.$$

Thus, $d_3 = 0$ is also true and

$$b_4(z) = \frac{A_4(z)}{\tilde{A}_4(z)} = d_0 + \frac{d_4}{z^4} + \cdots,$$

so that

$$\frac{A_4(z) - d_0 \tilde{A}_4(z)}{\tilde{A}_4(z)} = O(z^{-4}). \tag{4.A.5}$$

However, $\delta(\tilde{A}_4) = 4$ and the asymptotic behavior in (4.A.5) is achievable only if

$$A_4(z) - d_0 \tilde{A}_4(z) = \mu,$$

μ a constant. By inversion and elimination,

$$\tilde{A}_4(z) - d_0^* A_4(z) = \mu^* z^4$$

and

$$(1 - |d_0|^2)\tilde{A}_4(z) = \mu d_0^* + \mu^* z^4.$$

If $|d_0|^2 = 1$, $\mu = 0$ and $A_4(z) = \tilde{A}_4(z)e^{j\theta_0}$. But the only inversive strict Hurwitz polynomials are constants; hence, $\delta(A_4) = 0$, a contradiction, and $|d_0|^2 \neq 1$ is necessary. Actually, since all zeros of $\tilde{A}_4(z)$ lie in $|z| < 1$, $|d_0|^2 < 1$ and

$$A_4(z) = \eta(\mu + \mu^* d_0 z^4), \quad \tilde{A}_4(z) = \eta(\mu d_0^* + \mu^* z^4), \tag{4.A.6}$$

where $\eta = 1/(1 - |d_0|^2)$. To proceed further, we can make use of (3.42) and (3.39). Thus:

a) There exists a strict Hurwitz polynomial $B_4(z)$ of degree ≤ 4 such that

$$\tilde{B}_4(z)A_4(z) + B_4(z)\tilde{A}_4(z) = z^4. \tag{4.A.7}$$

b) Expressed in terms of $A_4(z)$, $B_4(z)$ and the bounded function $\rho_5(z)$,

$$Z(z) = 2\,\frac{B_4(z) + z\rho_5(z)\tilde{B}_4(z)}{A_4(z) - z\rho_5(z)\tilde{A}_4(z)}. \tag{4.A.8}$$

Substituting (4.A.6) into (4.A.7), we obtain

$$z^4 = \eta\left(\mu(B_4 d_0^* + \tilde{B}_4) + \mu^*(B_4 + d_0\tilde{B}_4)z^4\right),$$

and it is seen that z^4 divides $B_4(z)d_0^* + \tilde{B}_4(z)$. Therefore,

$$B_4(z)d_0^* + \tilde{B}_4(z) = \nu z^4,$$

ν a constant, or

$$\tilde{B}_4(z)d_0 + B_4(z) = \nu^*,$$

and, by elimination,

$$B_4(z) = \eta(\nu^* - \nu d_0 z^4), \quad \tilde{B}_4(z) = \eta(\nu z^4 - \nu^* d_0^*). \tag{4.A.9}$$

Equations (4.A.6), (4.A.9) and (4.A.8) combine to give

$$Z(z) = 2 \cdot \frac{\nu^* - \nu d_0 z^4 + z\rho_5(\nu z^4 - \nu^* d_0^*)}{\mu + \mu^* d_0 z^4 - z\rho_5(\mu d_0^* + \mu^* z^4)}. \tag{4.A.10}$$

In particular,

$$\frac{Z(z) - R_0}{Z(z) + R_0^*} = \frac{(z\rho_5 - d_0)z^4}{1 - z\rho_5 d_0^*}, \tag{4.A.11}$$

where $R_0 = 2\nu^*/\eta = Z(0)$. The boundedness of $\rho_5(z)$ and the inequality $|d_0| < 1$ imply, again by the maximum modulus theorem [26] (or from (4.96) directly), that

$$\left| \frac{z\rho_5(z) - d_0}{1 - z\rho_5(z)d_0^*} \right| \le 1, \quad |z| < 1.$$

It is now clear that the desired result

$$\left| \frac{Z(z) - Z(0)}{Z(z) + Z^*(0)} \right| \le |z|^4, \quad |z| < 1, \tag{4.A.12}$$

follows immediately from (4.A.11), Q.E.D.

Appendix 4.B

Diagonal Padé Approximations of e^{-p}

Consider the nonrational function

$$H(p) = e^{-p} \tag{4.B.1}$$

that represents a pure delay in the continuous (analog) domain $p = \sigma + j\omega$. Since $H(p)$ is analytic in the closed right-half plane and $|H(j\omega)| = 1$, by the maximum modulus theorem [26] $H(p)$ is a regular (nonrational) all-pass function. Extensive work has been reported on obtaining closed form Padé approximation solutions to this pure delay function in the sense of (4.143) along with a systematic study of the zeros/poles of such rational approximations [49, 95].

Since e^{-p} is a regular all-pass function, ideally it would be desirable for its rational approximations also to retain the all-pass property. However, since all regular rational all-pass functions are of the form (up to trivial factors)

$$H_r(p) = \frac{A_*(p)}{A(p)} = \frac{a_0^* - a_1^* p + \cdots + (-1)^n a_n^* p^n}{a_0 + a_1 p + \cdots + a_n p^n},$$

where $A(p)$ is a strict Hurwitz polynomial, it follows that only diagonal Padé approximations can possibly retain the all-pass nature of e^{-p}.

As Youla has shown in a rigorous manner [96], every diagonal Padé approximation of e^{-p} is a regular rational all-pass function! This interesting fact seems to have been overlooked in the literature, although, for several values of n, Euler has given the following approximations [97, 98]

$$\begin{aligned} e^{-p} &= \frac{2 - p}{2 + p} + O(p^3) \\ &= \frac{12 - 6p + p^2}{12 + 6p + p^2} + O(p^5) \\ &= \frac{120 - 60p + 12p^2 - p^3}{120 + 60p + 12p^2 + p^3} + O(p^7) \end{aligned} \tag{4.B.2}$$

that are regular rational all-pass functions.

Let $B_n(p)/A_n(p)$ represent the nth approximant to $H(p)$. Then in terms of the continued fraction expansion

$$H(p) = b_0 + \cfrac{a_1}{b_1 + \cfrac{a_2}{b_2 + \cfrac{a_3}{b_3 + \cdots + \cfrac{a_n}{b_n + \cdots}}}} , \qquad (4.B.3)$$

we have

$$\frac{B_n(p)}{A_n(p)} \triangleq b_0 + \cfrac{a_1}{b_1 + \cfrac{a_2}{b_2 + \cdots \cfrac{a_n}{b_n}}} . \qquad (4.B.4)$$

Clearly

$$\frac{B_0}{A_0} = \frac{b_0}{1} ; \quad \frac{B_1}{A_1} = b_0 + \frac{a_1}{b_1} = \frac{b_0 b_1 + a_1}{b_1}$$

and

$$\frac{B_2}{A_2} = b_0 + \cfrac{a_1}{b_1 + \cfrac{a_2}{b_2}} = b_0 + \frac{a_1 b_2}{b_1 b_2 + a_2} = \frac{b_2 B_1 + a_2 B_0}{b_2 A_1 + a_2 A_0} ,$$

and in general, using an induction argument, Wallis has first established that A_n and B_n satisfy the following recursions [99]

$$A_n = b_n A_{n-1} + a_n A_{n-2} \qquad (4.B.5)$$

$$B_n = b_n B_{n-1} + a_n B_{n-2} \qquad (4.B.6)$$

for $n \geq 2$. This follows easily from (4.B.4), since

$$\frac{B_{n+1}}{A_{n+1}} = b_0 + \cfrac{a_1}{b_1 + \cdots \cfrac{a_n}{b_n + \cfrac{a_{n+1}}{b_{n+1}}}} ,$$

which is the same as B_n/A_n with b_n replaced by $b_n + (a_{n+1}/b_{n+1})$. Thus

$$\frac{B_{n+1}}{A_{n+1}} = \frac{\left(b_n + \dfrac{a_{n+1}}{b_{n+1}}\right) B_{n-1} + a_n B_{n-2}}{\left(b_n + \dfrac{a_{n+1}}{b_{n+1}}\right) A_{n-1} + a_n B_{n-2}} = \frac{b_{n+1} B_n + a_{n+1} B_{n-1}}{b_{n+1} A_n + a_{n+1} A_{n+1}} ,$$

establishing the recursions (4.B.5)–(4.B.6) in the general case. Moreover, the difference between two such successive approximants is given by

$$\frac{B_n}{A_n} - \frac{B_{n-1}}{A_{n-1}} = \frac{B_n A_{n-1} - A_n B_{n-1}}{A_n A_{n-1}} = -a_n \frac{B_{n-1} A_{n-2} - A_{n-1} B_{n-2}}{A_n A_{n-1}}$$

$$= (-1)^{n-1} \frac{a_1 a_2 \cdots a_n}{A_n A_{n-1}} . \tag{4.B.7}$$

From (4.143), for these approximants to be optimal in the sense of Padé, we must have

$$H(p) - \frac{B_n(p)}{A_n(p)} = O(p^{N+M+1}), \tag{4.B.8}$$

where N and M represent the degrees of $A_n(p)$ and $B_n(p)$, respectively. Towards this purpose, it is best to make use of the following continued fraction expansion for e^{-p} due to Euler [98]:

$$e^{-p} = 1 - \cfrac{2p}{2 + p + \cfrac{p^2}{6 + \cfrac{p^2}{10 + \cfrac{p^2}{14 + \cdots + \cfrac{p^2}{2(2n-1) + \cdots}}}}} . \tag{4.B.9}$$

From the above expansion and (4.B.3), we have

$$b_0 = 1, \quad a_1 = -2p, \quad b_1 = 2 + p, \quad a_n = p^2, \quad b_n = 2(2n-1), \ n \geq 2, \tag{4.B.10}$$

and (4.B.5)–(4.B.6) reduces to

$$A_n(p) = 2(2n-1)A_{n-1}(p) + p^2 A_{n-2}(p) \tag{4.B.11}$$

$$B_n(p) = 2(2n-1)B_{n-1}(p) + p^2 B_{n-2}(p), \tag{4.B.12}$$

for $n \geq 2$. From (4.B.2)–(4.B.4) and (4.B.9), we also have

$$\frac{B_0(p)}{A_0(p)} = 1, \quad \frac{B_1(p)}{A_1(p)} = \frac{2-p}{2+p}, \quad \frac{B_2(p)}{A_2(p)} = \frac{12 - 6p + p^2}{12 + 6p + p^2} . \tag{4.B.13}$$

Clearly, $B_i(p)/A_i(p)$, $i = 0 \rightarrow 2$, are regular rational all-pass functions and $B_1(p) = A_1(-p)$ and $B_2(p) = A_2(-p)$. By induction on (4.B.11)–(4.B.12), it also follows that $A_n(p)$ and $B_n(p)$ are degree n polynomials and $B_n(p) = A_n(-p)$, since

$$B_n(-p) = 2(2n-1)B_{n-1}(-p) + p^2 B_{n-2}(-p)$$

$$= 2(2n-1)A_{n-1}(p) + p^2 A_{n-2}(p) = A_n(p) . \tag{4.B.14}$$

Consequently, there is only one iteration in (4.B.11)–(4.B.12). Moreover, using (4.B.10) in (4.B.7), we also have

$$\frac{B_{n+1}(p)}{A_{n+1}(p)} - \frac{B_n(p)}{A_n(p)} = (-1)^n \frac{2p^{2n+1}}{A_{n+1}(p)A_n(p)} = O(p^{2n+1})$$

and hence

$$e^{-p} - \frac{B_n(p)}{A_n(p)} = e^{-p} - \frac{A_n(-p)}{A_n(p)} = O(p^{2n+1}). \qquad (4.B.15)$$

From (4.B.8), it now follows that the all-pass functions $A_n(-p)/A_n(p)$, $n \geq 0$, indeed represent diagonal Padé approximations to e^{-p}, where $A_n(p)$ is given by (4.B.11) with $A_0(p) = 1$ and $A_1(p) = 2 + p$. To establish the claim that these diagonal Padé approximations of e^{-p} are also regular all-pass functions, from (4.B.11) and (4.B.15), it is enough to show that $A_n(p)$, $n \geq 2$, are strict Hurwitz polynomials. Following Youla [96], we will prove this using an induction argument and by employing elementary properties of positive-real (p.r.) functions.[14]

Consider the function [96]

$$Z_n(p) \triangleq \frac{A_n(p)}{pA_{n-1}(p)}, \quad n \geq 1. \qquad (4.B.16)$$

Using (4.B.11), we have

$$Z_n(p) = \frac{2(2n-1)}{p} + \frac{pA_{n-2}(p)}{A_{n-1}(p)} = \frac{2(2n-1)}{p} + \frac{1}{Z_{n-1}(p)}. \qquad (4.B.17)$$

Note that $2(2n-1)/p$ is clearly p.r. for every $n \geq 1$, and since

$$Z_1(p) = \frac{A_1(p)}{pA_0(p)} = \frac{2 + p}{p} = 1 + \frac{2}{p}$$

is also p.r., by induction on (4.B.17), it quickly follows that $Z_2(p)$, $Z_3(p)$, ..., and in general $Z_n(p)$ are p.r. functions. Since p.r. functions are free of zeros in the right-half plane, from (4.B.16), $A_n(p)$ is at least a Hurwitz polynomial.[15] To show that $A_n(p)$ is strict Hurwitz, assume the contrary, and suppose $p = j\omega_0$ is a zero of $A_n(p)$ on the $j\omega$-axis. Then using (4.B.11), since

$$\begin{aligned}
A_n(p)&A_{n-1*}(p) - A_{n*}(p)A_{n-1}(p) \\
&= -p^2 \Big(A_{n-1}(p)A_{n-2*}(p) - A_{n-1*}(p)A_{n-2}(p) \Big) \\
&= (-1)^{n-1}p^{2(n-1)} \Big(A_1(p)A_0(p) - A_{1*}(p)A_{0*}(p) \Big) \\
&= (-1)^{n-1}p^{2(n-1)} \big(2 + p - (2 - p) \big) \\
&= 2(-1)^{n-1}p^{2n-1},
\end{aligned}$$

---

[14]Recall that in the p-domain, a function $Z(p)$ is said to p.r., if it is analytic in $Re\, p > 0$ and $Re\, Z(p) \geq 0$ in $Re\, p > 0$. If $Z(p)$ is also rational, then it is also analytic on $p = j\omega$-axis except possibly for simple poles. The residues of such poles are positive.

[15]In (4.B.16), $A_n(p)$ and $A_{n-1}(p)$ cannot have any common zero cancellation, since in that case from (4.B.11), such a zero must be common to all $A_k(p)$, $k \geq 0$.

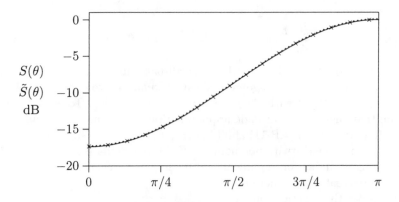

FIGURE 4.28. Diagonal Padé Approximations of e^{-z} and its power spectrum. Here $S(\theta) = e^{-2\cos\theta}$ (solid line), $\tilde{S}(\theta) = |A_n(-e^{j\theta})/A_n(e^{j\theta})|^2$, $n = 2$ and 3 with $A_n(z)$ as given in (4.B.11). Dotted line corresponds to $n = 2$ and crossed line corresponds to $n = 3$ case.

and for $p = j\omega_0$, this gives

$$0 = 2(-1)^{n-1}(j\omega_0)^{2(n-1)}.$$

Thus $p = j\omega_0 = 0$ is the only possible $j\omega$-axis zero of $A_n(p)$. However,

$$A_n(0) = 2(2n - 1)A_{n-1}(0) > 0,$$

since by induction $A_{n-1}(p)$ is strict Hurwitz and hence $A_n(p)$ has no zeros on the entire $j\omega$-axis. Thus $A_n(p)$ is a strict Hurwitz polynomial and this completes the proof.

From the above proof, more generally the iteration

$$A_n(p) = k(n)A_{n-1}(p) + p^2 A_{n-2}(p), \quad n \geq 2,$$

with $A_0(p) = 1$, $A_1(p) = k_1 + p$, $k_1 > 0$, will generate a series of strict Hurwitz polynomials for any $k(n) > 0$.

Interestingly, $A_n(z)$, $n \geq 1$, are also free of zeros in $|z| < 1$ and hence every diagonal Padé approximation to e^{-z} is also stable and is given by

$$e^{-z} = \frac{A_n(-z)}{A_n(z)} + O(z^{2n+1}), \tag{4.B.18}$$

where $A_n(z)$ is as given in (4.B.11) with p replaced by z. For $n = 2$ and 3, the associated rational spectra are shown in Fig. 4.28 along with the original spectrum (see also the Padé-like approximation in Fig. 4.16).

Problems

1. The first three autocorrelations of an unknown spectrum are given to be $r_0 = 2$, $r_1 = 0$ and $r_2 = -1$.

   (a) Is it possible to fit an ARMA$(1,1)$ model in this case?

   (b) Suppose it is given that the minimum-phase transfer function $H(z)$ also has a zero at $z = 2$. Find the $H(z)$ that has *minimum* numerator and denominator degrees.

2. Let $H(z)$ represent a rational stable system with a power series representation

$$H(z) = \frac{b_0 + b_1 z + \cdots + b_q z^q}{1 + a_1 z + \cdots + a_p z^p} = \sum_{k=0}^{\infty} h_k z^k$$

   that is valid in the entire z-plane (except at the poles of the system). Define the Hankel matrices

$$\mathbf{D}_k = \begin{bmatrix} h_0 & h_1 & \cdots & h_{k-1} \\ h_1 & h_2 & \cdots & h_k \\ \vdots & \vdots & \ddots & \vdots \\ h_{k-1} & h_k & \cdots & h_{2k-2} \end{bmatrix}, \quad \mathbf{H}_k = \begin{bmatrix} h_1 & h_2 & \cdots & h_k \\ h_2 & h_3 & \cdots & h_{k+1} \\ \vdots & \vdots & \ddots & \vdots \\ h_k & h_{k+1} & \cdots & h_{2k-1} \end{bmatrix}$$

   generated from the impulse response. Suppose $p \geq q$.

   (a) Show that

$$rank\, \mathbf{D}_{p-1} = rank\, \mathbf{D}_{p+k} = p, \quad k \geq 0, \tag{4.P.1}$$

   and

$$rank\, \mathbf{H}_p = rank\, \mathbf{H}_{p+k} = p, \quad k \geq 0, \tag{4.P.2}$$

   where p represents the degree of the rational system.

   (b) Conversely, suppose (4.P.1) and (4.P.2) hold for an arbitrary minimum-phase system $H(z) = \sum_{k=0}^{\infty} h_k z^k$, $|z| < 1$. Show that $H(z)$ is a rational system of degree p.

3. **ARMA(p, 1) Systems [100, 101]:** Let s_k, $k \geq 1$, represent the reflection coefficients of a stable ARMA$(p,1)$ system.

   (a) Show that [89]

$$|s_{k+1}| \leq \frac{|s_k|}{1 + |s_k|}, \quad k \geq p, \tag{4.P.3}$$

   i.e., For ARMA$(p, 1)$ systems, the reflection coefficients $\{s_k\}_{k=p}^{\infty}$ form a monotone decreasing sequence with $s_k \neq 0$, and $|s_{k+1}| \leq 1/2$, for every $k \geq p$.

(b) Show that the reflection coefficients satisfy the recursion [100]

$$s_{k+2} = \frac{s_{k+1}^2}{(1 - |s_{k+1}|^2)s_k}, \quad k \geq p. \tag{4.P.4}$$

Thus, in the case of real systems

$$sgn\,(s_p) = sgn\,(s_{p+2}) = sgn\,(s_{p+4}) = \cdots$$

and

$$sgn\,(s_{p+1}) = sgn\,(s_{p+3}) = sgn\,(s_{p+5}) = \cdots,$$

i.e., For ARMA$(p, 1)$ real systems, the signs of s_k, $k \geq p$ either alternate or are all equal.

(c) Show that the ratio $|s_k/s_{k+1}| > 1$ form a monotone nondecreasing sequence.

(*Hint*: From Fig. 4.2 and (4.100), after $p - 1$ line extractions, every ARMA$(p, 1)$ system can be represented in terms of a degree one bounded function as in (4.172).)

5

Multichannel System Identification

5.1 Introduction

A multichannel linear system has several inputs and several outputs, and all input signals interact linearly to generate every output signal. Thus in an m-input, m-output configuration, if $W_1(z), W_2(z), \ldots, W_m(z)$ represent the input signal transforms and $X_1(z), X_2(z), \ldots, X_m(z)$ the output signal transforms,[1] then

$$X_i(z) = \sum_{j=1}^{m} H_{ij}(z)W_j(z), \quad i = 1 \to m, \tag{5.1}$$

or in matrix form

$$\mathbf{X}(z) = \mathbf{H}(z)\mathbf{W}(z), \tag{5.2}$$

where

$$\mathbf{X}(z) \stackrel{\triangle}{=} [X_1(z), X_2(z), \ldots, X_m(z)]^T \tag{5.3}$$

$$\mathbf{W}(z) \stackrel{\triangle}{=} [W_1(z), W_2(z), \ldots, W_m(z)]^T. \tag{5.4}$$

Here $\mathbf{W}(z)$ and $\mathbf{X}(z)$ represent the input and output column vectors and

$$\mathbf{H}(z) \stackrel{\triangle}{=} \begin{bmatrix} H_{11}(z) & H_{12}(z) & \cdots & H_{1m}(z) \\ H_{21}(z) & H_{22}(z) & \cdots & H_{2m}(z) \\ \vdots & \vdots & \cdots & \vdots \\ H_{m1}(z) & H_{m2}(z) & \cdots & H_{mm}(z) \end{bmatrix}, \tag{5.5}$$

where $H_{ik}(z)$ represents the cross-transfer function between the ith output channel and the kth input channel. Physically, $H_{ik}(z)$ represents the contribution of the kth input signal to the ith output signal and $\mathbf{H}(z)$ represents the matrix transfer function of the system. If the system is rational, then all entries in (5.5) are rational functions, and (5.2) represents a Multichannel ARMA (or MARMA) system. In analogy to the single-channel case it is possible to define the order of a multichannel system in a consistent manner. However, it is not easily related to the orders (degrees) of the individual entries.

---

[1] The number of input and output channels does not have to be equal. But we will only consider that case here.

The degree of a rational matrix can be defined in several different but equivalent ways. The degree measures the true complexity of the system, and physically it represents the minimum number of delay elements required to implement that system. As the following example shows, the degree of $det\,\mathbf{H}(z)$ is not a proper choice for this purpose, since the matrix

$$\mathbf{H}_0(z) = \begin{bmatrix} 1 + z^2 & z \\ z & 1 \end{bmatrix}$$

has $det\,\mathbf{H}_0(z) = 1$, nevertheless, it requires two delay elements in its implementation. The McMillan degree of a rational matrix captures all these essentials and it is a true generalization of the ordinary degree concept for a scalar rational function [102, 103]. The McMillan degree of a rational matrix $\mathbf{H}(z)$ is defined as the sum of the McMillan degrees of its distinct poles. A pole of $\mathbf{H}(z)$ is, by definition, the pole of at least one entry of $\mathbf{H}(z)$, and the McMillan degree of a pole of $\mathbf{H}(z)$ equals the largest multiplicity with which it occurs as a pole of *any minor* of $\mathbf{H}(z)$.

The closely related concept of the order of a matrix turns out to be more useful in the present context. To start with, the order of a pole of a matrix is the largest multiplicity it possesses as a pole of *any entry* of $\mathbf{H}(z)$. The sum of the orders of distinct poles of $\mathbf{H}(z)$ is defined as the order of $\mathbf{H}(z)$. Thus, in particular, a matrix polynomial

$$\mathbf{A}(z) = \mathbf{A}_0 + \mathbf{A}_1 z + \cdots + \mathbf{A}_n z^n , \quad \mathbf{A}_n \not\equiv 0$$

has order n, since its only pole at $z = \infty$ must be of multiplicity n for at least one entry of $\mathbf{A}(z)$. To define the order of a rational matrix function in a useful manner it is best to make use of the left-coprime or the right-coprime polynomial decomposition of a rational matrix [104, 105]. Matrix polynomials

$$\mathbf{P}(z) = \mathbf{P}_0 + \mathbf{P}_1 z + \cdots + \mathbf{P}_p z^p \tag{5.6}$$

and

$$\mathbf{Q}(z) = \mathbf{Q}_0 + \mathbf{Q}_1 z + \cdots + \mathbf{Q}_q z^q \tag{5.7}$$

are said to be left-coprime if there exist two other $m \times m$ polynomial matrices $\mathbf{X}(z)$ and $\mathbf{Y}(z)$ such that[2]

$$\mathbf{P}(z)\mathbf{X}(z) + \mathbf{Q}(z)\mathbf{Y}(z) = \mathbf{I}_m . \tag{5.8}$$

Note that p and q in (5.6)–(5.7) represent the orders (or ordinary degrees) of the matrix polynomials $\mathbf{P}(z)$ and $\mathbf{Q}(z)$, respectively (if $\mathbf{P}_p \not\equiv 0$ and $\mathbf{Q}_q \not\equiv 0$), and we will denote them by $\delta_o(\mathbf{P}(z)) = p$, $\delta_o(\mathbf{Q}(z)) = q$. From (5.8), if the pair $\mathbf{P}(z)$, $\mathbf{Q}(z)$ can be represented as $\mathbf{P}(z) = \mathbf{E}(z)\mathbf{P}_a(z)$ and $\mathbf{Q}(z) = \mathbf{E}(z)\mathbf{Q}_a(z)$ for some polynomial matrices $\mathbf{E}(z)$, $\mathbf{P}_a(z)$ and $\mathbf{Q}_a(z)$,

---

[2]Equation (5.8) is the matrix Bezout identity [104].

then $\mathbf{E}(z)$ must be an elementary polynomial matrix.[3] Further, $\mathbf{P}^{-1}(z)\mathbf{Q}(z)$ is said to be a left-coprime decomposition to $\mathbf{H}(z)$ if

$$\mathbf{H}(z) = \mathbf{P}^{-1}(z)\mathbf{Q}(z) \tag{5.9}$$

and $\mathbf{P}(z)$ and $\mathbf{Q}(z)$ are left-coprime. Such a decomposition is only unique up to multiplication by an elementary polynomial matrix on the left [104]. This follows easily since if $\mathbf{H}(z) = \mathbf{P}_0^{-1}(z)\mathbf{Q}_0(z)$, then from (5.9), $\mathbf{P}_0(z) = \mathbf{R}(z)\mathbf{P}(z)$ and $\mathbf{Q}_0(z) = \mathbf{R}(z)\mathbf{Q}(z)$, where $\mathbf{R}(z)$ is some polynomial matrix. But $\mathbf{P}_0(z)$ and $\mathbf{Q}_0(z)$ are left-coprime, and from the above comment, $\mathbf{R}(z)$ must be an elementary polynomial matrix. However, since the product of two singular (nonzero) matrices can be identically equal to the zero matrix, the extra freedom present in (5.9) in terms of the elementary polynomial matrices makes it impossible to define the orders p and q uniquely in (5.6)–(5.7) and (5.9). This is because if $\mathbf{E}(z) = \sum_{k=0}^{m} \mathbf{E}_k z^k$ is an elementary polynomial matrix, then $\mathbf{E}_m$ is necessarily singular and if $\mathbf{P}_p$ in (5.6) is also singular, then since $\mathbf{E}_m\mathbf{P}_p$ can be identically zero, the highest term in $\mathbf{P}_0(z) = \mathbf{E}(z)\mathbf{P}(z)$ can be lower than z^{p+m} or, in principle, even lower than z^p. The same is the case with $\mathbf{Q}_0(z) = \mathbf{E}(z)\mathbf{Q}(z)$, and if $\mathbf{Q}_q$ is singular, the highest term in this case also can be lower than z^q. However, there is at least one representation in (5.9) for which the sum $p + q$ is less than or equal to that of any other such left-coprime representation, and we will refer to that particular pair p, q as the left-coprime order of the multichannel system $\mathbf{H}(z)$. Notice that if $\mathbf{P}_p$ and $\mathbf{Q}_q$ are both nonsingular, then p and q are both unique and they denote the left-coprime order of the representation that is free of any elementary polynomial factor.

Similarly the right-coprime polynomial decomposition has the form

$$\mathbf{H}(z) = \mathbf{Q}_1(z)\mathbf{P}_1^{-1}(z), \tag{5.10}$$

where

$$\mathbf{P}_1(z) = \tilde{\mathbf{P}}_0 + \tilde{\mathbf{P}}_1 z + \cdots \tilde{\mathbf{P}}_{p_1} z^{p_1} \tag{5.11}$$

$$\mathbf{Q}_1(z) = \tilde{\mathbf{Q}}_0 + \tilde{\mathbf{Q}}_1 z + \cdots + \tilde{\mathbf{Q}}_{q_1} z^{q_1} \tag{5.12}$$

and

$$\mathbf{Y}_1(z)\mathbf{Q}_1(z) + \mathbf{X}_1(z)\mathbf{P}_1(z) = \mathbf{I}_m. \tag{5.13}$$

Such a decomposition also is unique up to multiplication by an elementary polynomial matrix on the right. Once again, there exists a representation

---

[3]A square polynomial matrix $\mathbf{E}(z)$ is said to be elementary (or unimodular) if its determinant is a nonzero constant. Evidently the inverse of such a matrix is also a polynomial matrix, and moreover the coefficient matrix of the highest term of such a polynomial matrix is always singular and the constant term is always nonsingular.

for which the sum $p_1 + q_1$ is less than or equal to that of any other such right-coprime representation, and we will refer to that pair p_1, q_1 as the right-coprime order of the multichannel system $\mathbf{H}(z)$. From (5.9) and (5.10), we have

$$\mathbf{Q}(z)\mathbf{P}_1(z) = \mathbf{P}(z)\mathbf{Q}_1(z)$$

and if one of the matrices in each pair $\mathbf{P}_p$, $\tilde{\mathbf{Q}}_{q_1}$ and $\tilde{\mathbf{P}}_{p_1}$, $\mathbf{Q}_q$ is nonsingular, then necessarily

$$p - q = p_1 - q_1 . \tag{5.14}$$

A matrix function $\mathbf{H}(z)$ is said to be analytic in $|z| < 1$ if all its entries are analytic in $|z| < 1$. If, in addition, $\mathbf{H}(z)$ is also nonsingular in $|z| < 1$, i.e., $det\,\mathbf{H}(z) \neq 0$ in $|z| < 1$, it is said to be minimum-phase. In that case, $\mathbf{H}^{-1}(z)$ is also analytic in $|z| < 1$ and from (5.9)

$$det\,\mathbf{P}(z) \cdot det\,\mathbf{Q}(z) \neq 0, \quad \text{in } |z| < 1, \tag{5.15}$$

which, in particular, gives

$$det\,\mathbf{P}_0 \cdot det\,\mathbf{Q}_0 \neq 0. \tag{5.16}$$

By making use of (5.9) in (5.2), we can rewrite (5.2) as

$$\mathbf{P}(z)\mathbf{X}(z) = \mathbf{Q}(z)\mathbf{W}(z),$$

where $\mathbf{P}(z)$ and $\mathbf{Q}(z)$ are matrix polynomials that realize the left-coprime order (p, q). Or, represented in time domain,

$$\mathbf{P}_0\mathbf{x}(n) = -\sum_{k=1}^{p}\mathbf{P}_k\mathbf{x}(n - k) + \sum_{k=0}^{q}\mathbf{Q}_k\mathbf{w}(n - k), \tag{5.17}$$

where

$$\mathbf{x}(n) \overset{\triangle}{=} [\,x_1(n)\,, \, x_2(n)\,, \, \cdots, \, x_m(n)\,]^T$$

and

$$\mathbf{w}(n) \overset{\triangle}{=} [\,w_1(n)\,, \, w_2(n)\,, \, \cdots, \, w_m(n)\,]^T .$$

Clearly, (5.17) justifies the rational system in (5.2) and (5.9) to be represented as multichannel ARMA with left-coprime order (p, q) or simply as Left-ARMA (p, q) or L-ARMA(p, q) (see also Fig. 5.1). Similarly, the right-coprime representation is useful if a set of outputs $Y_1(z)$, $Y_2(z)$, ..., $Y_m(z)$ is related to the inputs $N_1(z)$, $N_2(z)$, ..., $N_m(z)$ as

$$\mathbf{Y}(z) = \mathbf{N}(z)\mathbf{H}(z), \tag{5.18}$$

where

$$\mathbf{Y}(z) = [\,Y_1(z),\, Y_2(z),\, \ldots,\, Y_m(z)\,] \tag{5.19}$$

and

$$\mathbf{N}(z) = [\,N_1(z),\, N_2(z),\, \ldots,\, N_m(z)\,] . \tag{5.20}$$

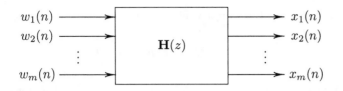

FIGURE 5.1. An $m \times m$ multichannel system. $X_i(z) = \sum_{j=1}^{m} H_{ij}(z)W_j(z)$, $i = 1 \to m$.

In that case using (5.10), the row vector (5.18) can be rewritten as

$$\mathbf{Y}(z)\mathbf{P}_1(z) = \mathbf{N}(z)\mathbf{Q}_1(z). \tag{5.21}$$

Represented in time domain, this reads

$$\mathbf{y}(n)\tilde{\mathbf{P}}_0 = -\sum_{k=1}^{p_1}\mathbf{y}(n-k)\tilde{\mathbf{P}}_k + \sum_{k=0}^{q_1}\mathbf{n}(n-k)\tilde{\mathbf{Q}}_k, \tag{5.22}$$

where

$$\mathbf{y}(n) \overset{\triangle}{=} [\, y_1(n), \, y_2(n), \, \cdots, \, y_m(n)\,]$$

and

$$\mathbf{n}(n) \overset{\triangle}{=} [\, n_1(n), \, n_2(n), \, \cdots, \, n_m(n)\,].$$

As before, (5.22) justifies the rational system $\mathbf{H}(z)$ in (5.18) to be represented as multichannel ARMA with right-coprime order (p_1, q_1) or simply as Right-ARMA(p_1, q_1) or R-ARMA(p_1, q_1). Notice that, the prefix in L-ARMA or R-ARMA clearly indicates the multichannel nature of the system. Define the $max\,(p, q)$ and $max\,(p_1, q_1)$ to represent respectively the left- and right-coprime orders (or ordinary degrees) of the multichannel rational function $\mathbf{H}(z)$. Thus

$$\delta_o(\mathbf{H}(z)) \overset{\triangle}{=} max\,(p, q) \quad \text{or} \quad max\,(p_1, q_1) \tag{5.23}$$

depending on whether we refer to the left-coprime form or the right-coprime form.

In (5.2), if the input excitation is a wide-sense stationary stochastic process, so is the output process, and if $\mathbf{S_W}(\theta)$ and $\mathbf{S_X}(\theta)$ represent the respective input and output power spectral density matrices, then

$$\mathbf{S_X}(\theta) = \mathbf{H}(e^{j\theta})\mathbf{S_W}(\theta)\mathbf{H}^*(e^{j\theta}). \tag{5.24}$$

Also

$$\mathbf{S_X}(\theta) = \sum_{k=-\infty}^{\infty} \mathbf{r}_k e^{jk\theta} \geq 0, \tag{5.25}$$

where

$$\mathbf{r}_k = E[\mathbf{x}(n+k)\mathbf{x}^*(n)] = \mathbf{r}^*_{-k}, \quad k = 0 \to \infty, \tag{5.26}$$

represents the kth output autocorrelation matrix. From (5.25), we also have

$$\mathbf{r}_k = \frac{1}{2\pi} \int_{-\pi}^{\pi} \mathbf{S}(\theta) e^{-jk\theta} d\theta, \quad k = 0 \to \infty. \tag{5.27}$$

The nonnegativity condition in (5.25) can be translated in terms of the nonnegativity of the Hermitian block Toeplitz matrices [6]

$$\mathbf{T}_k = \begin{bmatrix} \mathbf{r}_0 & \mathbf{r}_1 & \cdots & \mathbf{r}_k \\ \mathbf{r}_1^* & \mathbf{r}_0 & \cdots & \mathbf{r}_{k-1} \\ \vdots & \vdots & \ddots & \vdots \\ \mathbf{r}_k^* & \mathbf{r}_{k-1}^* & \cdots & \mathbf{r}_0 \end{bmatrix} \geq 0, \quad k = 0 \to \infty. \tag{5.28}$$

Thus

$$\mathbf{S}(\theta) \geq 0 \quad \Longleftrightarrow \quad \mathbf{T}_k \geq 0, \quad k = 0 \to \infty. \tag{5.29}$$

If the input process is also white noise that is channel to channel uncorrelated with unit variance on all channels, then

$$\mathbf{S_w}(\theta) = \mathbf{I}_m$$

and (5.24) becomes

$$\mathbf{S_x}(\theta) = \mathbf{H}(e^{j\theta})\mathbf{H}^*(e^{j\theta}). \tag{5.30}$$

Thus (5.30) represents the spectral density matrix associated with a system with transfer function $\mathbf{H}(z)$ that has been excited from right by a stationary white noise of unit spectral density. Similarly, if the input excitation in (5.18) is a wide-sense stationary process, so is the output process and with $\mathbf{S_n}(\theta)$ and $\mathbf{S_y}(\theta)$ representing the input and output power spectral density matrices, we have

$$\mathbf{S_y}(\theta) = \mathbf{H}^*(e^{j\theta})\mathbf{S_n}(\theta)\mathbf{H}(e^{j\theta}) = \sum_{k=-\infty}^{\infty} \tilde{\mathbf{r}}_k e^{jk\theta} \geq 0, \tag{5.31}$$

where

$$\tilde{\mathbf{r}}_k = E[\mathbf{y}^*(n+k)\mathbf{y}(n)] = \tilde{\mathbf{r}}_{-k}^*, \quad k = 0 \to \infty, \tag{5.32}$$

represents the kth output autocorrelation matrix of $\mathbf{y}(n)$. If the input process $\mathbf{n}(n)$ is also white noise that is channel to channel uncorrelated with unit variance, then

$$\mathbf{S_n}(\theta) = \mathbf{I}_m$$

and

$$\mathbf{S_y}(\theta) = \mathbf{H}^*(e^{j\theta})\mathbf{H}(e^{j\theta}). \tag{5.33}$$

Thus (5.33) represents the spectral density matrix associated with a system with transfer function $\mathbf{H}(z)$ that has been excited from the left by a stationary white noise process of unit spectral density.

The inverse problem of identifying the system from the power spectral density matrix is known as the spectral factorization problem, and several solutions are available in this context [7, 106]. From (5.25)–(5.33), every spectral density matrix (rational or nonrational) is necessarily Hermitian nonnegative definite

$$\mathbf{S}(\theta) = \mathbf{S}^*(\theta) \geq 0, \quad a.e.$$

If in addition $\mathbf{S}(\theta) \in L_1$, i.e., if its entries are absolutely integrable over $[-\pi, \pi]$, i.e.,

$$\frac{1}{2\pi} \int_{-\pi}^{\pi} |S_{ij}(\theta)| \, d\theta < \infty, \quad i, \, j = 1 \to m, \tag{5.34}$$

and if (Paley-Wiener or causality criterion)

$$\mathcal{H} \triangleq \frac{1}{2\pi} \int_{-\pi}^{\pi} \ln \det \mathbf{S}(\theta) \, d\theta > -\infty, \tag{5.35}$$

then every $\mathbf{T}_k$ in (5.28) can be shown to be positive definite [7], i.e.,

$$\frac{1}{2\pi} \int_{-\pi}^{\pi} \ln \det \mathbf{S}(\theta) d\theta > -\infty \implies \mathbf{T}_k > 0, \, k = 0 \to \infty. \tag{5.36}$$

Under these conditions, it is well known that [6, 7] there exists a unique function

$$\mathbf{H}(z) = \sum_{k=0}^{\infty} \mathbf{H}_k z^k \tag{5.37}$$

that is analytic in $|z| < 1$ such that

$$\det \mathbf{H}(z) \neq 0 \quad \text{in } |z| < 1 \tag{5.38}$$

$$\sum_{k=0}^{\infty} tr(\mathbf{H}_k \mathbf{H}_k^*) < \infty \tag{5.39}$$

and

$$\mathbf{S}(\theta) = \mathbf{H}(e^{j\theta}) \mathbf{H}^*(e^{j\theta}), \tag{5.40}$$

where

$$\mathbf{H}(e^{j\theta}) = \lim_{r \to 1-0} \mathbf{H}(re^{j\theta}). \tag{5.41}$$

Such a minimum-phase factor is unique up to multiplication on the right by an arbitrary constant unitary matrix and is known as the left-Wiener factor of $\mathbf{S}(\theta)$. Similarly the right-Wiener factor $\mathbf{B}(z) = \sum_{k=0}^{\infty} \mathbf{B}_k z^k$ of $\mathbf{S}(\theta)$ is also analytic in $|z| < 1$ such that $\det \mathbf{B}(z) \neq 0$ in $|z| < 1$,

$$\sum_{k=0}^{\infty} tr\, (\mathbf{B}_k^* \mathbf{B}_k) < \infty$$

and

$$\mathbf{S}(\theta) = \mathbf{B}^*(e^{j\theta})\mathbf{B}(e^{j\theta}), \tag{5.42}$$

where as before

$$\mathbf{B}(e^{j\theta}) = \lim_{r \to 1-0} \mathbf{B}(re^{j\theta}).$$

Again, the left- and right-Wiener factors are not related in an easy manner to each other. A multichannel power spectral density matrix $\mathbf{S}(\theta)$ is said to be rational if its Wiener factors are rational, and in that case these Wiener factors have left- and right-coprime representations as in (5.9) and (5.10). The identification problem then is to obtain the system parameters $\mathbf{P}_0 \to \mathbf{P}_p$ and $\mathbf{Q}_0 \to \mathbf{Q}_q$ (or $\tilde{\mathbf{P}}_0 \to \tilde{\mathbf{P}}_{p_1}$ and $\tilde{\mathbf{Q}}_0 \to \tilde{\mathbf{Q}}_{q_1}$) as well as the model orders p and q (or p_1 and q_1).

Since the spectral factorization as in (5.37)–(5.42) is unique iff $\mathbf{S}(\theta) \in L_1$ and satisfies the causality criterion in (5.35), we will assume that the power spectra associated with the systems under consideration satisfy these two requirements. In that case $\mathbf{H}(z)$ is minimum-phase and hence in particular $det\,\mathbf{P}_0 \neq 0$. Thus, for example, by multiplying by the inverse of $\mathbf{P}_0$ from the left-hand side, we have

$$\mathbf{H}(z) = \mathbf{P}^{-1}(z)\mathbf{Q}(z) = \mathbf{P}^{-1}(z)\mathbf{P}_0\mathbf{P}_0^{-1}\mathbf{Q}(z) = \left(\mathbf{P}_0^{-1}\mathbf{P}(z)\right)^{-1}\left(\mathbf{P}_0^{-1}\mathbf{Q}(z)\right)$$
$$= \left(\mathbf{I} + \mathbf{P}_1 z + \cdots + \mathbf{P}_p z^p\right)^{-1}\left(\mathbf{Q}_0 + \mathbf{Q}_1 z + \cdots + \mathbf{Q}_q z^q\right), \tag{5.43}$$

where $\mathbf{P}_k$, $k = 1 \to p$, and $\mathbf{Q}_k$, $k = 0 \to q$, are the *renormalized* $p + q + 1$ system parameters, and, hereonwards, we will assume such a representation. Thus an L-ARMA(p, q) system has $p+q+1$ matrix parameters, and in principle $p+q+1$ output autocorrelations should be able to identify these parameters. In fact, as shown in chapter 4, once the orders are known, certain rank properties of the covariance matrices can be utilized to identify these parameters. In what follows we review some standard results in this context and then develop a new technique that once again exploits rationality and makes use of a weak form of the multichannel version of Richards' theorem to determine the model order and the system parameters. However, the noncommutativity of the matrices brings in other subtleties, and these extensions are nontrivial.

Multichannel Rational Case: Exact Autocorrelations

To describe the exact relationship between the output autocorrelation matrices and the system parameters in the rational case, we start with the time domain equation (5.17) for a multichannel L-ARMA(p, q) system:

$$\mathbf{x}(n) = -\sum_{i=1}^{p} \mathbf{P}_i \mathbf{x}(n - i) + \sum_{i=0}^{q} \mathbf{Q}_i \mathbf{w}(n - i). \tag{5.44}$$

From (5.44), since the present system input $\mathbf{w}(n)$ is uncorrelated with its own past values and the past values of the system output, we have

$$E[\mathbf{w}(n)\,\mathbf{w}^*(n-k)] = \mathbf{0}\,, \quad k \geq 1\,, \tag{5.45}$$

and

$$E[\mathbf{w}(n)\,\mathbf{x}^*(n-k)] = \mathbf{0}\,, \quad k \geq q+1\,. \tag{5.46}$$

Multiplying (5.44) by $\mathbf{x}^*(n-k)$ on its right and taking expectation, we get

$$E[\mathbf{x}(n)\,\mathbf{x}^*(n-k)] = -\sum_{i=1}^{p} \mathbf{P}_i E[\mathbf{x}(n-i)\,\mathbf{x}^*(n-k)] + \sum_{i=0}^{q} \mathbf{Q}_i E[\mathbf{w}(n-i)\,\mathbf{x}^*(n-k)] \tag{5.47}$$

or

$$\mathbf{r}_k = -\sum_{i=1}^{p} \mathbf{P}_i \mathbf{r}_{k-i} + \sum_{i=1}^{q} \mathbf{Q}_i E[\mathbf{w}(n-i)\,\mathbf{x}^*(n-k)]\,. \tag{5.48}$$

Using (5.46) on (5.48), we have

$$\mathbf{r}_k + \sum_{i=1}^{p} \mathbf{P}_i \mathbf{r}_{k-i} \neq \mathbf{0}\,, \quad k \leq q\,, \tag{5.49}$$

and

$$\mathbf{r}_k + \sum_{i=1}^{p} \mathbf{P}_i \mathbf{r}_{k-i} = \mathbf{0}\,, \quad k \geq q+1\,. \tag{5.50}$$

Notice that the first p equations in (5.50) can be written as

$$[\mathbf{P}_1\,\mathbf{P}_2\,\cdots\,\mathbf{P}_p]\mathbf{R}(p,q) = -[\mathbf{r}_{q+1}\,\mathbf{r}_{q+2}\,\cdots\,\mathbf{r}_{q+p}]\,, \tag{5.51}$$

where

$$\mathbf{R}(p,q) \overset{\triangle}{=} \begin{bmatrix} \mathbf{r}_q & \mathbf{r}_{q+1} & \cdots & \mathbf{r}_{q+p-1} \\ \mathbf{r}_{q-1} & \mathbf{r}_q & \cdots & \mathbf{r}_{q+p-2} \\ \vdots & \vdots & \ddots & \vdots \\ \mathbf{r}_{q-p+1} & \mathbf{r}_{q-p+2} & \cdots & \mathbf{r}_q \end{bmatrix} \tag{5.52}$$

is a block matrix of size $mp \times mp$ whose upper left-hand corner block is $\mathbf{r}_q$. Since $\mathbf{P}_k$, $k = 1 \to p$, have a unique solution, $\mathbf{R}(p,q)$ in (5.52) must be nonsingular. Similarly, letting $\mathbf{P} = [\mathbf{I}\,\mathbf{P}_1\,\cdots\,\mathbf{P}_p]$ and using (5.52), the first $p+1$ equations in (5.50) give

$$\mathbf{P}\,\mathbf{R}(p+1,q+1) = \mathbf{0}\,. \tag{5.53}$$

Since $\mathbf{P} \not\equiv \mathbf{0}$, necessarily $\mathbf{R}(p+1,q+1)$ must be singular and hence the autocorrelation matrices of an L-ARMA(p,q) system satisfy

$$det\,\mathbf{R}(p,q) \neq 0 \tag{5.54}$$

$$det\,\mathbf{R}(p+1,q+1) = 0\,. \qquad (5.55)$$

Equation (5.55) also follows from certain block-linear dependencies inherent in (5.50). To see this, note that the first block row of $\mathbf{R}(p+1,q+1)$ is linearly dependent on its remaining p block rows. Thus, if $\mathbf{X}$ represents the last p block rows of $\mathbf{R}(p+1,q+1)$, then with $\mathbf{M}_1 = [\mathbf{P}_1\ \mathbf{P}_2\ \cdots\ \mathbf{P}_p]$,

$$\mathbf{R}(p+1,q+1) = \left[\frac{\mathbf{M}_1\mathbf{X}}{\mathbf{X}}\right] = \left[\frac{\mathbf{M}_1}{\mathbf{I}_{mp}}\right]\mathbf{X} \stackrel{\triangle}{=} \mathbf{D}\mathbf{X}\,. \qquad (5.56)$$

Since $rank\,\mathbf{X} \le mp$ and $rank\,\mathbf{D} \le mp$, from Sylvester's inequality [3], we have

$$rank\,\mathbf{R}(p+1,q+1) \le mp \qquad (5.57)$$

and hence

$$det\,\mathbf{R}(p+1,q+1) = 0\,.$$

More generally, for any L-ARMA(p,q) system,

$$det\,\mathbf{R}(p+k,q+i) = 0\,, \quad k > 0,\ i > 0\,. \qquad (5.58)$$

To establish this, first consider the case when $k = i$. In that case, from (5.50), the first k block rows of $\mathbf{R}(p+k,q+k)$ are linearly dependent on its last p block rows. Thus, if $\mathbf{Y}$ represents the last p block rows of $\mathbf{R}(p+k,q+k)$, then

$$\mathbf{R}(p+k,q+k) = \left[\frac{\mathbf{M}_2\mathbf{Y}}{\mathbf{Y}}\right] = \left[\frac{\mathbf{M}_2}{\mathbf{I}_{mp}}\right]\mathbf{Y} \stackrel{\triangle}{=} \mathbf{E}\mathbf{Y}\,, \qquad (5.59)$$

where $\mathbf{M}_2$ is an $mk \times mp$ matrix. Since $rank\,\mathbf{Y} \le mp$ and $rank\,\mathbf{E} \le mp$, again from Sylvester's inequality,

$$rank\,\mathbf{R}(p+k,q+k) \le mp\,, \quad k > 0,\ i > 0\,, \qquad (5.60)$$

from which (5.58) follows for $k = i$.

When $k < i$, the first i block rows of $\mathbf{R}(p+k,q+i)$ are linearly dependent on the rows of the above $\mathbf{Y}$ matrix, i.e.,

$$\mathbf{R}(p+k,q+i) = \left[\frac{\mathbf{M}_3}{\mathbf{L}}\right]\mathbf{Y}\,, \qquad (5.61)$$

where $\mathbf{M}_3$ is an $mi \times mp$ matrix and $\mathbf{L}$ is an $m(p+k-i) \times mp$ matrix given by

$$\mathbf{L} = \left[\ \mathbf{I}_{m(p+k-i)}\ \big|\ \mathbf{0}_{m(p+k-i),m(i-k)}\ \right]\,.$$

Since $rank\,\mathbf{Y} \le mp$, again (5.58) holds. Similarly, (5.58) can be established for $k > i$ by considering its columns.

When $k < 0$ or $i < 0$, no such linear dependency can be proved among the rows or columns of $\mathbf{R}(p+k,q+i)$, and hence, in general, $\mathbf{R}(p+k,q+i)$

will be nonsingular if $k < 0$ or $i < 0$. To summarize, for an L-ARMA(p, q) system,

$$det\, \mathbf{R}(p + k, q + i) = \begin{cases} = 0\,, & \text{if } k > 0 \text{ and } i > 0 \\ \neq 0\,, & \text{if } k = 0 \text{ and/or } i = 0 \\ \text{generally } \neq 0\,, & \text{if } k < 0 \text{ or } i < 0 \end{cases} \quad (5.62)$$

Assuming $p > q$, from (5.62), we can derive a determinantal test that does not involve q. Let $k = 0$ and $i = p - q > 0$. Then (5.62) reads

$$det\, \mathbf{R}(p, p) \neq 0\,. \quad (5.63)$$

If we let $k > 0$ and $i = p + k - q > 0$, then

$$det\, \mathbf{R}(p + k, p + k) = 0\,. \quad (5.64)$$

Let

$$\mathbf{H}_p \overset{\triangle}{=} \begin{bmatrix} \mathbf{0} & \cdots & \mathbf{0} & \mathbf{I} \\ \mathbf{0} & \cdots & \mathbf{I} & \mathbf{0} \\ \vdots & \cdots & \vdots & \vdots \\ \mathbf{I} & \cdots & \mathbf{0} & \mathbf{0} \end{bmatrix} \quad \mathbf{R}(p, p) = \begin{bmatrix} \mathbf{r}_1 & \mathbf{r}_2 & \cdots & \mathbf{r}_p \\ \mathbf{r}_2 & \mathbf{r}_3 & \cdots & \mathbf{r}_{p+1} \\ \vdots & \vdots & \cdots & \vdots \\ \mathbf{r}_p & \mathbf{r}_{p+1} & \cdots & \mathbf{r}_{2p-1} \end{bmatrix} \quad (5.65)$$

represent the $mp \times mp$ block Hankel matrix. Then, for $\mathbf{H}_{p+k}$, $k \geq 1$, from (5.50), the last k block rows (columns) are linearly dependent on the previous p block rows (columns), and hence equations (5.63)–(5.65) take the classical form:

$$rank\, \mathbf{H}_p = rank\, \mathbf{H}_{p+k} = mp\,, \quad k \geq 0\,. \quad (5.66)$$

As pointed out before, an L-ARMA(p, q) system has $p + q + 1$ unknown parameters and hence the first $p + q + 1$ autocorrelation matrices $\mathbf{r}_0$, $\mathbf{r}_1$, ..., $\mathbf{r}_{p+q}$ should be able to identify these system parameters and their order. To make further progress, it is best to describe the classical solution to the class of all admissible spectral extensions that interpolate $\mathbf{r}_0 \to \mathbf{r}_{p+q}$ and invoke rationality and other ordinary-degree (order) constraints to complete the identification problem.

5.2 Multichannel Admissible Spectral Extensions

To develop techniques similar to those in Chapter 3, it is necessary to review some standard classical results on positive matrix functions and bounded matrix functions. Together with the Schur algorithm, this will enable us to exhibit the class of all admissible extensions that interpolate the given autocorrelations [25]. Once again the ordinary-degree reduction conditions imposed by the Richards' theorem in the rational case can be profitably used to identify the multichannel system order and its matrix system parameters. We start with the following standard definition [25].

Definition: A matrix $\mathbf{Z}(z)$ is said to be positive if

  (i) $\mathbf{Z}(z)$ is analytic in $|z| < 1$

and

  (ii) its Hermitian part $\mathbf{R}(z) \stackrel{\triangle}{=} \mathbf{Z}(z) + \mathbf{Z}^*(z) \geq 0$ in $|z| < 1$.

Since a positive matrix function is analytic in $|z| < 1$, it has a power series expansion given by

$$\mathbf{Z}(z) = \mathbf{r}_0 + 2 \sum_{k=1}^{\infty} \mathbf{r}_k z^k, \quad |z| < 1. \tag{5.67}$$

The next classic theorem due to Schur translates the positivity condition in terms of the positivity of certain Hermitian Toeplitz matrices.

Schur's Theorem: The power series

$$\mathbf{Z}(z) = \mathbf{r}_0 + 2 \sum_{k=1}^{\infty} \mathbf{r}_k z^k, \quad |z| < 1, \tag{5.68}$$

represents a positive matrix function iff every

$$\mathbf{T}_k = \begin{bmatrix} \mathbf{r}_0 & \mathbf{r}_1 & \cdots & \mathbf{r}_k \\ \mathbf{r}_1^* & \mathbf{r}_0 & \cdots & \mathbf{r}_{k-1} \\ \vdots & \vdots & \ddots & \vdots \\ \mathbf{r}_k^* & \mathbf{r}_{k-1}^* & \cdots & \mathbf{r}_0 \end{bmatrix} \geq 0, \quad k = 0 \to \infty. \tag{5.69}$$

Proof: See section 2.2.

The boundary values of $\mathbf{Z}(z)$ are given by the interior radial limits [27]

$$\mathbf{Z}(e^{j\theta}) = \lim_{r \to 1-0} \mathbf{Z}(re^{j\theta}), \tag{5.70}$$

and they exist almost everywhere. In that case by definition

$$\mathbf{Z}(e^{j\theta}) + \mathbf{Z}^*(e^{j\theta}) \geq 0$$

and using (5.67)

$$\mathbf{S}(\theta) \stackrel{\triangle}{=} \frac{\mathbf{Z}(e^{j\theta}) + \mathbf{Z}^*(e^{j\theta})}{2} = \sum_{k=-\infty}^{\infty} \mathbf{r}_k e^{jk\theta} \geq 0 \tag{5.71}$$

where we have made use of the notation $\mathbf{r}_{-k} = \mathbf{r}_k^*$, $k \geq 0$. Comparing (5.25)–(5.29) with (5.67)–(5.71), the connection between power spectra and positive matrix functions is obvious, i.e., every positive matrix function (5.67) corresponds to a power spectrum through the relations (5.68)–(5.71), and conversely on comparing (5.25)–(5.29) with (5.67), every power spectrum generates a positive matrix function.

As in the single-channel case, it is possible to define bounded matrix functions associated with every positive matrix function. Recall that in the single-channel case a function $d(z)$, that is analytic in $|z| < 1$, is said to be bounded if

$$1 - |d(z)|^2 \geq 0 \quad \text{in } |z| < 1, \tag{5.72}$$

and for every R with $\mathrm{Re}\, R > 0$,

$$d(z) \stackrel{\triangle}{=} \frac{Z(z) - R}{Z(z) + R^*} \tag{5.73}$$

is a bounded function if and only if $Z(z)$ is a positive function. Following the definition in (5.72), a matrix $\mathbf{d}(z)$ that is analytic in $|z| < 1$ is said to be bounded if[4]

$$\mathbf{I} - \mathbf{d}(z)\mathbf{d}^*(z) \geq 0 \quad \text{in } |z| < 1. \tag{5.74}$$

However, the transformation from positive to bounded as in (5.73) for the single-channel case does not easily translate into the matrix case, and it requires proper renormalization or initialization [107]. To see this, let $\mathbf{R}$ be any matrix with a positive Hermitian part, i.e.,

$$\mathbf{K} \stackrel{\triangle}{=} \frac{\mathbf{R} + \mathbf{R}^*}{2} > 0 \tag{5.75}$$

and let $\mathbf{K}^{1/2}$ represent *any* Hermitian square root of $\mathbf{K}$. Obviously, $\mathbf{K}^{1/2} = (\mathbf{K}^{1/2})^*$ and $(\mathbf{K}^{1/2})^2 = \mathbf{K}$. Notice that such a square root can be made unique by requiring it to be also positive definite. Then for every positive matrix function $\mathbf{Z}(z)$, the function

$$\mathbf{d}(z) \stackrel{\triangle}{=} \mathbf{K}^{-1/2}(\mathbf{Z}(z) - \mathbf{R})(\mathbf{Z}(z) + \mathbf{R}^*)^{-1}\mathbf{K}^{1/2} \tag{5.76}$$

represents a bounded matrix[5] [25, 107].

To prove this, notice that (5.76) can be rewritten as

$$\begin{aligned}
\mathbf{d}(z) &\stackrel{\triangle}{=} \mathbf{K}^{-1/2}(\mathbf{Z}(z) - \mathbf{R})(\mathbf{Z}(z) + \mathbf{R}^*)^{-1}\mathbf{K}^{1/2} \\
&= \mathbf{K}^{-1/2}(\mathbf{Z}(z) + \mathbf{R}^* - \mathbf{R}^* - \mathbf{R})(\mathbf{Z}(z) + \mathbf{R}^*)^{-1}\mathbf{K}^{1/2} \\
&= \mathbf{I} - \mathbf{K}^{-1/2}(\mathbf{R} + \mathbf{R}^*)(\mathbf{Z}(z) + \mathbf{R}^*)^{-1}\mathbf{K}^{1/2} \\
&= \mathbf{I} - 2\mathbf{K}^{1/2}(\mathbf{Z}(z) + \mathbf{R}^*)^{-1}\mathbf{K}^{1/2}.
\end{aligned}$$

---

[4]Or equivalently, $\mathbf{I} - \mathbf{d}^*(z)\mathbf{d}(z) \geq 0$ in $|z| < 1$. Since $\mathbf{d}(z)\mathbf{d}^*(z)$ and $\mathbf{d}^*(z)\mathbf{d}(z)$ have the same set of eigenvalues, both these definitions are equivalent.

[5]Note that if $\mathbf{Z}(z)$ is a positive matrix function, then for an $\mathbf{R}$ with positive Hermitian part, $(\mathbf{Z}(z) - \mathbf{R})(\mathbf{Z}(z) + \mathbf{R}^*)^{-1}$ is *not* a bounded matrix function unless $\mathbf{R} = \mathbf{I}$.

Thus,

$$\mathbf{I} - \mathbf{d}(z)\mathbf{d}^*(z)$$

$$= \mathbf{I} - \left\{\mathbf{I} - 2\mathbf{K}^{1/2}(\mathbf{Z}(z) + \mathbf{R}^*)^{-1}\mathbf{K}^{1/2}\right\}\left\{\mathbf{I} - 2\mathbf{K}^{1/2}(\mathbf{Z}^*(z) + \mathbf{R})^{-1}\mathbf{K}^{1/2}\right\}$$

$$= 2\mathbf{K}^{1/2}(\mathbf{Z}(z) + \mathbf{R}^*)^{-1}\mathbf{K}^{1/2} + 2\mathbf{K}^{1/2}(\mathbf{Z}^*(z) + \mathbf{R})^{-1}\mathbf{K}^{1/2}$$

$$\quad - 4\mathbf{K}^{1/2}(\mathbf{Z}(z) + \mathbf{R}^*)^{-1}\mathbf{K}(\mathbf{Z}^*(z) + \mathbf{R})^{-1}\mathbf{K}^{1/2}$$

$$= 2\mathbf{K}^{1/2}(\mathbf{Z}(z) + \mathbf{R}^*)^{-1}\left\{(\mathbf{Z}^*(z) + \mathbf{R}) + (\mathbf{Z}(z) + \mathbf{R}^*) - 2\mathbf{K}\right\}$$

$$\quad \times (\mathbf{Z}^*(z) + \mathbf{R})^{-1}\mathbf{K}^{1/2}$$

$$= 2\mathbf{K}^{1/2}(\mathbf{Z}(z) + \mathbf{R}^*)^{-1}(\mathbf{Z}(z) + \mathbf{Z}^*(z))(\mathbf{Z}^*(z) + \mathbf{R})^{-1}\mathbf{K}^{1/2} \geq 0, \quad (5.77)$$

in $|z| < 1$. Thus $\mathbf{d}(z)$ is analytic in $|z| < 1$ and from (5.77) it represents a bounded matrix function. Q.E.D.

Solving for $\mathbf{Z}(z)$ in terms of $\mathbf{d}(z)$, from (5.76) we have

$$\mathbf{K}^{1/2}\mathbf{d}(z)\mathbf{K}^{-1/2} = (\mathbf{Z}(z) - \mathbf{R})(\mathbf{Z}(z) + \mathbf{R}^*)^{-1} \tag{5.78}$$

$$\mathbf{K}^{1/2}\mathbf{d}(z)\mathbf{K}^{-1/2}(\mathbf{Z}(z) + \mathbf{R}^*) = \mathbf{Z}(z) - \mathbf{R}$$

$$(\mathbf{I} - \mathbf{K}^{1/2}\mathbf{d}(z)\mathbf{K}^{-1/2})\mathbf{Z}(z) = \mathbf{K}^{1/2}\mathbf{d}(z)\mathbf{K}^{-1/2}\mathbf{R}^* + \mathbf{R}$$

$$\mathbf{K}^{1/2}(\mathbf{I} - \mathbf{d}(z))\mathbf{K}^{-1/2}\mathbf{Z}(z) = \mathbf{K}^{1/2}\mathbf{d}(z)\mathbf{K}^{-1/2}\mathbf{R}^* + \mathbf{R} \tag{5.79}$$

or

$$(\mathbf{I} - \mathbf{d}(z))\mathbf{K}^{-1/2}\mathbf{Z}(z) = \mathbf{d}(z)\mathbf{K}^{-1/2}\mathbf{R}^* + \mathbf{K}^{-1/2}\mathbf{R}$$

$$= \mathbf{d}(z)\mathbf{K}^{-1/2}(\mathbf{R} + \mathbf{R}^* - \mathbf{R} + \mathbf{R}^*)/2 + \mathbf{K}^{-1/2}(\mathbf{R} + \mathbf{R}^* + \mathbf{R} - \mathbf{R}^*)/2$$

$$= \mathbf{d}(z)\mathbf{K}^{-1/2}(2\mathbf{K} - \mathbf{R} + \mathbf{R}^*)/2 + \mathbf{K}^{-1/2}(2\mathbf{K} + \mathbf{R} - \mathbf{R}^*)/2$$

$$= \mathbf{d}(z)\mathbf{K}^{1/2} + \mathbf{K}^{1/2} - (\mathbf{d}(z)\mathbf{K}^{-1/2} - \mathbf{K}^{-1/2})(\mathbf{R} - \mathbf{R}^*)/2$$

$$= (\mathbf{I} + \mathbf{d}(z))\mathbf{K}^{1/2} + (\mathbf{I} - \mathbf{d}(z))\mathbf{K}^{-1/2}(\mathbf{R} - \mathbf{R}^*)/2. \tag{5.80}$$

Thus,

$$\mathbf{K}^{-1/2}\mathbf{Z}(z) = (\mathbf{I} - \mathbf{d}(z))^{-1}(\mathbf{I} + \mathbf{d}(z))\mathbf{K}^{1/2} + \mathbf{K}^{-1/2}\frac{(\mathbf{R} - \mathbf{R}^*)}{2}$$

and hence

$$\mathbf{Z}(z) = \mathbf{K}^{1/2}(\mathbf{I} - \mathbf{d}(z))^{-1}(\mathbf{I} + \mathbf{d}(z))\mathbf{K}^{1/2} + \frac{(\mathbf{R} - \mathbf{R}^*)}{2} \tag{5.81}$$

or

$$\mathbf{Z}(z) = \mathbf{K}^{1/2}(\mathbf{I} + \mathbf{d}(z))(\mathbf{I} - \mathbf{d}(z))^{-1}\mathbf{K}^{1/2} + \frac{(\mathbf{R} - \mathbf{R}^*)}{2}, \tag{5.82}$$

a relation that proves to be useful later on. More interestingly, starting from (5.82), and rewriting it as

$$\mathbf{Z}(z)\mathbf{K}^{-1/2}(\mathbf{I} - \mathbf{d}(z)) = \mathbf{K}^{1/2}(\mathbf{I} + \mathbf{d}(z)) + \left(\frac{\mathbf{R} - \mathbf{R}^*}{2}\right)\mathbf{K}^{-1/2}(\mathbf{I} - \mathbf{d}(z)) \tag{5.83}$$

and proceeding backwards as in (5.80)–(5.78), we also obtain

$$\mathbf{d}(z) = \mathbf{K}^{1/2}(\mathbf{Z}(z) + \mathbf{R}^*)^{-1}(\mathbf{Z}(z) - \mathbf{R})\mathbf{K}^{-1/2}. \tag{5.84}$$

Notice that (5.76) together with (5.84) constitute two alternate representations of the same bounded matrix function $\mathbf{d}(z)$, and they represent the direct generalization of the single-channel relation (5.73). Interestingly, from (5.74), (5.76) and (5.84), with $\mathbf{U}$ and $\mathbf{V}$ representing any two constant unitary matrices, the function

$$\rho(z) \stackrel{\triangle}{=} \mathbf{U}\mathbf{d}(z)\mathbf{V} \tag{5.85}$$

also defines a bounded matrix function, and the extra freedom that exists in (5.85) in terms of the unitary matrices $\mathbf{U}$ and $\mathbf{V}$ can be utilized for any subsequent renormalization.

5.2.1 MULTICHANNEL SCHUR ALGORITHM AND A WEAK FORM OF RICHARDS' THEOREM

In particular, consider the special case when

$$\mathbf{R} = \mathbf{Z}(0) \stackrel{\triangle}{=} \mathbf{R}_0 \tag{5.86}$$

in (5.75)–(5.76) and (5.84). Using (5.68), we have $\mathbf{R}_0 = \mathbf{r}_0$, and since the power spectral density matrix $\mathbf{S}(\theta)$ satisfies the Paley-Wiener criterion in (5.35), from (5.36), $\mathbf{r}_0 > 0$ and hence

$$\mathbf{K}_0 = \frac{\mathbf{R}_0 + \mathbf{R}_0^*}{2} > 0. \tag{5.87}$$

As before, let $\mathbf{K}_0^{1/2}$ represent any Hermitian square root of $\mathbf{K}_0$. Then

$$\begin{aligned}
\mathbf{d}_0(z) &= \mathbf{K}_0^{-1/2}(\mathbf{Z}_0(z) - \mathbf{R}_0)(\mathbf{Z}_0(z) + \mathbf{R}_0^*)^{-1}\mathbf{K}_0^{1/2}\\
&= \mathbf{K}_0^{1/2}(\mathbf{Z}_0(z) + \mathbf{R}_0^*)^{-1}(\mathbf{Z}_0(z) - \mathbf{R}_0)\mathbf{K}_0^{-1/2}
\end{aligned} \tag{5.88}$$

is a bounded matrix function where

$$\mathbf{Z}_0(z) \equiv \mathbf{Z}(z). \tag{5.89}$$

Clearly $\mathbf{d}_0(0) \equiv \mathbf{0}$ in (5.88) and hence the function

$$\begin{aligned}
\mathbf{d}_1(z) \stackrel{\triangle}{=} \frac{1}{z}\mathbf{d}_0(z) &= \frac{1}{z}\mathbf{K}_0^{-1/2}(\mathbf{Z}_0(z) - \mathbf{R}_0)(\mathbf{Z}_0(z) + \mathbf{R}_0^*)^{-1}\mathbf{K}_0^{1/2}\\
&= \frac{1}{z}\mathbf{K}_0^{1/2}(\mathbf{Z}_0(z) + \mathbf{R}_0^*)^{-1}(\mathbf{Z}_0(z) - \mathbf{R}_0)\mathbf{K}_0^{-1/2}
\end{aligned} \tag{5.90}$$

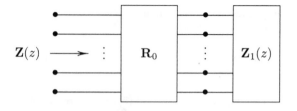

FIGURE 5.2. Multichannel line extraction.

is also bounded [25]. This follows by noticing that the only possible simple pole of $\mathbf{d}_1(z)$ at $z = 0$ is cancelled by the zero of $\mathbf{d}_0(z)$ at $z = 0$, and since $\mathbf{d}_1(z)$ is also bounded by $1/r$ for every $|z| = r < 1$, using a matrix version of the Schwartz' Lemma, $\mathbf{d}_1(z)$ represents a bounded matrix function. Notice that $\mathbf{d}_1(z)$ is also normalized with respect to $\mathbf{R}_0$. Let $\mathbf{Z}_1(z)$ represent the positive matrix function corresponding to this bounded matrix function. Then from (5.76) and (5.84),

$$\begin{aligned}
\mathbf{d}_1(z) &= \mathbf{K}_0^{-1/2}(\mathbf{Z}_1(z) - \mathbf{R}_0)(\mathbf{Z}_1(z) + \mathbf{R}_0^*)^{-1}\mathbf{K}_0^{1/2} \\
&= \mathbf{K}_0^{1/2}(\mathbf{Z}_1(z) + \mathbf{R}_0^*)^{-1}(\mathbf{Z}_1(z) - \mathbf{R}_0)\mathbf{K}_0^{-1/2}.
\end{aligned} \tag{5.91}$$

This procedure of extracting a new positive matrix function $\mathbf{Z}_1(z)$ from a given positive matrix function $\mathbf{Z}_0(z)$ represents the matrix version of the Schur algorithm described in chapter 2 and can be given a similar physical interpretation as in the single-channel case (refer to Fig. 2.7) by making use of the incident and reflected wave vectors $\mathbf{a}(z)$, $\mathbf{b}(z)$ together with an ideal delay line. Referring to Fig. 5.2, a multiline extraction from the input positive matrix function $\mathbf{Z}(z)$ has resulted in a new positive matrix function $\mathbf{Z}_1(z)$. Notice that the above line extraction is possible only if the normalization is carried out with respect to $\mathbf{R}_0 = \mathbf{Z}(0)$ in (5.88)–(5.91), and hence the quantity $\mathbf{R}_0$ that is characteristic to the line is known as its characteristic impedance matrix.

Under what conditions will the new bounded matrix function $\mathbf{d}_1(z)$ or the positive matrix function $\mathbf{Z}_1(z)$ be 'simpler' than $\mathbf{Z}(z)$? In the rational case, this question is answered by the matrix version of a weak form of Richards' theorem[6] in terms of the ordinary-degree of the positive matrix function.

A Weak Form of the Multichannel Richard's Theorem

Let $\mathbf{Z}_0(z)$ represent a rational positive matrix function. Define $\mathbf{d}_0(z)$ and

---

[6]The true generalization of Richards' theorem to the multichannel case involves the McMillan degree reduction and, as Youla has first shown, such is the case iff the para-conjugate Hermitian matrix $(\mathbf{Z}(z)+\mathbf{Z}_*(z))^{-1}$ has a pole at $z = 0$ [108]. Of course, (5.92) is sufficient, but obviously not necessary in that case.

$\mathbf{d}_1(z)$ as in (5.88)–(5.91). Then $\mathbf{d}_1(z)$ represents a bounded rational matrix function, and if $\mathbf{Z}_1(z)$ represents the associated positive matrix function normalized to $\mathbf{R}_0$ as in (5.91) then $\delta_0(\mathbf{Z}_1(z)) \leq \delta_0(\mathbf{Z}(z))$. Moreover $\delta_0(\mathbf{Z}_1(z)) = \delta_0(\mathbf{Z}(z)) - 1$ if

$$\mathbf{Z}(z) + \mathbf{Z}_*(z)|_{z=0} = \mathbf{0}. \tag{5.92}$$

Proof: Let $\mathbf{Z}_0(z)$ be a rational matrix with ordinary degree n as in (5.23), with a left-coprime representation

$$\mathbf{Z}_0(z) = \mathbf{A}^{-1}(z)\mathbf{B}(z), \tag{5.93}$$

where, in general,

$$\mathbf{A}(z) = \mathbf{A}_0 + \mathbf{A}_1 z + \cdots + \mathbf{A}_n z^n$$

and

$$\mathbf{B}(z) = \mathbf{B}_0 + \mathbf{B}_1 z + \cdots + \mathbf{B}_n z^n.$$

Substituting this into the later form in (5.90), we obtain

$$\mathbf{d}_1(z) = \mathbf{K}_0^{1/2} \mathbf{A}_0^* \left[(\mathbf{A}_0 \mathbf{B}_0^* + \mathbf{B}_0 \mathbf{A}_0^*) + \cdots + (\mathbf{A}_n \mathbf{B}_0^* + \mathbf{B}_n \mathbf{A}_0^*) z^n \right]^{-1}$$
$$\times \left[(\mathbf{B}_1 - \mathbf{A}_1 \mathbf{A}_0^{-1} \mathbf{B}_0) + \cdots + (\mathbf{B}_n - \mathbf{A}_n \mathbf{A}_0^{-1} \mathbf{B}_0) z^{n-1} \right] \mathbf{K}_0^{-1/2}. \tag{5.94}$$

From (5.94), clearly the ordinary degree of $\mathbf{d}_1(z)$ is n and degree reduction happens iff the matrix coefficient of z^n goes to zero there. Thus

$$\delta_0(\mathbf{d}_1(z)) = n - 1 \quad \Longleftrightarrow \quad \mathbf{A}_n \mathbf{B}_0^* + \mathbf{B}_n \mathbf{A}_0^* = \mathbf{0}$$

which is the same as

$$\mathbf{B}_0 \mathbf{A}_n^* + \mathbf{A}_0 \mathbf{B}_n^* = \mathbf{0}. \tag{5.95}$$

Using (1.9)

$$\mathbf{Z}(z) + \mathbf{Z}_*(z) = \mathbf{A}^{-1}(z)\mathbf{B}(z) + \mathbf{B}_*(z)\mathbf{A}_*^{-1}(z)$$
$$= \mathbf{A}^{-1}(z) \left(\mathbf{B}(z)\mathbf{A}_*(z) + \mathbf{A}(z)\mathbf{B}_*(z) \right) \mathbf{A}_*^{-1}(z)$$

or

$$\mathbf{A}(z) \left(\mathbf{Z}(z) + \mathbf{Z}_*(z) \right) \mathbf{A}_*(z) = \mathbf{B}(z)\mathbf{A}_*(z) + \mathbf{A}(z)\mathbf{B}_*(z).$$

Multiplying by z^n on both sides of the above equation and making use of the reciprocal matrix polynomials $\tilde{\mathbf{A}}(z) = z^n \mathbf{A}_*(z) = \mathbf{A}_n^* + \mathbf{A}_{n-1}^* z + \cdots + \mathbf{A}_0^* z^n$, and $\tilde{\mathbf{B}}(z) = z^n \mathbf{B}_*(z) = \mathbf{B}_n^* + \mathbf{B}_{n-1}^* z + \cdots + \mathbf{B}_0^* z^n$ there, we get

$$\mathbf{A}(z) \left(\mathbf{Z}(z) + \mathbf{Z}_*(z) \right) \tilde{\mathbf{A}}(z) = \mathbf{B}(z)\tilde{\mathbf{A}}(z) + \mathbf{A}(z)\tilde{\mathbf{B}}(z).$$

Finally, letting $z \to 0$ on both sides of the above expression, we obtain

$$\mathbf{A}_0 \left(\lim_{z \to 0} \left(\mathbf{Z}(z) + \mathbf{Z}_*(z) \right) \right) \mathbf{A}_n^* = \mathbf{B}_0 \mathbf{A}_n^* + \mathbf{A}_0 \mathbf{B}_n^*. \tag{5.96}$$

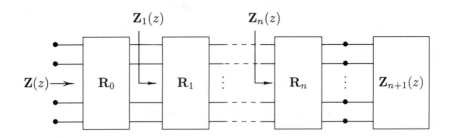

FIGURE 5.3. Multichannel cascade of equi-delay lines.

Since $\mathbf{A}_0$ is nonsingular, from (5.95), ordinary degree reduction happens in (5.90) iff

$$\left(\mathbf{Z}(z) + \mathbf{Z}_*(z)\right)\big|_{z=0}\, \mathbf{A}_n^* = \mathbf{0}\,. \tag{5.97}$$

and clearly (5.92) is sufficient in that case. However, if $\mathbf{A}_n$ is also non-singular, from (5.95)–(5.97), the above degree reduction condition reduces to

$$\mathbf{Z}(z) + \mathbf{Z}_*(z)\big|_{z=0} = \mathbf{0}\,, \tag{5.98}$$

the familiar 'even part' condition in (5.92) (see also (4.86)). Q.E.D.

When (5.97) is satisfied, using (5.94)–(5.95), (5.91) simplifies to

$$\mathbf{Z}_1(z) = \mathbf{A}_1^{-1}(z)\mathbf{B}_1(z)\,, \tag{5.99}$$

where $\delta_0(\mathbf{A}_1(z)) = n-1$ and $\delta_0(\mathbf{B}_1(z)) = n-1$.

More generally, rational or otherwise, this procedure for generating a new positive matrix function $\mathbf{Z}_1(z)$ from a given positive matrix function $\mathbf{Z}_0(z)$ by extracting a line of characteristic impedance $\mathbf{R}_0 = \mathbf{Z}(0)$ can be repeated, and continuing this process, by proceeding sequentially and pulling out lines of characteristic impedances $\mathbf{R}_0 = \mathbf{Z}(0)$, $\mathbf{R}_1 = \mathbf{Z}_1(0)$, ..., $\mathbf{R}_n = \mathbf{Z}_n(0)$, we obtain the situation shown in Fig. 5.3.

In general, with

$$\mathbf{R}_n = \mathbf{Z}_n(0) \tag{5.100}$$

and[7]

$$\mathbf{K}_n = \frac{\mathbf{R}_n + \mathbf{R}_n^*}{2} > 0\,, \tag{5.101}$$

as before

$$\mathbf{K}_n^{-1/2}\left(\mathbf{Z}_n(z) - \mathbf{R}_n\right)\left(\mathbf{Z}_n(z) + \mathbf{R}_n^*\right)^{-1}\mathbf{K}_n^{1/2}$$

---

[7]The Paley-Wiener criterion guarantees $\mathbf{K}_n > 0$. See (5.239).

is a bounded matrix function with a zero at $z = 0$. Consequently

$$\mathbf{d}_{n+1}(z) \triangleq \frac{1}{z} \mathbf{K}_n^{-1/2} \big(\mathbf{Z}_n(z) - \mathbf{R}_n\big)\big(\mathbf{Z}_n(z) + \mathbf{R}_n^*\big)^{-1} \mathbf{K}_n^{1/2} \qquad (5.102)$$

represents a bounded matrix function as in (5.90). Physically, it is related to the positive matrix function $\mathbf{Z}_{n+1}(z)$, that is obtained from $\mathbf{Z}_n(z)$ by extracting the next multiline with characteristic impedance matrix $\mathbf{R}_n$, through the relation (see (5.91))

$$\mathbf{d}_{n+1}(z) = \mathbf{K}_n^{-1/2} \big(\mathbf{Z}_{n+1}(z) - \mathbf{R}_n\big)\big(\mathbf{Z}_{n+1}(z) + \mathbf{R}_n^*\big)^{-1} \mathbf{K}_n^{1/2} \qquad (5.103)$$

for every $n \geq 0$. From (5.103) with $n + 1$ replaced by n we have

$$\mathbf{d}_n(z) = \mathbf{K}_{n-1}^{-1/2} \big(\mathbf{Z}_n(z) - \mathbf{R}_{n-1}\big)\big(\mathbf{Z}_n(z) + \mathbf{R}_{n-1}^*\big)^{-1} \mathbf{K}_{n-1}^{1/2}. \qquad (5.104)$$

More generally, since (5.76) and (5.84) constitute two equivalent representations of $\mathbf{d}(z)$, from (5.104), we also obtain

$$\mathbf{d}_n(z) = \mathbf{K}_{n-1}^{1/2} \big(\mathbf{Z}_n(z) + \mathbf{R}_{n-1}^*\big)^{-1} \big(\mathbf{Z}_n(z) - \mathbf{R}_{n-1}\big)\mathbf{K}_{n-1}^{-1/2}. \qquad (5.105)$$

Finally, solving for $\mathbf{Z}_n(z)$ using (5.104) together with (5.81)–(5.82) gives

$$\mathbf{Z}_n(z) = \mathbf{K}_{n-1}^{1/2} \big(\mathbf{I} - \mathbf{d}_n(z)\big)^{-1} \big(\mathbf{I} + \mathbf{d}_n(z)\big)\mathbf{K}_{n-1}^{1/2} + \frac{(\mathbf{R}_{n-1} - \mathbf{R}_{n-1}^*)}{2} \qquad (5.106)$$

or

$$\mathbf{Z}_n(z) = \mathbf{K}_{n-1}^{1/2} \big(\mathbf{I} + \mathbf{d}_n(z)\big)\big(\mathbf{I} - \mathbf{d}_n(z)\big)^{-1} \mathbf{K}_{n-1}^{1/2} + \frac{(\mathbf{R}_{n-1} - \mathbf{R}_{n-1}^*)}{2}. \qquad (5.107)$$

Similarly (5.102) gives

$$\mathbf{Z}_n(z) = \mathbf{K}_n^{1/2} \big(\mathbf{I} - z\mathbf{d}_{n+1}(z)\big)^{-1} \big(\mathbf{I} + z\mathbf{d}_{n+1}(z)\big)\mathbf{K}_n^{1/2} + \frac{(\mathbf{R}_n - \mathbf{R}_n^*)}{2} \qquad (5.108)$$

or

$$\mathbf{Z}_n(z) = \mathbf{K}_n^{1/2} \big(\mathbf{I} + z\mathbf{d}_{n+1}(z)\big)\big(\mathbf{I} - z\mathbf{d}_{n+1}(z)\big)^{-1} \mathbf{K}_n^{1/2} + \frac{(\mathbf{R}_n - \mathbf{R}_n^*)}{2}. \qquad (5.109)$$

Thus, in the multichannel case, two distinct representations are possible for the positive matrix function $\mathbf{Z}_n(z)$ in Fig. 5.3 as shown by the pair (5.106)–(5.107) or (5.108)–(5.109). For each pair, the first and second representations are known as the left- and right-inverse forms.

The next step is to express $\mathbf{d}_n(z)$ in terms of $\mathbf{d}_{n+1}(z)$ and to complete the iteration so that $\mathbf{d}_0(z)$ is expressible in terms of $\mathbf{d}_{n+1}(z)$. This will allow the associated power spectral density matrix to be expressed in terms of an arbitrary bounded matrix function. Even though (5.106) or (5.107) can be equated with either (5.108) or (5.109) for this purpose, because of the noncommutativity of matrices, a little investigation will show that it is best to equate (5.107) and (5.108) or (5.106) and (5.109). For reasons that will become obvious later, we will refer to these two formulations as the left-inverse form and the right-inverse form, respectively.

5.2.2   LEFT-INVERSE FORM

Equating (5.107) and (5.108), we obtain

$$\mathbf{K}_{n-1}^{1/2}\big(\mathbf{I} + \mathbf{d}_n(z)\big)\big(\mathbf{I} - \mathbf{d}_n(z)\big)^{-1}\mathbf{K}_{n-1}^{1/2} + (\mathbf{R}_{n-1} - \mathbf{R}_{n-1}^*)/2$$
$$= \mathbf{K}_n^{1/2}\big(\mathbf{I} - z\mathbf{d}_{n+1}(z)\big)^{-1}\big(\mathbf{I} + z\mathbf{d}_{n+1}(z)\big)\mathbf{K}_n^{1/2} + (\mathbf{R}_n - \mathbf{R}_n^*)/2$$

or

$$\mathbf{K}_{n-1}^{1/2}\big(\mathbf{I} + \mathbf{d}_n(z)\big) + (\mathbf{R}_{n-1} - \mathbf{R}_{n-1}^*)\mathbf{K}_{n-1}^{-1/2}\big(\mathbf{I} - \mathbf{d}_n(z)\big)/2$$
$$= \mathbf{K}_n^{1/2}\big(\mathbf{I} - z\mathbf{d}_{n+1}(z)\big)^{-1}\big(\mathbf{I} + z\mathbf{d}_{n+1}(z)\big)\mathbf{K}_n^{1/2}\mathbf{K}_{n-1}^{-1/2}\big(\mathbf{I} - \mathbf{d}_n(z)\big)$$
$$+ (\mathbf{R}_n - \mathbf{R}_n^*)\mathbf{K}_{n-1}^{-1/2}\big(\mathbf{I} - \mathbf{d}_n(z)\big)/2$$

or

$$\big(\mathbf{I} - z\mathbf{d}_{n+1}(z)\big)\mathbf{K}_n^{-1/2}\mathbf{K}_{n-1}^{1/2}\big(\mathbf{I} + \mathbf{d}_n(z)\big)$$
$$+ \big(\mathbf{I} - z\mathbf{d}_{n+1}(z)\big)\mathbf{K}_n^{-1/2}(\mathbf{R}_{n-1} - \mathbf{R}_{n-1}^*)\mathbf{K}_{n-1}^{-1/2}\big(\mathbf{I} - \mathbf{d}_n(z)\big)/2$$
$$= \big(\mathbf{I} + z\mathbf{d}_{n+1}(z)\big)\mathbf{K}_n^{1/2}\mathbf{K}_{n-1}^{-1/2}\big(\mathbf{I} - \mathbf{d}_n(z)\big)$$
$$+ \big(\mathbf{I} - z\mathbf{d}_{n+1}(z)\big)\mathbf{K}_n^{-1/2}(\mathbf{R}_n - \mathbf{R}_n^*)\mathbf{K}_{n-1}^{-1/2}\big(\mathbf{I} - \mathbf{d}_n(z)\big)/2$$

and hence a simple rearrangement gives

$$\Big\{\big(\mathbf{I} - z\mathbf{d}_{n+1}(z)\big)\mathbf{K}_n^{-1/2}\mathbf{K}_{n-1}^{1/2}$$
$$- \big(\mathbf{I} - z\mathbf{d}_{n+1}(z)\big)\mathbf{K}_n^{-1/2}(\mathbf{R}_{n-1} - \mathbf{R}_{n-1}^*)\mathbf{K}_{n-1}^{-1/2}/2$$
$$+ \big(\mathbf{I} + z\mathbf{d}_{n+1}(z)\big)\mathbf{K}_n^{1/2}\mathbf{K}_{n-1}^{-1/2}$$
$$+ \big(\mathbf{I} - z\mathbf{d}_{n+1}(z)\big)\mathbf{K}_n^{-1/2}(\mathbf{R}_n - \mathbf{R}_n^*)\mathbf{K}_{n-1}^{-1/2}/2\Big\}\mathbf{d}_n(z)$$
$$= \Big\{\big(\mathbf{I} + z\mathbf{d}_{n+1}(z)\big)\mathbf{K}_n^{1/2}\mathbf{K}_{n-1}^{-1/2}$$
$$+ \big(\mathbf{I} - z\mathbf{d}_{n+1}(z)\big)\mathbf{K}_n^{-1/2}(\mathbf{R}_n - \mathbf{R}_n^*)\mathbf{K}_{n-1}^{-1/2}/2$$
$$- \big(\mathbf{I} - z\mathbf{d}_{n+1}(z)\big)\mathbf{K}_n^{-1/2}\mathbf{K}_{n-1}^{1/2}$$
$$- \big(\mathbf{I} - z\mathbf{d}_{n+1}(z)\big)\mathbf{K}_n^{-1/2}(\mathbf{R}_{n-1} - \mathbf{R}_{n-1}^*)\mathbf{K}_{n-1}^{-1/2}/2\Big\}. \quad (5.110)$$

Equation (5.110) is of the form

$$\big(\mathbf{C}_1 + z\mathbf{d}_{n+1}(z)\mathbf{C}_2\big)\mathbf{d}_n(z) = \mathbf{C}_3 + z\mathbf{d}_{n+1}(z)\mathbf{C}_4, \quad (5.111)$$

where

$$\mathbf{C}_1 = \mathbf{K}_n^{-1/2}\mathbf{K}_{n-1}^{1/2} - \mathbf{K}_n^{-1/2}(\mathbf{R}_{n-1} - \mathbf{R}_{n-1}^*)\mathbf{K}_{n-1}^{-1/2}/2$$
$$+ \mathbf{K}_n^{1/2}\mathbf{K}_{n-1}^{-1/2} + \mathbf{K}_n^{-1/2}(\mathbf{R}_n - \mathbf{R}_n^*)\mathbf{K}_{n-1}^{-1/2}/2$$
$$= \frac{1}{2}\mathbf{K}_n^{-1/2}\big\{2\mathbf{K}_{n-1} - \mathbf{R}_{n-1} + \mathbf{R}_{n-1}^* + 2\mathbf{K}_n + \mathbf{R}_n - \mathbf{R}_n^*\big\}\mathbf{K}_{n-1}^{-1/2}$$
$$= \mathbf{K}_n^{-1/2}\big(\mathbf{R}_n + \mathbf{R}_{n-1}^*\big)\mathbf{K}_{n-1}^{-1/2} \quad (5.112)$$

$$\mathbf{C}_2 = -\mathbf{K}_n^{-1/2}\mathbf{K}_{n-1}^{1/2} + \mathbf{K}_n^{-1/2}(\mathbf{R}_{n-1} - \mathbf{R}_{n-1}^*)\mathbf{K}_{n-1}^{-1/2}/2$$
$$+\mathbf{K}_n^{1/2}\mathbf{K}_{n-1}^{-1/2} - \mathbf{K}_n^{-1/2}(\mathbf{R}_n - \mathbf{R}_n^*)\mathbf{K}_{n-1}^{-1/2}/2$$
$$= \frac{1}{2}\mathbf{K}_n^{-1/2}\left\{-2\mathbf{K}_{n-1} + \mathbf{R}_{n-1} - \mathbf{R}_{n-1}^* + 2\mathbf{K}_n - \mathbf{R}_n + \mathbf{R}_n^*\right\}\mathbf{K}_{n-1}^{-1/2}$$
$$= \mathbf{K}_n^{-1/2}\left(\mathbf{R}_n^* - \mathbf{R}_{n-1}^*\right)\mathbf{K}_{n-1}^{-1/2} \tag{5.113}$$

$$\mathbf{C}_3 = \mathbf{K}_n^{1/2}\mathbf{K}_{n-1}^{-1/2} + \mathbf{K}_n^{-1/2}(\mathbf{R}_n - \mathbf{R}_n^*)\mathbf{K}_{n-1}^{-1/2}/2$$
$$-\mathbf{K}_n^{-1/2}\mathbf{K}_{n-1}^{1/2} - \mathbf{K}_n^{-1/2}(\mathbf{R}_{n-1} - \mathbf{R}_{n-1}^*)\mathbf{K}_{n-1}^{-1/2}/2$$
$$= \frac{1}{2}\mathbf{K}_n^{-1/2}\left\{2\mathbf{K}_{n-1} + \mathbf{R}_n - \mathbf{R}_n^* - 2\mathbf{K}_{n-1} - \mathbf{R}_{n-1} + \mathbf{R}_{n-1}^*\right\}\mathbf{K}_{n-1}^{-1/2}$$
$$= \mathbf{K}_n^{-1/2}\left(\mathbf{R}_n - \mathbf{R}_{n-1}\right)\mathbf{K}_{n-1}^{-1/2} \tag{5.114}$$

and

$$\mathbf{C}_4 = \mathbf{K}_n^{1/2}\mathbf{K}_{n-1}^{-1/2} - \mathbf{K}_n^{-1/2}(\mathbf{R}_n - \mathbf{R}_n^*)\mathbf{K}_{n-1}^{-1/2}/2$$
$$+\mathbf{K}_n^{1/2}\mathbf{K}_{n-1}^{1/2} + \mathbf{K}_n^{-1/2}(\mathbf{R}_{n-1} - \mathbf{R}_{n-1}^*)\mathbf{K}_{n-1}^{-1/2}/2$$
$$= \frac{1}{2}\mathbf{K}_n^{-1/2}\left\{2\mathbf{K}_n - \mathbf{R}_n + \mathbf{R}_n^* + 2\mathbf{K}_{n-1} + \mathbf{R}_{n-1} - \mathbf{R}_{n-1}^*\right\}\mathbf{K}_{n-1}^{-1/2}$$
$$= \mathbf{K}_n^{-1/2}\left(\mathbf{R}_n^* + \mathbf{R}_{n-1}\right)\mathbf{K}_{n-1}^{-1/2}. \tag{5.115}$$

Thus, (5.110)–(5.111) reads

$$\left\{\mathbf{K}_n^{-1/2}\left(\mathbf{R}_n + \mathbf{R}_{n-1}^*\right)\mathbf{K}_{n-1}^{-1/2} + z\mathbf{d}_{n+1}(z)\mathbf{K}_n^{-1/2}\left(\mathbf{R}_n^* - \mathbf{R}_{n-1}^*\right)\mathbf{K}_{n-1}^{-1/2}\right\}\mathbf{d}_n(z)$$
$$= \left\{\mathbf{K}_n^{-1/2}\left(\mathbf{R}_n - \mathbf{R}_{n-1}\right)\mathbf{K}_{n-1}^{-1/2} + z\mathbf{d}_{n+1}(z)\mathbf{K}_n^{-1/2}\left(\mathbf{R}_n^* + \mathbf{R}_{n-1}\right)\mathbf{K}_{n-1}^{-1/2}\right\}. \tag{5.116}$$

Equation (5.116) can be further simplified by introducing the parameter

$$\mathbf{S}_n \overset{\triangle}{=} \mathbf{d}_n(0) = \mathbf{K}_{n-1}^{-1/2}\left(\mathbf{R}_n - \mathbf{R}_{n-1}\right)\left(\mathbf{R}_n + \mathbf{R}_{n-1}^*\right)^{-1}\mathbf{K}_{n-1}^{1/2} \tag{5.117}$$

where we have made use of (5.104) as well as (5.100). Alternatively from (5.105) or (5.116) with $z = 0$, we also obtain

$$\mathbf{S}_n = \mathbf{d}_n(0) = \mathbf{K}_{n-1}^{1/2}\left(\mathbf{R}_n + \mathbf{R}_{n-1}^*\right)^{-1}\left(\mathbf{R}_n - \mathbf{R}_{n-1}\right)\mathbf{K}_{n-1}^{-1/2}. \tag{5.118}$$

Note that (5.117) and (5.118) represent two equivalent representations of $\mathbf{S}_n$. Further, since $\mathbf{d}_n(z)$ is analytic in $|z| < 1$, clearly $\mathbf{S}_n$ satisfies $\mathbf{I} - \mathbf{S}_n\mathbf{S}_n^* \geq 0$. By making use of (5.117) and (5.118) on the second term of the left- and right-hand sides of (5.116) respectively, we obtain

$$\left\{\mathbf{K}_n^{-1/2}\left(\mathbf{R}_n + \mathbf{R}_{n-1}^*\right)\mathbf{K}_{n-1}^{-1/2} + z\mathbf{d}_{n+1}(z)\mathbf{K}_n^{-1/2}\left(\mathbf{R}_n^* + \mathbf{R}_{n-1}\right)\mathbf{K}_{n-1}^{-1/2}\mathbf{S}_n^*\right\}\mathbf{d}_n(z)$$
$$= \left\{\mathbf{K}_n^{-1/2}\left(\mathbf{R}_n + \mathbf{R}_{n-1}^*\right)\mathbf{K}_{n-1}^{-1/2}\mathbf{S}_n + z\mathbf{d}_{n+1}(z)\mathbf{K}_n^{-1/2}\left(\mathbf{R}_n^* + \mathbf{R}_{n-1}\right)\mathbf{K}_{n-1}^{-1/2}\right\}$$

or

$$\{\mathbf{A} + z\mathbf{d}_{n+1}(z)\mathbf{BS}_n^*\}\, \mathbf{d}_n(z) = \{\mathbf{AS}_n + z\mathbf{d}_{n+1}(z)\mathbf{B}\}$$

or

$$\mathbf{d}_n(z) = \{\mathbf{A} + z\mathbf{d}_{n+1}(z)\mathbf{BS}_n^*\}^{-1}\{\mathbf{AS}_n + z\mathbf{d}_{n+1}(z)\mathbf{B}\}\,, \tag{5.119}$$

where for later convenience, we define

$$\mathbf{A} \triangleq \frac{\mathbf{C}_1}{2} = \mathbf{K}_n^{-1/2}\left(\frac{\mathbf{R}_n + \mathbf{R}_{n-1}^*}{2}\right)\mathbf{K}_{n-1}^{-1/2} \tag{5.120}$$

and

$$\mathbf{B} \triangleq \frac{\mathbf{C}_4}{2} = \mathbf{K}_n^{-1/2}\left(\frac{\mathbf{R}_n^* + \mathbf{R}_{n-1}}{2}\right)\mathbf{K}_{n-1}^{-1/2}\,. \tag{5.121}$$

Notice that $\mathbf{A}$ and $\mathbf{B}$ are functions of n. Interestingly as Appendix 5.A shows, at every stage, $\mathbf{A}$ and $\mathbf{B}$ are related to the junction reflection coefficient matrix $\mathbf{S}_n$ through the relations[8]

$$\mathbf{A}^{-1}(\mathbf{A}^{-1})^* = \mathbf{I} - \mathbf{S}_n\mathbf{S}_n^* \tag{5.122}$$

and

$$\mathbf{B}^{-1}(\mathbf{B}^{-1})^* = \mathbf{I} - \mathbf{S}_n^*\mathbf{S}_n\,, \tag{5.123}$$

and, as shown later, through proper renormalization all parameters in (5.119) are expressible in terms of a single parameter. Notice that (5.119) represents the left-inverse form of $\mathbf{d}_n(z)$, and hence in particular with $n = 1$ in (5.119), we get

$$\mathbf{d}_1(z) = \left(\mathbf{A}_1 + z\mathbf{d}_2(z)\mathbf{B}_1\mathbf{S}_1^*\right)^{-1}\left(\mathbf{A}_1\mathbf{S}_1 + z\mathbf{d}_2(z)\mathbf{B}_1\right)\,. \tag{5.124}$$

Notice that (5.90) also can also be represented in this form, since $\mathbf{S}_0 = \mathbf{d}_0(0) = \mathbf{0}$ and hence

$$\mathbf{d}_0(z) = z\mathbf{d}_1(z) = \left(\mathbf{I} + z\mathbf{d}_1(z)\mathbf{S}_0^*\right)^{-1}\left(\mathbf{S}_0 + z\mathbf{d}_1(z)\right)\,. \tag{5.125}$$

Substituting (5.124) into (5.125) and simplifying, we obtain

$$\begin{aligned}\mathbf{d}_0(z) = &\left\{\mathbf{A}_1\left(\mathbf{I} + z\mathbf{S}_1\mathbf{S}_0^*\right) + z\mathbf{d}_2(z)\mathbf{B}_1\left(z\mathbf{S}_0^* + \mathbf{S}_1^*\right)\right\}^{-1}\\ &\times \left\{\mathbf{A}_1\left(\mathbf{S}_0 + z\mathbf{S}_1\right) + z\mathbf{d}_2(z)\mathbf{B}_1\left(\mathbf{S}_1^*\mathbf{S}_0 + z\mathbf{I}\right)\right\}\end{aligned} \tag{5.126}$$

and repeating this iteration for $(n-2)$ more steps with the help of (5.119), we get

$$\mathbf{d}_0(z) = \left\{\mathbf{G}_{n-1}(z) + z\mathbf{d}_n(z)\tilde{\mathbf{L}}_{n-1}(z)\right\}^{-1}\left\{\mathbf{H}_{n-1}(z) + z\mathbf{d}_n(z)\tilde{\mathbf{J}}_{n-1}(z)\right\}\,, \tag{5.127}$$

---

[8]As shown in (5.237)–(5.238), the invertibility of $\mathbf{A}$ and $\mathbf{B}$ follows from the Paley-Wiener criterion.

where $\mathbf{G}_{n-1}(z)$, $\mathbf{H}_{n-1}(z)$, $\mathbf{J}_{n-1}(z)$ and $\mathbf{L}_{n-1}(z)$ are matrix polynomials of ordinary degree $n - 1$, that are yet to be determined. Here $\tilde{\mathbf{L}}_{n-1}(z) \overset{\triangle}{=} z^{n-1}\mathbf{L}_{n-1}^*(1/z^*)$ and $\tilde{\mathbf{J}}_{n-1}(z) \overset{\triangle}{=} z^{n-1}\mathbf{J}_{n-1}^*(1/z^*)$ represent the matrix polynomials reciprocal to $\mathbf{L}_{n-1}(z)$ and $\mathbf{J}_{n-1}(z)$, respectively. Once again, updating (5.127) with the help of (5.119), we obtain

$$
\begin{aligned}
\mathbf{d}_0(z) &= \left(\mathbf{G}_{n-1} + z\left\{\mathbf{A} + z\mathbf{d}_{n+1}\mathbf{BS}_n^*\right\}^{-1}\left\{\mathbf{AS}_n + z\mathbf{d}_{n+1}\mathbf{B}\right\}\tilde{\mathbf{L}}_{n-1}\right)^{-1} \\
&\quad \times \left(\mathbf{H}_{n-1} + z\left\{\mathbf{A} + z\mathbf{d}_{n+1}\mathbf{BS}_n^*\right\}^{-1}\left\{\mathbf{AS}_n + z\mathbf{d}_{n+1}\mathbf{B}\right\}\tilde{\mathbf{J}}_{n-1}\right) \\
&= \left(\left\{\mathbf{A} + z\mathbf{d}_{n+1}\mathbf{BS}_n^*\right\}\mathbf{G}_{n-1} + z\left\{\mathbf{AS}_n + z\mathbf{d}_{n+1}\mathbf{B}\right\}\tilde{\mathbf{L}}_{n-1}\right)^{-1} \\
&\quad \times \left(\left\{\mathbf{A} + z\mathbf{d}_{n+1}\mathbf{BS}_n^*\right\}\mathbf{H}_{n-1} + z\left\{\mathbf{AS}_n + z\mathbf{d}_{n+1}\mathbf{B}\right\}\tilde{\mathbf{J}}_{n-1}\right) \\
&= \left(\mathbf{A}\left\{\mathbf{G}_{n-1} + z\mathbf{S}_n\tilde{\mathbf{L}}_{n-1}\right\} + z\mathbf{d}_{n+1}\mathbf{B}\left\{\mathbf{S}_n^*\mathbf{G}_{n-1} + z\tilde{\mathbf{L}}_{n-1}\right\}\right)^{-1} \\
&\quad \times \left(\mathbf{A}\left\{\mathbf{H}_{n-1} + z\mathbf{S}_n\tilde{\mathbf{J}}_{n-1}\right\} + z\mathbf{d}_{n+1}\mathbf{B}\left\{\mathbf{S}_n^*\mathbf{H}_{n-1} + z\tilde{\mathbf{J}}_{n-1}\right\}\right) \\
&\overset{\triangle}{=} \left(\mathbf{G}_n(z) + z\mathbf{d}_{n+1}(z)\tilde{\mathbf{L}}_n(z)\right)^{-1}\left(\mathbf{H}_n(z) + z\mathbf{d}_{n+1}(z)\tilde{\mathbf{J}}_n(z)\right), (5.128)
\end{aligned}
$$

the left-inverse form of $\mathbf{d}_0(z)$, where

$$\mathbf{G}_n(z) \overset{\triangle}{=} \mathbf{A}\left(\mathbf{G}_{n-1}(z) + z\mathbf{S}_n\tilde{\mathbf{L}}_{n-1}(z)\right) \tag{5.129}$$

$$\tilde{\mathbf{L}}_n(z) \overset{\triangle}{=} \mathbf{B}\left(\mathbf{S}_n^*\mathbf{G}_{n-1}(z) + z\tilde{\mathbf{L}}_{n-1}(z)\right) \tag{5.130}$$

$$\mathbf{H}_n(z) \overset{\triangle}{=} \mathbf{A}\left(\mathbf{H}_{n-1}(z) + z\mathbf{S}_n\tilde{\mathbf{J}}_{n-1}(z)\right) \tag{5.131}$$

$$\tilde{\mathbf{J}}_n(z) \overset{\triangle}{=} \mathbf{B}\left(\mathbf{S}_n^*\mathbf{H}_{n-1}(z) + z\tilde{\mathbf{J}}_{n-1}(z)\right) \tag{5.132}$$

and hence

$$\mathbf{L}_n(z) \overset{\triangle}{=} z^n\tilde{\mathbf{L}}_n^*(1/z^*) = \left(\mathbf{L}_{n-1}(z) + z\tilde{\mathbf{G}}_{n-1}(z)\mathbf{S}_n\right)\mathbf{B}^* \tag{5.133}$$

$$\mathbf{J}_n(z) \overset{\triangle}{=} z^n\tilde{\mathbf{J}}_n^*(1/z^*) = \left(\mathbf{J}_{n-1}(z) + z\tilde{\mathbf{H}}_{n-1}(z)\mathbf{S}_n\right)\mathbf{B}^*. \tag{5.134}$$

Notice that the above matrix polynomials $\mathbf{G}_n(z)$ and $\mathbf{H}_n(z)$ are the counterparts of $g_n(z)$ and $h_n(z)$ in the single-channel case. However, unlike the single-channel case, in addition to $\mathbf{G}_n(z)$, $\mathbf{H}_n(z)$ we also have the new pair $\mathbf{J}_n(z)$ and $\mathbf{L}_n(z)$. To analyze their interrelationship, note that from (5.125) the above iteration starts with

$$\mathbf{G}_0(z) = \mathbf{I}, \quad \mathbf{H}_0(z) = \mathbf{0} \tag{5.135}$$

$$\mathbf{J}_0(z) = \mathbf{I}, \quad \mathbf{L}_0(z) = \mathbf{0}. \tag{5.136}$$

Further, as shown below, these matrix polynomials are related to each other through the nested relations [25]

$$\mathbf{G}_n(z)\mathbf{G}_{n*}(z) - \mathbf{H}_n(z)\mathbf{H}_{n*}(z) = \mathbf{I} \tag{5.137}$$

$$\mathbf{J}_n(z)\mathbf{J}_{n*}(z) - \mathbf{H}_{n*}(z)\mathbf{H}_n(z) = \mathbf{I} \tag{5.138}$$

$$\mathbf{L}_n(z) = \mathbf{G}_n^{-1}(z)\mathbf{H}_n(z)\mathbf{J}_n(z) = \mathbf{G}_{n*}(z)\mathbf{H}_n(z)\mathbf{J}_{n*}^{-1}(z) \tag{5.139}$$

and finally

$$\mathbf{J}_{n*}(z)\mathbf{J}_n(z) - \mathbf{L}_{n*}(z)\mathbf{L}_n(z) = \mathbf{I}. \tag{5.140}$$

From (5.135)–(5.136), since the above relations are obviously true for $n = 0$, to establish them for an arbitrary n, it is best to proceed through induction. Thus assume (5.137)–(5.140) are true for $n = m-1$, and using (5.129) and (5.131) for $n = m$, the left-hand side of (5.137) reduces to

$$
\begin{aligned}
\mathbf{G}_m&(z)\mathbf{G}_{m*}(z) - \mathbf{H}_m(z)\mathbf{H}_{m*}(z) \\
&= \mathbf{A}\big(\mathbf{G}_{m-1} + z^m\mathbf{S}_m\mathbf{L}_{m-1*}\big)\big(\mathbf{G}_{m-1*} + z^{-m}\mathbf{L}_{m-1}\mathbf{S}_m^*\big)\mathbf{A}^* \\
&\quad -\mathbf{A}\big(\mathbf{H}_{m-1} + z^m\mathbf{S}_m\mathbf{J}_{m-1*}\big)\big(\mathbf{H}_{m-1*} + z^{-m}\mathbf{J}_{m-1}\mathbf{S}_m^*\big)\mathbf{A}^* \\
&= \mathbf{A}\big(\mathbf{G}_{m-1}\mathbf{G}_{m-1*} - \mathbf{H}_{m-1}\mathbf{H}_{m-1*}\big)\mathbf{A}^* \\
&\quad -\mathbf{A}\mathbf{S}_m\big(\mathbf{J}_{m-1*}\mathbf{J}_{m-1} - \mathbf{L}_{m-1*}\mathbf{L}_{m-1}\big)\mathbf{S}_m^*\mathbf{A}^* \\
&\quad +z^m\mathbf{A}\mathbf{S}_m\big(\mathbf{L}_{m-1*}\mathbf{G}_{m-1*} - \mathbf{J}_{m-1*}\mathbf{H}_{m-1*}\big)\mathbf{A}^* \\
&\quad +z^{-m}\mathbf{A}\big(\mathbf{G}_{m-1}\mathbf{L}_{m-1} - \mathbf{H}_{m-1}\mathbf{J}_{m-1}\big)\mathbf{S}_m^*\mathbf{A}^* \\
&= \mathbf{A}\mathbf{A}^* - \mathbf{A}\mathbf{S}_m\mathbf{S}_m^*\mathbf{A}^* = \mathbf{A}\big(\mathbf{I} - \mathbf{S}_m\mathbf{S}_m^*\big)\mathbf{A}^* \\
&= \mathbf{A}\big(\mathbf{A}^{-1}(\mathbf{A}^{-1})^*\big)\mathbf{A}^* = \mathbf{I} \tag{5.141}
\end{aligned}
$$

where we have made use of the induction on m and the last step follows from Appendix 5.A. Similarly with $n = m$, (5.138) simplifies to

$$
\begin{aligned}
\mathbf{J}_m&(z)\mathbf{J}_{m*}(z) - \mathbf{H}_{m*}(z)\mathbf{H}_m(z) \\
&= \big(\mathbf{J}_{m-1} + z^m\mathbf{H}_{m-1*}\mathbf{S}_m\big)\mathbf{B}^*\mathbf{B}\big(\mathbf{J}_{m-1*} + z^{-m}\mathbf{S}_m^*\mathbf{H}_{m-1}\big) \\
&\quad -\big(\mathbf{H}_{m-1*} + z^{-m}\mathbf{J}_{m-1}\mathbf{S}_m^*\big)\mathbf{A}^*\mathbf{A}\big(\mathbf{H}_{m-1} + z^m\mathbf{S}_m\mathbf{J}_{m-1*}\big) \\
&= \mathbf{J}_{m-1}\big(\mathbf{B}^*\mathbf{B} - \mathbf{S}_m^*\mathbf{A}^*\mathbf{A}\mathbf{S}_m\big)\mathbf{J}_{m-1*} \\
&\quad -\mathbf{H}_{m-1*}\big(\mathbf{A}^*\mathbf{A} - \mathbf{S}_m\mathbf{B}^*\mathbf{B}\mathbf{S}_m^*\big)\mathbf{H}_{m-1} \\
&\quad +z^m\mathbf{H}_{m-1*}\big(\mathbf{S}_m\mathbf{B}^*\mathbf{B} - \mathbf{A}^*\mathbf{A}\mathbf{S}_m\big)\mathbf{J}_{m-1*} \\
&\quad +z^{-m}\mathbf{J}_{m-1}\big(\mathbf{B}^*\mathbf{B}\mathbf{S}_m^* - \mathbf{S}_m^*\mathbf{A}^*\mathbf{A}\big)\mathbf{H}_{m-1}. \tag{5.142}
\end{aligned}
$$

Since

$$\mathbf{B}^*\mathbf{B} = \big(\mathbf{I} - \mathbf{S}_m^*\mathbf{S}_m\big)^{-1} \quad \text{and} \quad \mathbf{A}^*\mathbf{A} = \big(\mathbf{I} - \mathbf{S}_m\mathbf{S}_m^*\big)^{-1},$$

$$
\begin{aligned}
\mathbf{B}^*\mathbf{B}\mathbf{S}_m^* &= \big((\mathbf{S}_m^*)^{-1}(\mathbf{I} - \mathbf{S}_m^*\mathbf{S}_m)\big)^{-1} = \big((\mathbf{I} - \mathbf{S}_m\mathbf{S}_m^*)(\mathbf{S}_m^*)^{-1}\big)^{-1} \\
&= \mathbf{S}_m^*\big(\mathbf{I} - \mathbf{S}_m\mathbf{S}_m^*\big)^{-1} = \mathbf{S}_m^*\mathbf{A}^*\mathbf{A} \tag{5.143}
\end{aligned}
$$

or
$$\mathbf{S}_m \mathbf{B}^* \mathbf{B} = \mathbf{A}^* \mathbf{A} \mathbf{S}_m \tag{5.144}$$

and hence
$$\mathbf{S}_m \mathbf{B}^* \mathbf{B} \mathbf{S}_m^* = \mathbf{S}_m \mathbf{S}_m^* \mathbf{A}^* \mathbf{A} = \mathbf{A}^* \mathbf{A} \mathbf{S}_m \mathbf{S}_m^* \tag{5.145}$$

$$\mathbf{S}_m^* \mathbf{A}^* \mathbf{A} \mathbf{S}_m = \mathbf{S}_m^* \mathbf{S}_m \mathbf{B}^* \mathbf{B} = \mathbf{B}^* \mathbf{B} \mathbf{S}_m^* \mathbf{S}_m \,. \tag{5.146}$$

Thus
$$\mathbf{B}^* \mathbf{B} - \mathbf{S}_m^* \mathbf{A}^* \mathbf{A} \mathbf{S}_m = \mathbf{B}^* \mathbf{B} \big(\mathbf{I} - \mathbf{S}_m^* \mathbf{S}_m \big) = \mathbf{I} \tag{5.147}$$

and
$$\mathbf{A}^* \mathbf{A} - \mathbf{S}_m \mathbf{B}^* \mathbf{B} \mathbf{S}_m^* = \mathbf{A}^* \mathbf{A} \big(\mathbf{I} - \mathbf{S}_m \mathbf{S}_m^* \big) = \mathbf{I} \,. \tag{5.148}$$

Using (5.143)–(5.148) in (5.142), we obtain

$$\mathbf{J}_m(z) \mathbf{J}_{m*}(z) - \mathbf{H}_{m*}(z) \mathbf{H}_m(z)$$
$$= \mathbf{J}_{m-1}(z) \mathbf{J}_{m-1*}(z) - \mathbf{H}_{m-1*}(z) \mathbf{H}_{m-1}(z) = \mathbf{I} \,,$$

by induction. Similarly, to establish (5.139), it is enough to show that

$$\mathbf{G}_n(z) \mathbf{L}_n(z) = \mathbf{H}_n(z) \mathbf{J}_n(z) \tag{5.149}$$

and
$$\mathbf{L}_n(z) \mathbf{J}_{n*}(z) = \mathbf{G}_{n*}(z) \mathbf{H}_n(z) \,. \tag{5.150}$$

With $n = m$, (5.150) becomes

$$\mathbf{G}_m(z) \mathbf{L}_m(z) - \mathbf{H}_m(z) \mathbf{J}_m(z)$$
$$= \mathbf{A}(\mathbf{G}_{m-1} + z^m \mathbf{S}_m \mathbf{L}_{m-1*})(\mathbf{L}_{m-1} + z^m \mathbf{G}_{m-1*} \mathbf{S}_m) \mathbf{B}^*$$
$$- \mathbf{A}(\mathbf{H}_{m-1} + z^m \mathbf{S}_m \mathbf{J}_{m-1*})(\mathbf{J}_{m-1} + z^m \mathbf{H}_{m-1*} \mathbf{S}_m) \mathbf{B}^*$$
$$= \mathbf{A}(\mathbf{G}_{m-1} \mathbf{L}_{m-1} - \mathbf{H}_{m-1} \mathbf{J}_{m-1}) \mathbf{B}^*$$
$$+ z^{2m} \mathbf{A} \mathbf{S}_m(\mathbf{L}_{m-1*} \mathbf{G}_{m-1*} - \mathbf{J}_{m-1*} \mathbf{H}_{m-1*}) \mathbf{B}^*$$
$$+ z^m \mathbf{A}(\mathbf{G}_{m-1} \mathbf{G}_{m-1*} - \mathbf{H}_{m-1} \mathbf{H}_{m-1*}) \mathbf{S}_m \mathbf{B}^*$$
$$- z^m \mathbf{A} \mathbf{S}_m(\mathbf{J}_{m-1*} \mathbf{J}_{m-1} - \mathbf{L}_{m-1*} \mathbf{L}_{m-1}) \mathbf{B}^* = 0$$

and

$$\mathbf{L}_m(z) \mathbf{J}_{m*}(z) - \mathbf{G}_{m*} \mathbf{H}_m(z)$$
$$= (\mathbf{L}_{m-1} + z^m \mathbf{G}_{m-1*} \mathbf{S}_m) \mathbf{B}^* \mathbf{B}(\mathbf{J}_{m-1*} + z^{-m} \mathbf{S}_m^* \mathbf{H}_{m-1})$$
$$- (\mathbf{G}_{m-1} + z^{-m} \mathbf{L}_{m-1} \mathbf{S}_m^*) \mathbf{A}^* \mathbf{A}(\mathbf{H}_{m-1} + z^m \mathbf{S}_m \mathbf{J}_{m-1*})$$
$$= \mathbf{L}_{m-1}(\mathbf{B}^* \mathbf{B} - \mathbf{S}_m^* \mathbf{A}^* \mathbf{A} \mathbf{S}_m) \mathbf{J}_{m-1*}$$
$$- \mathbf{G}_{m-1*}(\mathbf{A}^* \mathbf{A} - \mathbf{S}_m \mathbf{B}^* \mathbf{B} \mathbf{S}_m) \mathbf{H}_{m-1}$$
$$+ z^m \mathbf{G}_{m-1*}(\mathbf{S}_m \mathbf{B}^* \mathbf{B} - \mathbf{A}^* \mathbf{A} \mathbf{S}_m) \mathbf{J}_{m-1*}$$
$$+ z^{-m} \mathbf{L}_{m-1}(\mathbf{B}^* \mathbf{B} \mathbf{S}_m^* - \mathbf{S}_m^* \mathbf{A}^* \mathbf{A}) \mathbf{H}_{m-1}$$
$$= \mathbf{L}_{m-1} \mathbf{J}_{m-1*} - \mathbf{G}_{m-1*} \mathbf{H}_{m-1} = 0 \,.$$

Finally with $n = m$, (5.140) becomes

$$
\begin{aligned}
\mathbf{J}_{m*}&(z)\mathbf{J}_m(z) - \mathbf{L}_{m*}(z)\mathbf{L}_m(z) \\
&= \mathbf{B}(\mathbf{J}_{m-1*} + z^{-m}\mathbf{S}_m^*\mathbf{H}_{m-1})(\mathbf{J}_{m-1} + z^m\mathbf{H}_{m-1*}\mathbf{S}_m)\mathbf{B}^* \\
&\quad -\mathbf{B}(\mathbf{L}_{m-1*} + z^{-m}\mathbf{S}_m^*\mathbf{G}_{m-1})(\mathbf{L}_{m-1} + z^m\mathbf{G}_{m-1*}\mathbf{S}_m)\mathbf{B}^* \\
&= \mathbf{B}(\mathbf{J}_{m-1*}\mathbf{J}_{m-1} - \mathbf{L}_{m-1*}\mathbf{L}_{m-1})\mathbf{B}^* \\
&\quad -\mathbf{B}\mathbf{S}_m^*(\mathbf{G}_{m-1}\mathbf{G}_{m-1*} - \mathbf{H}_{m-1}\mathbf{H}_{m-1*})\mathbf{S}_m\mathbf{B}^* \\
&\quad +z^{-m}\mathbf{B}\mathbf{S}_m^*(\mathbf{H}_{m-1}\mathbf{J}_{m-1} - \mathbf{G}_{m-1}\mathbf{L}_{m-1})\mathbf{B}^* \\
&\quad +z^m\mathbf{B}(\mathbf{J}_{m-1*}\mathbf{H}_{m-1*} - \mathbf{L}_{m-1*}\mathbf{G}_{m-1*})\mathbf{S}_m\mathbf{B}^* \\
&= \mathbf{B}(\mathbf{I} - \mathbf{S}_m^*\mathbf{S}_m)\mathbf{B}^* = \mathbf{B}(\mathbf{B}^*\mathbf{B})^{-1}\mathbf{B}^* = \mathbf{I},
\end{aligned}
$$

thus completing the proof of the above nested relations. From (5.131) and (5.135)

$$
\mathbf{H}_n(0) = \mathbf{A}\mathbf{H}_{n-1}(0) = \mathbf{H}_0(0) = \mathbf{0}
$$

and hence $\mathbf{H}_n(z)$ must be of the form

$$
\mathbf{H}_n(z) = z(\mathbf{H}_1 + \mathbf{H}_2 z + \cdots + \mathbf{H}_n z^{n-1}) \tag{5.151}
$$

and from (5.137) and (5.138), both $\mathbf{G}_n(z)$ and $\mathbf{J}_n(z)$ are at most of degree $n - 1$, i.e.,

$$
\mathbf{G}_n(z) = \mathbf{G}_0 + \mathbf{G}_1 z + \cdots + \mathbf{G}_{n-1} z^{n-1} \tag{5.152}
$$

and

$$
\mathbf{J}_n(z) = \mathbf{J}_0 + \mathbf{J}_1 z + \cdots + \mathbf{J}_{n-1} z^{n-1}. \tag{5.153}
$$

An easy induction argument on (5.129), (5.134), together with (5.137)–(5.138) also shows that

$$
det\,\mathbf{G}_n(z) \cdot det\,\mathbf{J}_n(z) \neq 0, \quad |z| \leq 1, \tag{5.154}
$$

i.e., $\mathbf{G}_n(z)$ and $\mathbf{J}_n(z)$ are both minimum-phase matrix polynomials.

Equations (5.137)–(5.140) also provide an alternate interpretation to the polynomial matrices introduced there. Referring back to (5.137)–(5.138), $\mathbf{G}_n(z)$ and $\mathbf{J}_n(z)$ are seen to be the left-Wiener factor of $\mathbf{I} + \mathbf{H}_n(z)\mathbf{H}_{n*}(z)$ and $\mathbf{I} + \mathbf{H}_{n*}(z)\mathbf{H}_n(z)$, respectively. Having determined $\mathbf{G}_n(z)$ and $\mathbf{J}_n(z)$ in this manner, $\mathbf{L}(z)$ is defined through (5.139). But the right-hand side of (5.139) is analytic in $|z| \geq 1$ and the left-hand side is analytic in $|z| \leq 1$ and hence $\mathbf{L}(z)$ is analytic in the entire finite z-plane. Thus $\mathbf{L}(z)$ is a matrix polynomial [25].

Equation (5.128) together with (5.129)–(5.134) represent the left-inverse form of $\mathbf{d}_0(z)$ and its recursion. Similarly, its right-inverse form can be obtained by starting with (5.106) and (5.109).

5.2.3 RIGHT-INVERSE FORM

Equating (5.106) and (5.109), we obtain

$$\mathbf{K}_{n-1}^{1/2}(\mathbf{I} - \mathbf{d}_n(z))^{-1}(\mathbf{I} + \mathbf{d}_n(z))\mathbf{K}_{n-1}^{1/2} + (\mathbf{R}_{n-1} - \mathbf{R}_{n-1}^*)/2$$
$$= \mathbf{K}_n^{1/2}(\mathbf{I} + z\mathbf{d}_{n+1}(z))(\mathbf{I} - z\mathbf{d}_{n+1}(z))^{-1}\mathbf{K}_n^{1/2} + (\mathbf{R}_n - \mathbf{R}_n^*)/2$$

or

$$(\mathbf{I} + \mathbf{d}_n(z))\mathbf{K}_{n-1}^{1/2} + (\mathbf{I} - \mathbf{d}_n(z))\mathbf{K}_{n-1}^{-1/2}(\mathbf{R}_{n-1} - \mathbf{R}_{n-1}^*)/2$$
$$= (\mathbf{I} - \mathbf{d}_n(z))\mathbf{K}_{n-1}^{-1/2}\mathbf{K}_n^{1/2}(\mathbf{I} + z\mathbf{d}_{n+1}(z))(\mathbf{I} - z\mathbf{d}_{n+1}(z))^{-1}\mathbf{K}_n^{1/2}$$
$$+(\mathbf{I} - \mathbf{d}_n(z))\mathbf{K}_{n-1}^{-1/2}(\mathbf{R}_n - \mathbf{R}_n^*)/2$$

or

$$(\mathbf{I} + \mathbf{d}_n(z))\mathbf{K}_{n-1}^{1/2}\mathbf{K}_n^{-1/2}(\mathbf{I} - z\mathbf{d}_{n+1}(z))$$
$$+(\mathbf{I} - \mathbf{d}_n(z))\mathbf{K}_{n-1}^{-1/2}(\mathbf{R}_{n-1} - \mathbf{R}_{n-1}^*)\mathbf{K}_n^{-1/2}(\mathbf{I} - z\mathbf{d}_{n+1}(z))/2$$
$$= (\mathbf{I} - \mathbf{d}_n(z))\mathbf{K}_{n-1}^{-1/2}\mathbf{K}_n^{1/2}(\mathbf{I} + z\mathbf{d}_{n+1}(z))$$
$$+(\mathbf{I} - \mathbf{d}_n(z))\mathbf{K}_{n-1}^{-1/2}(\mathbf{R}_n - \mathbf{R}_n^*)\mathbf{K}_n^{-1/2}(\mathbf{I} - z\mathbf{d}_{n+1}(z))/2$$

and hence

$$\mathbf{d}_n(z)\left\{\mathbf{K}_{n-1}^{1/2}\mathbf{K}_n^{-1/2}(\mathbf{I} - z\mathbf{d}_{n+1}(z))\right.$$
$$-\mathbf{K}_{n-1}^{-1/2}(\mathbf{R}_{n-1} - \mathbf{R}_{n-1}^*)\mathbf{K}_n^{-1/2}(\mathbf{I} - z\mathbf{d}_{n+1}(z))/2$$
$$+\mathbf{K}_{n-1}^{-1/2}\mathbf{K}_n^{1/2}(\mathbf{I} + z\mathbf{d}_{n+1}(z))$$
$$\left.+ \mathbf{K}_{n-1}^{-1/2}(\mathbf{R}_n - \mathbf{R}_n^*)\mathbf{K}_n^{-1/2}(\mathbf{I} - z\mathbf{d}_{n+1}(z))/2\right\}$$
$$= \left\{\mathbf{K}_{n-1}^{-1/2}\mathbf{K}_n^{1/2}(\mathbf{I} + z\mathbf{d}_{n+1}(z))\right.$$
$$+\mathbf{K}_{n-1}^{-1/2}(\mathbf{R}_n - \mathbf{R}_n^*)\mathbf{K}_n^{-1/2}(\mathbf{I} - z\mathbf{d}_{n+1}(z))/2$$
$$-\mathbf{K}_{n-1}^{1/2}\mathbf{K}_n^{-1/2}(\mathbf{I} - z\mathbf{d}_{n+1}(z))$$
$$\left.- \mathbf{K}_{n-1}^{-1/2}(\mathbf{R}_{n-1} - \mathbf{R}_{n-1}^*)\mathbf{K}_n^{-1/2}(\mathbf{I} - z\mathbf{d}_{n+1}(z))/2\right\}. \quad (5.155)$$

Equation (5.155) is of the form

$$\mathbf{d}_n(z)(\mathbf{C}_4^* + z\mathbf{C}_3^*\mathbf{d}_{n+1}(z)) = \mathbf{C}_2^* + z\mathbf{C}_1^*\mathbf{d}_{n+1}(z),$$

where $\mathbf{C}_1$, $\mathbf{C}_2$, $\mathbf{C}_3$ and $\mathbf{C}_4$ are as defined in (5.112)–(5.115). Thus, (5.155) reads

$$\mathbf{d}_n(z)\left\{\mathbf{K}_{n-1}^{-1/2}(\mathbf{R}_n + \mathbf{R}_{n-1}^*)\mathbf{K}_n^{-1/2} + z\mathbf{K}_{n-1}^{-1/2}(\mathbf{R}_n^* - \mathbf{R}_{n-1}^*)\mathbf{K}_n^{-1/2}\mathbf{d}_{n+1}(z)\right\}$$
$$= \left\{\mathbf{K}_{n-1}^{-1/2}(\mathbf{R}_n - \mathbf{R}_{n-1})\mathbf{K}_n^{-1/2} + z\mathbf{K}_{n-1}^{-1/2}(\mathbf{R}_n^* + \mathbf{R}_{n-1})\mathbf{K}_n^{-1/2}\mathbf{d}_{n+1}(z)\right\}.$$

Making use of $\mathbf{S}_n$ defined in (5.117)–(5.118), the above equation simplifies to

$$\mathbf{d}_n(z)\left\{\mathbf{K}_{n-1}^{-1/2}(\mathbf{R}_n + \mathbf{R}_{n-1}^*)\mathbf{K}_n^{-1/2}\right.$$
$$\left. + z\mathbf{S}_n^*\mathbf{K}_{n-1}^{-1/2}(\mathbf{R}_n^* + \mathbf{R}_{n-1})\mathbf{K}_n^{-1/2}\mathbf{d}_{n+1}(z)\right\}$$
$$= \left\{\mathbf{S}_n\mathbf{K}_{n-1}^{-1/2}(\mathbf{R}_n + \mathbf{R}_{n-1}^*)\mathbf{K}_n^{-1/2}\right.$$
$$\left. + z\mathbf{K}_{n-1}^{-1/2}(\mathbf{R}_n^* + \mathbf{R}_{n-1})\mathbf{K}_n^{-1/2}\mathbf{d}_{n+1}(z)\right\}$$

or

$$\mathbf{d}_n(z)\left\{\mathbf{B}^* + z\mathbf{S}_n^*\mathbf{A}^*\mathbf{d}_{n+1}(z)\right\} = \left\{\mathbf{S}_n\mathbf{B}^* + z\mathbf{A}^*\mathbf{d}_{n+1}(z)\right\}$$

or

$$\mathbf{d}_n(z) = \left\{\mathbf{S}_n\mathbf{B}^* + z\mathbf{A}^*\mathbf{d}_{n+1}(z)\right\}\left\{\mathbf{B}^* + z\mathbf{S}_n^*\mathbf{A}^*\mathbf{d}_{n+1}(z)\right\}^{-1}, \qquad (5.156)$$

where $\mathbf{A}$ and $\mathbf{B}$ are as defined in (5.120)–(5.121). Once again, proceeding as in (5.124)–(5.127), we get

$$\mathbf{d}_0(z) = \left\{\mathbf{L}_{n-1}(z) + z\tilde{\mathbf{G}}_{n-1}(z)\mathbf{d}_n(z)\right\}\left\{\mathbf{J}_{n-1}(z) + z\tilde{\mathbf{H}}_{n-1}(z)\mathbf{d}_n(z)\right\}^{-1}$$
$$(5.157)$$

where $\mathbf{G}_{n-1}(z)$, $\mathbf{H}_{n-1}(z)$, $\mathbf{J}_{n-1}(z)$ and $\mathbf{L}_{n-1}(z)$ are matrix polynomials to be determined. Updating (5.157) with the help of (5.156), we get

$$\mathbf{d}_0(z)$$
$$= \left(\mathbf{L}_{n-1} + z\tilde{\mathbf{G}}_{n-1}\left\{\mathbf{S}_n\mathbf{B}^* + z\mathbf{A}^*\mathbf{d}_{n+1}\right\}\left\{\mathbf{B}^* + z\mathbf{S}_n^*\mathbf{A}^*\mathbf{d}_{n+1}\right\}^{-1}\right)$$
$$\times \left(\mathbf{J}_{n-1} + z\tilde{\mathbf{H}}_{n-1}\left\{\mathbf{S}_n\mathbf{B}^* + z\mathbf{A}^*\mathbf{d}_{n+1}\right\}\left\{\mathbf{B}^* + z\mathbf{S}_n^*\mathbf{A}^*\mathbf{d}_{n+1}\right\}^{-1}\right)^{-1}$$
$$= \left(\mathbf{L}_{n-1}\left\{\mathbf{B}^* + z\mathbf{S}_n^*\mathbf{A}^*\mathbf{d}_{n+1}\right\} + z\tilde{\mathbf{G}}_{n-1}\left\{\mathbf{S}_n\mathbf{B}^* + z\mathbf{A}^*\mathbf{d}_{n+1}\right\}\right)$$
$$\times \left(\mathbf{J}_{n-1}\left\{\mathbf{B}^* + z\mathbf{S}_n^*\mathbf{A}^*\mathbf{d}_{n+1}\right\} + z\tilde{\mathbf{H}}_{n-1}\left\{\mathbf{S}_n\mathbf{B}^* + z\mathbf{A}^*\mathbf{d}_{n+1}\right\}\right)^{-1}$$
$$= \left(\left\{\mathbf{L}_{n-1} + z\tilde{\mathbf{G}}_{n-1}\mathbf{S}_n\right\}\mathbf{B}^* + z\left\{\mathbf{L}_{n-1}\mathbf{S}_n^* + z\tilde{\mathbf{G}}_{n-1}\right\}\mathbf{A}^*\mathbf{d}_{n+1}\right)$$
$$\times \left(\left\{\mathbf{J}_{n-1} + z\tilde{\mathbf{H}}_{n-1}\mathbf{S}_n\right\}\mathbf{B}^* + z\left\{\mathbf{J}_{n-1}\mathbf{S}_n^* + z\tilde{\mathbf{H}}_{n-1}\right\}\mathbf{A}^*\mathbf{d}_{n+1}\right)^{-1}$$
$$\overset{\triangle}{=} \left\{\mathbf{L}_n(z) + z\tilde{\mathbf{G}}_n(z)\mathbf{d}_{n+1}(z)\right\}\left\{\mathbf{J}_n(z) + z\tilde{\mathbf{H}}_n(z)\mathbf{d}_{n+1}(z)\right\}^{-1}, \quad (5.158)$$

where

$$\mathbf{G}_n(z) = \mathbf{A}\left(\mathbf{G}_{n-1}(z) + z\mathbf{S}_n\tilde{\mathbf{L}}_{n-1}(z)\right) \qquad (5.159)$$

$$\mathbf{H}_n(z) = \mathbf{A}\left(\mathbf{H}_{n-1}(z) + z\mathbf{S}_n\tilde{\mathbf{J}}_{n-1}(z)\right) \qquad (5.160)$$

$$\mathbf{J}_n(z) = \left(\mathbf{J}_{n-1}(z) + z\tilde{\mathbf{H}}_{n-1}(z)\mathbf{S}_n \right) \mathbf{B}^* \tag{5.161}$$

and

$$\mathbf{L}_n(z) = \left(\mathbf{L}_{n-1}(z) + z\tilde{\mathbf{G}}_{n-1}(z)\mathbf{S}_n \right) \mathbf{B}^* . \tag{5.162}$$

Further, since $\mathbf{S}_0 = \mathbf{0}$, from (5.90) we have

$$\mathbf{d}_0(z) = (\mathbf{S}_0 + z\mathbf{d}_1(z)) \left(\mathbf{I} + z\mathbf{S}_0^*\mathbf{d}_1(z) \right)^{-1}$$

and hence these iterations also start with

$$\mathbf{G}_0(z) = \mathbf{I}, \quad \mathbf{H}_0(z) = \mathbf{0} \tag{5.163}$$

and

$$\mathbf{J}_0(z) = \mathbf{I}, \quad \mathbf{L}_0(z) = \mathbf{0} . \tag{5.164}$$

On comparing (5.159)–(5.164) with (5.129)–(5.136), we conclude that these two sets of matrix polynomials are identical and hence the *same* set of matrix polynomials take part in both the left-inverse form and the right-inverse form for $\mathbf{d}_0(z)$.

Thus, (5.128) and (5.158) represent the left-inverse form and the right-inverse form of the input reflection coefficient matrix $\mathbf{d}_0(z)$ and hence

$$\mathbf{d}_0(z) = \mathbf{P}_1^{-1}(z)\mathbf{P}_2(z) \tag{5.165}$$

or

$$\mathbf{d}_0(z) = \mathbf{P}_3(z)\mathbf{P}_4^{-1}(z) , \tag{5.166}$$

where

$$\mathbf{P}_1(z) = \mathbf{G}_n(z) + z\mathbf{d}_{n+1}(z)\tilde{\mathbf{L}}_n(z)$$

$$\mathbf{P}_2(z) = \mathbf{H}_n(z) + z\mathbf{d}_{n+1}(z)\tilde{\mathbf{J}}_n(z)$$

$$\mathbf{P}_3(z) = \mathbf{L}_n(z) + z\tilde{\mathbf{G}}_n(z)\mathbf{d}_{n+1}(z)$$

and

$$\mathbf{P}_4(z) = \mathbf{J}_n(z) + z\tilde{\mathbf{H}}_n(z)\mathbf{d}_{n+1}(z) .$$

Since $\mathbf{R}_0 = \mathbf{R}_0^*$ and $\mathbf{K}_0 = \mathbf{R}_0$, from (5.75)–(5.76) and (5.81), the associated input positive matrix function is given by

$$\mathbf{Z}_0(z) = \mathbf{R}_0^{1/2}(\mathbf{I} - \mathbf{d}_0(z))^{-1}(\mathbf{I} + \mathbf{d}_0(z))\mathbf{R}_0^{1/2} \tag{5.167}$$

or, equivalently, from (5.82)

$$\mathbf{Z}_0(z) = \mathbf{R}_0^{1/2}(\mathbf{I} + \mathbf{d}_0(z))(\mathbf{I} - \mathbf{d}_0(z))^{-1}\mathbf{R}_0^{1/2} . \tag{5.168}$$

Using (5.165) in (5.167), we get the left-inverse form of $\mathbf{Z}_0(z)$ to be

$$
\begin{aligned}
\mathbf{Z}_0(z) &= \mathbf{R}_0^{1/2}(\mathbf{P}_1(z) - \mathbf{P}_2(z))^{-1}(\mathbf{P}_1(z) + \mathbf{P}_2(z))\mathbf{R}_0^{1/2} \\
&= \mathbf{R}_0^{1/2}\left((\mathbf{G}_n - \mathbf{H}_n) - z\mathbf{d}_{n+1}(\tilde{\mathbf{J}}_n - \tilde{\mathbf{L}}_n)\right)^{-1} \\
&\quad \times \left((\mathbf{G}_n + \mathbf{H}_n) + z\mathbf{d}_{n+1}(\tilde{\mathbf{J}}_n + \tilde{\mathbf{L}}_n)\right)\mathbf{R}_0^{1/2} \\
&= 2\left(\mathbf{A}_n(z) - z\mathbf{d}_{n+1}(z)\tilde{\mathbf{C}}_n(z)\right)^{-1}\left(\mathbf{B}_n(z) + z\mathbf{d}_{n+1}(z)\tilde{\mathbf{D}}_n(z)\right),
\end{aligned}
$$

(5.169)

where by definition

$$
\mathbf{A}_n(z) = (\mathbf{G}_n(z) - \mathbf{H}_n(z))\mathbf{R}_0^{-1/2} \tag{5.170}
$$

$$
\mathbf{B}_n(z) = \frac{1}{2}(\mathbf{G}_n(z) + \mathbf{H}_n(z))\mathbf{R}_0^{1/2} \tag{5.171}
$$

$$
\mathbf{C}_n(z) = \mathbf{R}_0^{-1/2}(\mathbf{J}_n(z) - \mathbf{L}_n(z)) \tag{5.172}
$$

and

$$
\mathbf{D}_n(z) = \frac{1}{2}\mathbf{R}_0^{1/2}(\mathbf{J}_n(z) + \mathbf{L}_n(z)). \tag{5.173}
$$

Similarly, using (5.166) in (5.168), the right-inverse form of $\mathbf{Z}_0(z)$ turns out to be

$$
\begin{aligned}
\mathbf{Z}_0(z) &= \mathbf{R}_0^{1/2}(\mathbf{P}_4(z) + \mathbf{P}_3(z))(\mathbf{P}_4(z) - \mathbf{P}_3(z))^{-1}\mathbf{R}_0^{1/2} \\
&= \mathbf{R}_0^{1/2}\left((\mathbf{J}_n + \mathbf{L}_n) + z(\tilde{\mathbf{G}}_n + \tilde{\mathbf{H}}_n)\mathbf{d}_{n+1}\right) \\
&\quad \times \left((\mathbf{J}_n - \mathbf{L}_n) - z(\tilde{\mathbf{G}}_n - \tilde{\mathbf{H}}_n)\mathbf{d}_{n+1}\right)^{-1}\mathbf{R}_0^{1/2} \\
&= 2\left(\mathbf{D}_n(z) + z\tilde{\mathbf{B}}_n(z)\mathbf{d}_{n+1}(z)\right)\left(\mathbf{C}_n(z) - z\tilde{\mathbf{A}}_n(z)\mathbf{d}_{n+1}(z)\right)^{-1}.
\end{aligned}
$$

(5.174)

Evidently (5.169) and (5.174) are two equivalent representations of the same positive matrix function $\mathbf{Z}_0(z)$. Thus

$$
\begin{aligned}
\mathbf{Z}_0(z) &= 2\left(\mathbf{A}_n(z) - z\mathbf{d}_{n+1}(z)\tilde{\mathbf{C}}_n(z)\right)^{-1}\left(\mathbf{B}_n(z) + z\mathbf{d}_{n+1}(z)\tilde{\mathbf{D}}_n(z)\right) \\
&= 2\left(\mathbf{D}_n(z) + z\tilde{\mathbf{B}}_n(z)\mathbf{d}_{n+1}(z)\right)\left(\mathbf{C}_n(z) - z\tilde{\mathbf{A}}_n(z)\mathbf{d}_{n+1}(z)\right)^{-1}
\end{aligned}
$$

(5.175)

and, alternately, (5.175) represents the input positive matrix function at the input of an $(n+1)$ multichannel cascade with junction reflection coefficient matrices $\mathbf{S}_1, \mathbf{S}_2, \ldots, \mathbf{S}_n$ (or equivalently, characteristic impedance matrices

$\mathbf{R}_0$, $\mathbf{R}_1$, ..., $\mathbf{R}_n$) that has been terminated upon an arbitrary bounded matrix function $\mathbf{d}_{n+1}(z)$ (see Fig. 5.3).

In particular, $\mathbf{d}_{n+1}(z) \equiv \mathbf{0}$ in (5.175) yields

$$\mathbf{Z}_0^{(0)}(z) = 2\mathbf{A}_n^{-1}(z)\mathbf{B}_n(z) = 2\mathbf{D}_n(z)\mathbf{C}_n^{-1}(z). \tag{5.176}$$

Thus $\mathbf{A}_n(z)$, $\mathbf{B}_n(z)$, $\mathbf{C}_n(z)$ and $\mathbf{D}_n(z)$ are minimum-phase polynomial matrices and a direct calculation shows

$$\begin{aligned}
\mathbf{Z}_0&(z) - \mathbf{Z}_0^{(0)}(z) \\
&= \mathbf{Z}_0(z) - 2\mathbf{D}_n(z)\mathbf{C}_n^{-1}(z) \\
&= 2\left(\mathbf{A}_n(z) - z\mathbf{d}_{n+1}(z)\tilde{\mathbf{C}}_n(z)\right)^{-1}\left(\mathbf{B}_n(z) + z\mathbf{d}_{n+1}(z)\tilde{\mathbf{D}}_n(z)\right) \\
&\quad -2\mathbf{D}_n(z)\mathbf{C}_n^{-1}(z) \\
&= 2\left(\mathbf{A}_n(z) - z\mathbf{d}_{n+1}(z)\tilde{\mathbf{C}}_n(z)\right)^{-1}\left\{\left(\mathbf{B}_n(z) + z\mathbf{d}_{n+1}(z)\tilde{\mathbf{D}}_n(z)\right)\mathbf{C}_n(z)\right. \\
&\quad \left. - \left(\mathbf{A}_n(z) - z\mathbf{d}_{n+1}(z)\tilde{\mathbf{C}}_n(z)\right)\mathbf{D}_n(z)\right\}\mathbf{C}_n^{-1}(z) \\
&= 2\left(\mathbf{A}_n(z) - z\mathbf{d}_{n+1}(z)\tilde{\mathbf{C}}_n(z)\right)^{-1}\left\{\left(\mathbf{B}_n(z)\mathbf{C}_n(z) - \mathbf{A}_n(z)\mathbf{D}_n(z)\right)\right. \\
&\quad \left. +z^{n+1}\mathbf{d}_{n+1}(z)\left(\mathbf{C}_{n*}(z)\mathbf{D}_n(z) + \mathbf{D}_{n*}(z)\mathbf{C}_n(z)\right)\right\}\mathbf{C}_n^{-1}(z). \tag{5.177}
\end{aligned}$$

But from (5.139)–(5.140),

$$\begin{aligned}
\mathbf{C}_{n*}&(z)\mathbf{D}_n(z) + \mathbf{D}_{n*}(z)\mathbf{C}_n(z) \\
&= \frac{1}{2}\left(\left(\mathbf{J}_{n*}(z) - \mathbf{L}_{n*}(z)\right)\left(\mathbf{J}_n(z) + \mathbf{L}_n(z)\right)\right. \\
&\quad \left. +\left(\mathbf{J}_{n*}(z) + \mathbf{L}_{n*}(z)\right)\left(\mathbf{J}_n(z) - \mathbf{L}_n(z)\right)\right) \\
&= \mathbf{J}_{n*}(z)\mathbf{J}_n(z) - \mathbf{L}_{n*}(z)\mathbf{L}_n(z) = \mathbf{I} \tag{5.178}
\end{aligned}$$

and

$$\begin{aligned}
\mathbf{B}_n&(z)\mathbf{C}_n(z) - \mathbf{A}_n(z)\mathbf{D}_n(z) \\
&= \frac{1}{2}\left(\left(\mathbf{G}_n(z) + \mathbf{H}_n(z)\right)\left(\mathbf{J}_n(z) - \mathbf{L}_n(z)\right)\right. \\
&\quad \left. -\left(\mathbf{G}_n(z) - \mathbf{H}_n(z)\right)\left(\mathbf{J}_n(z) + \mathbf{L}_n(z)\right)\right) \\
&= \mathbf{H}_n(z)\mathbf{J}_n(z) - \mathbf{G}_n(z)\mathbf{L}_n(z) = \mathbf{0} \tag{5.179}
\end{aligned}$$

and hence (5.176)–(5.177) reduces to

$$\begin{aligned}
\mathbf{Z}_0(z) - 2\mathbf{A}_n^{-1}(z)\mathbf{B}_n(z) &= \mathbf{Z}_0(z) - 2\mathbf{D}_n(z)\mathbf{C}_n^{-1}(z) \\
&= 2z^{n+1}\left(\mathbf{A}_n(z) - z\mathbf{d}_{n+1}(z)\tilde{\mathbf{C}}_n(z)\right)^{-1}\mathbf{d}_{n+1}(z)\mathbf{C}_n^{-1}(z) \\
&= 2\sum_{k=n+1}^{\infty} \mathbf{c}_k z^k = O(z^{n+1}), \quad |z| < 1, \tag{5.180}
\end{aligned}$$

i.e., the power series expansion of $\mathbf{Z}_0(z)$ and $\mathbf{Z}_0^{(0)}(z) = 2\mathbf{A}_n^{-1}(z)\mathbf{B}_n(z) = 2\mathbf{D}_n(z)\mathbf{C}_n^{-1}(z)$ agree for the first $(n+1)$ terms in $|z| < 1$. Since $\mathbf{Z}_0(z)$ contains an arbitrary bounded matrix function $\mathbf{d}_{n+1}(z)$, this implies that the first $(n+1)$ terms in the expansion of $\mathbf{Z}_0(z)$ *do not* depend on $\mathbf{d}_{n+1}(z)$, and from (5.176) and (5.180), they only depend on the pair $\mathbf{C}_n(z)$ and $\mathbf{D}_n(z)$ or equivalently on the pair $\mathbf{A}_n(z)$ and $\mathbf{B}_n(z)$. Thus

$$\mathbf{Z}_0(z) = \mathbf{r}_0 + 2\sum_{k=1}^{n}\mathbf{r}_k z^k + O(z^{n+1})$$

irrespective of the choice of the bounded matrix function $\mathbf{d}_{n+1}(z)$.

Finally, to obtain an explicit expression for the power spectral density matrices associated with (5.175), we can make use of (5.71). Since

$$\mathbf{S}(\theta) = \frac{\mathbf{Z}_0(e^{j\theta}) + \mathbf{Z}_0^*(e^{j\theta})}{2} = \left.\frac{\mathbf{Z}_0(z) + \mathbf{Z}_{0*}(z)}{2}\right|_{z=e^{j\theta}},$$

using the left-inverse form in (5.175), we get

$$\frac{1}{2}(\mathbf{Z}_0(z) + \mathbf{Z}_{0*}(z))$$

$$= (\mathbf{Q}_1^{-1}\mathbf{Q}_2 + \mathbf{Q}_{2*}\mathbf{Q}_{1*}^{-1}) = \mathbf{Q}_1^{-1}(\mathbf{Q}_2\mathbf{Q}_{1*} + \mathbf{Q}_1\mathbf{Q}_{2*})\mathbf{Q}_{1*}^{-1}$$

$$= \mathbf{Q}_1^{-1}\left\{\left(\mathbf{B}_n + z^{n+1}\mathbf{d}_{n+1}\mathbf{D}_{n*}\right)\left(\mathbf{A}_{n*} - z^{-(n+1)}\mathbf{C}_n\mathbf{d}_{n+1*}\right)\right.$$

$$\left. + \left(\mathbf{A}_n - z^{n+1}\mathbf{d}_{n+1}\mathbf{C}_{n*}\right)\left(\mathbf{B}_{n*} + z^{-(n+1)}\mathbf{D}_n\mathbf{d}_{n+1*}\right)\right\}\mathbf{Q}_{1*}^{-1}$$

$$= \mathbf{Q}_1^{-1}\left\{(\mathbf{A}_n\mathbf{B}_{n*} + \mathbf{B}_n\mathbf{A}_{n*}) + z^{n+1}\mathbf{d}_{n+1}(\mathbf{D}_{n*}\mathbf{A}_{n*} - \mathbf{C}_{n*}\mathbf{B}_{n*})\right.$$

$$+ z^{-(n+1)}\left(\mathbf{A}_n\mathbf{D}_n - \mathbf{B}_n\mathbf{C}_n\right)\mathbf{d}_{n+1*}$$

$$\left.-\mathbf{d}_{n+1}\left(\mathbf{D}_{n*}\mathbf{C}_n + \mathbf{C}_{n*}\mathbf{D}_n\right)\mathbf{d}_{n+1*}\right\}\mathbf{Q}_{1*}^{-1}, \qquad (5.181)$$

where

$$\mathbf{Q}_1(z) \triangleq \mathbf{A}_n(z) - z\mathbf{d}_{n+1}(z)\tilde{\mathbf{C}}_n(z) \qquad (5.182)$$

and

$$\mathbf{Q}_2(z) \triangleq \mathbf{B}_n(z) + z\mathbf{d}_{n+1}(z)\tilde{\mathbf{D}}_n(z).$$

But using (5.170)–(5.173), a direct calculation also gives

$$\mathbf{A}_n(z)\mathbf{B}_{n*}(z) + \mathbf{B}_n(z)\mathbf{A}_{n*}(z) = \mathbf{G}_n(z)\mathbf{G}_{n*}(z) - \mathbf{H}_n(z)\mathbf{H}_{n*}(z) = \mathbf{I}, \qquad (5.183)$$

and together with (5.178)–(5.179), equation (5.181) simplifies to

$$\frac{\mathbf{Z}_0(z) + \mathbf{Z}_{0*}(z)}{2} = \mathbf{Q}_1^{-1}(z)\left(\mathbf{I} - \mathbf{d}_{n+1}(z)\mathbf{d}_{n+1*}(z)\right)\mathbf{Q}_{1*}^{-1}(z). \qquad (5.184)$$

Similarly the right-inverse form in (5.175) gives

$$\frac{\mathbf{Z}_0(z) + \mathbf{Z}_{0*}(z)}{2} = (\mathbf{Q}_3\mathbf{Q}_4^{-1} + \mathbf{Q}_{4*}^{-1}\mathbf{Q}_{3*})$$

$$= \mathbf{Q}_{4*}^{-1}(\mathbf{Q}_{4*}\mathbf{Q}_3 + \mathbf{Q}_{3*}\mathbf{Q}_4)\mathbf{Q}_4^{-1}$$

$$= \mathbf{Q}_{4*}^{-1}(z)(\mathbf{I} - \mathbf{d}_{n+1*}(z)\mathbf{d}_{n+1}(z))\mathbf{Q}_4^{-1}(z), \quad (5.185)$$

where

$$\mathbf{Q}_3(z) \triangleq \mathbf{D}_n(z) + z\tilde{\mathbf{B}}_n(z)\mathbf{d}_{n+1}(z)$$

and

$$\mathbf{Q}_4(z) \triangleq \mathbf{C}_n(z) - z\tilde{\mathbf{A}}_n(z)\mathbf{d}_{n+1}(z). \quad (5.186)$$

Thus

$$\mathbf{S}(\theta) = \mathbf{Q}_1^{-1}(e^{j\theta})\Big(\mathbf{I} - \mathbf{d}_{n+1}(e^{j\theta})\mathbf{d}_{n+1}^*(e^{j\theta})\Big)\mathbf{Q}_1^{-1*}(e^{j\theta})$$

$$= \mathbf{Q}_4^{-1*}(e^{j\theta})\Big(\mathbf{I} - \mathbf{d}_{n+1}^*(e^{j\theta})\mathbf{d}_{n+1}(e^{j\theta})\Big)\mathbf{Q}_4^{-1}(e^{j\theta}). \quad (5.187)$$

For $\mathbf{S}(\theta)$ to satisfy the Paley-Wiener criterion, from (5.35), clearly the arbitrary bounded matrix function $\mathbf{d}_{n+1}(z)$ must satisfy

$$\frac{1}{2\pi}\int_{-\pi}^{\pi} \ln \det \left(\mathbf{I} - \mathbf{d}_{n+1}(e^{j\theta})\mathbf{d}_{n+1}^*(e^{j\theta})\right) d\theta > -\infty \quad (5.188)$$

and in that case let $\mathbf{\Gamma}_l(z)$ and $\mathbf{\Gamma}_r(z)$ represent the left- and right-Wiener factor of $\mathbf{I} - \mathbf{d}_{n+1}(e^{j\theta})\mathbf{d}_{n+1}^*(e^{j\theta})$ and $\mathbf{I} - \mathbf{d}_{n+1}^*(e^{j\theta})\mathbf{d}_{n+1}(e^{j\theta})$, respectively. Thus,

$$\mathbf{I} - \mathbf{d}_{n+1}(e^{j\theta})\mathbf{d}_{n+1}^*(e^{j\theta}) = \mathbf{\Gamma}_l(e^{j\theta})\mathbf{\Gamma}_l^*(e^{j\theta})$$

and

$$\mathbf{I} - \mathbf{d}_{n+1}^*(e^{j\theta})\mathbf{d}_{n+1}(e^{j\theta}) = \mathbf{\Gamma}_r^*(e^{j\theta})\mathbf{\Gamma}_r(e^{j\theta})$$

and (5.187) simplifies to

$$\mathbf{S}(\theta) = \Big(\mathbf{Q}_1^{-1}(e^{j\theta})\mathbf{\Gamma}_l(e^{j\theta})\Big)\Big(\mathbf{Q}_1^{-1}(e^{j\theta})\mathbf{\Gamma}_l(e^{j\theta})\Big)^* \triangleq \mathbf{B}_l(z)\mathbf{B}_l^*(z)|_{z=e^{j\theta}}$$

$$= \Big(\mathbf{\Gamma}_r(e^{j\theta})\mathbf{Q}_4^{-1}(e^{j\theta})\Big)^*\Big(\mathbf{\Gamma}_r(e^{j\theta})\mathbf{Q}_4^{-1}(e^{j\theta})\Big) \triangleq \mathbf{B}_r^*(z)\mathbf{B}_r(z)|_{z=e^{j\theta}}$$

$$= \sum_{k=-n}^{n} \mathbf{r}_k e^{jk\theta} + O(e^{j(n+1)\theta}). \quad (5.189)$$

Clearly

$$\mathbf{B}_l(z) = \mathbf{Q}_1^{-1}(z)\mathbf{\Gamma}_l(z) = \left\{\mathbf{A}_n(z) - z\mathbf{d}_{n+1}(z)\tilde{\mathbf{C}}_n(z)\right\}^{-1}\mathbf{\Gamma}_l(z) \quad (5.190)$$

and

$$\mathbf{B}_r(z) = \boldsymbol{\Gamma}_r(z)\mathbf{Q}_4^{-1}(z) = \boldsymbol{\Gamma}_r(z)\left\{\mathbf{C}_n(z) - z\tilde{\mathbf{A}}_n(z)\mathbf{d}_{n+1}(z)\right\}^{-1} \quad (5.191)$$

represent the left- and right-inverse Wiener factors (that are unique up to multiplication by a unitary constant matrix from right and left in (5.190) and (5.191), respectively) of $\mathbf{S}(\theta)$ that match the first $(n+1)$ autocorrelation matrices. It can be shown that, in the rational case, (5.190)–(5.191) exhibit the left- and right-coprime representations of the respective Wiener factors. From (5.189), (5.71) and (5.180), the first $(n + 1)$ matrix coefficients $\mathbf{r}_0$, $\mathbf{r}_1$, ..., $\mathbf{r}_n$ in the Fourier expansion of $\mathbf{S}(\theta)$ are independent of the choice of the arbitrary bounded matrix function $\mathbf{d}_{n+1}(z)$ in (5.189)–(5.191) and hence they are solely determined from $\mathbf{A}_n(z)$ and $\mathbf{C}_n(z)$. To establish their explicit interdependence, it is best to analyze the recursions for $\mathbf{A}_n(z)$, $\mathbf{B}_n(z)$, $\mathbf{C}_n(z)$ and $\mathbf{D}_n(z)$.

Using (5.170)–(5.173) together with (5.129)–(5.134), we obtain the recursion

$$\mathbf{A}_n(z) = \mathbf{A}(\mathbf{A}_{n-1}(z) - z\mathbf{S}_n\tilde{\mathbf{C}}_{n-1}(z)) \quad (5.192)$$

$$\mathbf{B}_n(z) = \mathbf{A}(\mathbf{B}_{n-1}(z) + z\mathbf{S}_n\tilde{\mathbf{D}}_{n-1}(z)) \quad (5.193)$$

$$\mathbf{C}_n(z) = (\mathbf{C}_{n-1}(z) - z\tilde{\mathbf{A}}_{n-1}(z)\mathbf{S}_n)\mathbf{B}^* \quad (5.194)$$

and

$$\mathbf{D}_n(z) = (\mathbf{D}_{n-1}(z) + z\tilde{\mathbf{B}}_{n-1}(z)\mathbf{S}_n)\mathbf{B}^* . \quad (5.195)$$

The matrices $\mathbf{A}$ and $\mathbf{B}$ required in the above updating procedure are given by (5.120)–(5.121), and they satisfy the identities (5.A.4)–(5.A.5). Although $\mathbf{A}$ and $\mathbf{B}$ are 'square root factors' of $(\mathbf{I}-\mathbf{S}_n\mathbf{S}_n^*)^{-1}$ and $(\mathbf{I}-\mathbf{S}_n^*\mathbf{S}_n)^{-1}$, respectively, their complicated forms in (5.A.1)–(5.A.2) involve the characteristic impedance matrices $\mathbf{R}_{n-1}$, $\mathbf{R}_n$, and these forms are not particularly convenient for easy evaluation. However, this can be rectified by making use of the extra freedom present in the factorizations (5.A.4)–(5.A.5) in terms of unitary matrices. As shown in Appendix 5.B, by employing a specific sequence of unitary matrices, a new set of bounded matrix functions can be defined, whereby the equivalent $\mathbf{A}$ and $\mathbf{B}$ can be chosen to be lower triangular matrices with positive diagonal entries, and consequently they can be immediately obtained by performing the above 'square root' factorizations.

5.2.4   Multichannel Spectral Extension — Youla's Representation

To be specific, given $\mathbf{A}_i$, $\mathbf{B}_i$, $i \geq 1$, as in (5.A.1)–(5.A.2), let $(\mathbf{U}_1, \mathbf{V}_1)$, $(\mathbf{U}_2, \mathbf{V}_2)$, ..., $(\mathbf{U}_n, \mathbf{V}_n)$ represent the unique set of unitary matrix pairs generated sequentially through the lower triangular Schur factorization

$$\mathbf{A}_1^{-1} = \mathbf{M}_1^*\mathbf{U}_1 , \quad \mathbf{B}_1^{-1} = \mathbf{N}_1^*\mathbf{V}_1$$

$$\mathbf{U}_1 \mathbf{A}_2^{-1} \mathbf{U}_1^* = \mathbf{M}_2^* \mathbf{U}_2, \quad \mathbf{V}_1 \mathbf{B}_2^{-1} \mathbf{V}_1^* = \mathbf{N}_2^* \mathbf{V}_2$$

$$\vdots \tag{5.196}$$

$$\mathbf{U}_{n-1} \cdots \mathbf{U}_2 \mathbf{U}_1 \mathbf{A}_n^{-1} \mathbf{U}_1^* \mathbf{U}_2^* \cdots \mathbf{U}_{n-1}^* = \mathbf{M}_n^* \mathbf{U}_n$$

$$\mathbf{V}_{n-1} \cdots \mathbf{V}_2 \mathbf{V}_1 \mathbf{B}_n^{-1} \mathbf{V}_1^* \mathbf{V}_2^* \cdots \mathbf{V}_{n-1}^* = \mathbf{N}_n^* \mathbf{V}_n \,.$$

Here $\mathbf{M}_i$, $\mathbf{N}_i$, $i \geq 1$, are unique lower triangular matrices with positive diagonal entries. Define a new set of bounded matrix functions

$$\boldsymbol{\rho}_0(z) = \mathbf{d}_0(z), \quad \boldsymbol{\rho}_1(z) = \mathbf{d}_1(z),$$

$$\boldsymbol{\rho}_{n+1}(z) \overset{\triangle}{=} \mathbf{U}_n \mathbf{U}_{n-1} \cdots \mathbf{U}_1 \mathbf{d}_{n+1}(z) \mathbf{V}_1^* \mathbf{V}_2^* \cdots \mathbf{V}_n^*, \quad n \geq 1, \tag{5.197}$$

with $\mathbf{d}_n(z)$ as in (5.119) or (5.156). As shown in Appendix 5.B, these modified bounded matrix functions satisfy the Schur recursions

$$\boldsymbol{\rho}_n(z) = \left(\mathbf{I} + z\mathbf{M}_n^* \boldsymbol{\rho}_{n+1}(z) \mathbf{N}_n^{*-1} \mathbf{s}_n^*\right)^{-1} \left(\mathbf{s}_n + z\mathbf{M}_n^* \boldsymbol{\rho}_{n+1}(z) \mathbf{N}_n^{*-1}\right) \tag{5.198}$$

or, equivalently,

$$\boldsymbol{\rho}_n(z) = \left(\mathbf{s}_n + z\mathbf{M}_n^{-1} \boldsymbol{\rho}_{n+1}(z) \mathbf{N}_n\right) \left(\mathbf{I} + z\mathbf{s}_n^* \mathbf{M}_n^{-1} \boldsymbol{\rho}_{n+1}(z) \mathbf{N}_n\right)^{-1}, \tag{5.199}$$

where

$$\mathbf{s}_n \overset{\triangle}{=} \boldsymbol{\rho}_n(0) = \mathbf{U}_{n-1} \cdots \mathbf{U}_2 \mathbf{U}_1 \mathbf{S}_n \mathbf{V}_1^* \mathbf{V}_2^* \cdots \mathbf{V}_{n-1}^* \,.$$

Further, the above lower triangular matrices $\mathbf{M}_n$ and $\mathbf{N}_n$ with positive diagonal entries satisfy the identities

$$\mathbf{M}_n^* \mathbf{M}_n = \mathbf{I} - \mathbf{s}_n \mathbf{s}_n^* > 0, \quad \mathbf{N}_n^* \mathbf{N}_n = \mathbf{I} - \mathbf{s}_n^* \mathbf{s}_n > 0, \tag{5.200}$$

and hence given $\mathbf{s}_n$, they can be computed directly from (5.200). In that case, (5.192)–(5.195) simplifies into the standard form [25]

$$\mathbf{M}_n^* \mathbf{A}_n(z) = \mathbf{A}_{n-1}(z) - z\mathbf{s}_n \tilde{\mathbf{C}}_{n-1}(z) \tag{5.201}$$

$$\mathbf{M}_n^* \mathbf{B}_n(z) = \mathbf{B}_{n-1}(z) + z\mathbf{s}_n \tilde{\mathbf{D}}_{n-1}(z) \tag{5.202}$$

$$\mathbf{C}_n(z) \mathbf{N}_n = \mathbf{C}_{n-1}(z) - z\tilde{\mathbf{A}}_{n-1}(z) \mathbf{s}_n \tag{5.203}$$

and

$$\mathbf{D}_n(z) \mathbf{N}_n = \mathbf{D}_{n-1}(z) + z\tilde{\mathbf{B}}_{n-1}(z) \mathbf{s}_n \,. \tag{5.204}$$

From (5.135)–(5.136) and (5.170)–(5.173), these recursions start with

$$\mathbf{A}_0(z) = \mathbf{R}_0^{-1/2}, \quad \mathbf{B}_0(z) = \frac{1}{2} \mathbf{R}_0^{1/2} \tag{5.205}$$

$$\mathbf{C}_0(z) = \mathbf{R}_0^{-1/2}, \quad \mathbf{D}_0(z) = \frac{1}{2} \mathbf{R}_0^{1/2} \,. \tag{5.206}$$

Clearly, under the above renormalization, the input positive matrix function in (5.175) takes the form (replace $\mathbf{d}_{n+1}(z)$ by $\boldsymbol{\rho}_{n+1}(z)$)

$$\mathbf{Z}_0(z) = 2\left(\mathbf{A}_n(z) - z\boldsymbol{\rho}_{n+1}(z)\tilde{\mathbf{C}}_n(z)\right)^{-1}\left(\mathbf{B}_n(z) + z\boldsymbol{\rho}_{n+1}(z)\tilde{\mathbf{D}}_n(z)\right)$$

$$= 2\left(\mathbf{D}_n(z) + z\tilde{\mathbf{B}}_n(z)\boldsymbol{\rho}_{n+1}(z)\right)\left(\mathbf{C}_n(z) - z\tilde{\mathbf{A}}_n(z)\boldsymbol{\rho}_{n+1}(z)\right)^{-1}$$

$$= \mathbf{r}_0 + 2\sum_{k=1}^{n}\mathbf{r}_k z^k + O(z^{n+1}), \tag{5.207}$$

where $\mathbf{A}_n(z)$, $\mathbf{B}_n(z)$, $\mathbf{C}_n(z)$ and $\mathbf{D}_n(z)$ are as given in (5.201)–(5.206). Similarly, the class of all multichannel spectral extension formula in (5.189)–(5.191) reads [25]

$$\mathbf{S}(\theta) = \mathbf{B}_l(z)\mathbf{B}_l^*(z)|_{z=e^{j\theta}} = \mathbf{B}_r^*(z)\mathbf{B}_r(z)|_{z=e^{j\theta}}$$

$$= \sum_{k=-n}^{n}\mathbf{r}_k e^{jk\theta} + O(e^{j(n+1)\theta}) \geq 0, \tag{5.208}$$

where the left- and right-coprime Wiener factors are given by

$$\mathbf{B}_l(z) = \left(\mathbf{A}_n(z) - z\boldsymbol{\rho}_{n+1}(z)\tilde{\mathbf{C}}_n(z)\right)^{-1}\boldsymbol{\Gamma}_l(z) \tag{5.209}$$

and

$$\mathbf{B}_r(z) = \boldsymbol{\Gamma}_r(z)\left(\mathbf{C}_n(z) - z\tilde{\mathbf{A}}_n(z)\boldsymbol{\rho}_{n+1}(z)\right)^{-1}, \tag{5.210}$$

respectively. Here $\boldsymbol{\Gamma}_l(z)$ and $\boldsymbol{\Gamma}_r(z)$ represent the left- and right-Wiener factors in the factorization

$$\mathbf{I} - \boldsymbol{\rho}_{n+1}(e^{j\theta})\boldsymbol{\rho}_{n+1}^*(e^{j\theta}) = \boldsymbol{\Gamma}_l(e^{j\theta})\boldsymbol{\Gamma}_l^*(e^{j\theta}) \tag{5.211}$$

and

$$\mathbf{I} - \boldsymbol{\rho}_{n+1}^*(e^{j\theta})\boldsymbol{\rho}_{n+1}(e^{j\theta}) = \boldsymbol{\Gamma}_r^*(e^{j\theta})\boldsymbol{\Gamma}_r(e^{j\theta}), \tag{5.212}$$

and they are guaranteed to exist under the condition

$$\frac{1}{2\pi}\int_{-\pi}^{\pi} \ln \det\left(\mathbf{I} - \boldsymbol{\rho}_{n+1}(e^{j\theta})\boldsymbol{\rho}_{n+1}^*(e^{j\theta})\right) d\theta > -\infty. \tag{5.213}$$

The above Wiener factors are unique up to multiplication by an arbitrary unitary matrix from right in (5.209) and left in (5.210), and $\mathbf{S}(\theta)$ in (5.208) matches the first $(n+1)$ autocorrelation matrices for any $\boldsymbol{\rho}_{n+1}(z)$.

Reflection Coefficient Matrices

Equations (5.201)–(5.206) represent the Levinson recursion algorithm in the multichannel case. Notice that these matrix polynomials can be computed recursively once their initialization parameter

$$\mathbf{R}_0 = \mathbf{Z}(0) = \mathbf{r}_0 = \frac{1}{2\pi}\int_{-\pi}^{\pi}\mathbf{S}(\theta)d\theta = \mathbf{r}_0^* = \mathbf{R}_0^* \tag{5.214}$$

and the reflection coefficient matrices s_1, s_2, ..., s_n are known.

To complete this discussion, it is necessary to obtain the relationship between the reflection coefficient matrices s_1, s_2, ..., s_n and the given autocorrelation matrices r_0, r_1, ..., r_n. This can be easily derived from (5.207) with $\rho_{n+1}(z) \equiv 0$. In that case (see (5.207))

$$Z_0^{(0)}(z) = r_0 + 2\sum_{k=1}^{n} r_k z^k + O(z^{n+1}) = 2A_n^{-1}(z)B_n(z) = 2D_n(z)C_n^{-1}(z).$$

Thus using (5.178) and (5.183), we get

$$\frac{Z_0^{(0)}(z) + Z_{0*}^{(0)}(z)}{2} = \left(r_0 + \sum_{k=1}^{n} r_k z^k + \sum_{k=1}^{n} r_k^* z^{-k} + O(z^{\pm(n+1)})\right)$$

$$= A_n^{-1}(z)A_{n*}^{-1}(z) = C_{n*}^{-1}(z)C_n^{-1}(z), \qquad (5.215)$$

which can be rewritten as

$$\left(r_0 + \sum_{k=1}^{n} r_k z^k + \sum_{k=1}^{n} r_k^* z^{-k} + O(z^{\pm(n+1)})\right)\tilde{A}_n(z) = z^n A_n^{-1}(z) \qquad (5.216)$$

and

$$\tilde{C}_n(z)\left(r_0 + \sum_{k=1}^{n} r_k z^k + \sum_{k=1}^{n} r_k^* z^{-k} + O(z^{\pm(n+1)})\right) = z^n C_n^{-1}(z). \qquad (5.217)$$

Let

$$A_n(z) = A_0 + A_1 z + \cdots + A_n z^n \qquad (5.218)$$

and

$$C_n(z) = C_0 + C_1 z + \cdots + C_n z^n \qquad (5.219)$$

represent the minimum-phase matrix polynomials $A_n(z)$ and $C_n(z)$. Thus $A_n^{-1}(z)$ and $C_n^{-1}(z)$ can be expanded in power series form in $|z| < 1$, and comparing coefficients of z^n, z^{n-1}, ..., z, 1 on both sides of (5.216) and (5.217), we obtain

$$r_0 A_0^* + r_1 A_1^* + \cdots + r_n A_n^* = A_0^{-1} \qquad (5.220)$$
$$r_1^* A_0^* + r_0 A_1^* + \cdots + r_{n-1} A_n^* = 0 \qquad (5.221)$$

$$\vdots \quad \vdots$$

$$r_n^* A_0^* + r_{n-1}^* A_1^* + \cdots + r_0 A_n^* = 0 \qquad (5.222)$$

and

$$C_0^* r_0 + C_1^* r_1 + \cdots + C_n^* r_n = C_0^{-1} \qquad (5.223)$$
$$C_0^* r_1^* + C_1^* r_0 + \cdots + C_n^* r_{n-1} = 0 \qquad (5.224)$$

$$\vdots \quad \vdots$$

$$C_0^* r_n^* + C_1^* r_{n-1}^* + \cdots + C_n^* r_0 = 0. \qquad (5.225)$$

From (5.222),

$$\mathbf{A}_n \mathbf{r}_0 = -[\mathbf{A}_0, \ \mathbf{A}_1, \ \ldots, \ \mathbf{A}_{n-1}] \begin{bmatrix} \mathbf{r}_n \\ \mathbf{r}_{n-1} \\ \vdots \\ \mathbf{r}_1 \end{bmatrix}. \tag{5.226}$$

To make further progress, we can make use of the recursion in (5.201) for $\mathbf{A}_n(z)$ by letting

$$\mathbf{A}_{n-1}(z) = \mathbf{X}_0 + \mathbf{X}_1 z + \cdots + \mathbf{X}_{n-1} z^{n-1}$$

and

$$\mathbf{C}_{n-1}(z) = \mathbf{Y}_0 + \mathbf{Y}_1 z + \cdots + \mathbf{Y}_{n-1} z^{n-1}.$$

Thus (5.201) gives

$$\mathbf{M}_n^*[\mathbf{A}_0, \ \mathbf{A}_1, \ \ldots, \ \mathbf{A}_{n-1}] =$$

$$[\mathbf{X}_0, \ \mathbf{X}_1, \ \ldots, \ \mathbf{X}_{n-1}] - \mathbf{s}_n[\mathbf{0}, \ \mathbf{Y}_{n-1}^*, \ \mathbf{Y}_{n-2}^*, \ \ldots, \ \mathbf{Y}_1^*] \tag{5.227}$$

and

$$\mathbf{M}_n^* \mathbf{A}_n = -\mathbf{s}_n \mathbf{Y}_0^*. \tag{5.228}$$

With (5.227) in (5.226), we get

$$\mathbf{A}_n \mathbf{r}_0 = -(\mathbf{M}_n^*)^{-1} \left\{ [\mathbf{X}_0, \ \mathbf{X}_1, \ \ldots, \ \mathbf{X}_{n-1}] \begin{bmatrix} \mathbf{r}_n \\ \mathbf{r}_{n-1} \\ \vdots \\ \mathbf{r}_1 \end{bmatrix} \right.$$

$$\left. -\mathbf{s}_n[\mathbf{Y}_{n-1}^*, \ \mathbf{Y}_{n-2}^*, \ \ldots, \ \mathbf{Y}_1^*] \begin{bmatrix} \mathbf{r}_{n-1} \\ \mathbf{r}_{n-2} \\ \vdots \\ \mathbf{r}_1 \end{bmatrix} \right\}$$

or

$$\mathbf{M}_n^* \mathbf{A}_n \mathbf{r}_0 = - \left(\sum_{k=0}^{n-1} \mathbf{X}_k \mathbf{r}_{n-k} \right) + \mathbf{s}_n \left(\sum_{k=1}^{n-1} \mathbf{Y}_{n-k}^* \mathbf{r}_{n-k} \right). \tag{5.229}$$

But from (5.223) with n replaced by $(n-1)$, we get

$$\sum_{k=1}^{n-1} \mathbf{Y}_{n-k}^* \mathbf{r}_{n-k} = \mathbf{Y}_0^{-1} - \mathbf{Y}_0^* \mathbf{r}_0$$

and using this in (5.229), we have

$$\mathbf{M}_n^* \mathbf{A}_n \mathbf{r}_0 = - \sum_{k=0}^{n-1} \mathbf{X}_k \mathbf{r}_{n-k} + \mathbf{s}_n \left\{ \mathbf{Y}_0^{-1} - \mathbf{Y}_0^* \mathbf{r}_0 \right\}$$

$$= - \sum_{k=0}^{n-1} \mathbf{X}_k \mathbf{r}_{n-k} + \mathbf{s}_n \mathbf{Y}_0^{-1} - \mathbf{s}_n \mathbf{Y}_0^* \mathbf{r}_0 . \qquad (5.230)$$

Finally, with the help of (5.228), (5.230) simplifies into the classic expression [25]

$$\mathbf{s}_n = \left(\sum_{k=0}^{n-1} \mathbf{X}_k \mathbf{r}_{n-k} \right) \mathbf{Y}_0 = \left\{ \mathbf{A}_{n-1}(z) \sum_{k=1}^{n} \mathbf{r}_k z^k \right\}_n \mathbf{C}_{n-1}(0) , \qquad (5.231)$$

where $\{ \ \}_n$ represents, as before, the coefficient of z^n in $\{ \ \}$. Notice that the reflection coefficient matrix $\mathbf{s}_n$ has been expressed in terms of the coefficients of the Levinson polynomial matrices at the previous stage along with $\mathbf{r}_0$, $\mathbf{r}_1$, ..., $\mathbf{r}_n$, and (5.231) can be computed quite efficiently. Finally, (5.220)–(5.222) can be expressed in matrix form as

$$[\, \mathbf{A}_0, \ \mathbf{A}_1, \ \ldots, \ \mathbf{A}_n \,] \begin{bmatrix} \mathbf{r}_0 & \mathbf{r}_1 & \cdots & \mathbf{r}_n \\ \mathbf{r}_1^* & \mathbf{r}_0 & \cdots & \mathbf{r}_{n-1} \\ \vdots & \vdots & \ddots & \vdots \\ \mathbf{r}_n^* & \mathbf{r}_{n-1}^* & \cdots & \mathbf{r}_0 \end{bmatrix} = [\, (\mathbf{A}_0^*)^{-1}, \ \mathbf{0}, \ \ldots, \ \mathbf{0} \,]$$

$$(5.232)$$

and since $\mathbf{r}_k^* = \mathbf{r}_{-k}$, similarly (5.223)–(5.225) can be expressed as

$$\begin{bmatrix} \mathbf{r}_0 & \mathbf{r}_{-1} & \cdots & \mathbf{r}_{-n} \\ \mathbf{r}_1 & \mathbf{r}_0 & \cdots & \mathbf{r}_{-(n-1)} \\ \vdots & \vdots & \ddots & \vdots \\ \mathbf{r}_n & \mathbf{r}_{n-1} & \cdots & \mathbf{r}_0 \end{bmatrix} \begin{bmatrix} \mathbf{C}_0 \\ \mathbf{C}_1 \\ \vdots \\ \mathbf{C}_n \end{bmatrix} = \begin{bmatrix} (\mathbf{C}_0^*)^{-1} \\ \mathbf{0} \\ \vdots \\ \mathbf{0} \end{bmatrix} . \qquad (5.233)$$

It is easy to show that (5.232) has exactly the same form as the equation representing the best one-step predictor that makes use of n data samples from its left (forward predictor) and the complex conjugate transpose of the unknowns in (5.233) represents the best one-step predictor that makes use of n data samples from its right (backward predictor). As a result $\mathbf{A}_n(z)$ and $\mathbf{C}_n(z)$ are respectively known as the left (forward) and right (backward) Levinson polynomial matrices of the first kind. Similarly $\mathbf{B}_n(z)$ and $\mathbf{D}_n(z)$ are known as the left and right Levinson polynomial matrices of the second kind, respectively. From (5.232), to obtain the left polynomial matrix $\mathbf{A}_n(z)$, the Hermitian block Toeplitz matrix $\mathbf{T}_n$ is involved, whereas to obtain the right polynomial matrix $\mathbf{C}_n(z)$, the Hermitian block Toeplitz matrix $\mathbf{T}_{-n}$ given by

$$\mathbf{T}_{-n} \triangleq \begin{bmatrix} \mathbf{r}_0 & \mathbf{r}_{-1} & \cdots & \mathbf{r}_{-n} \\ \mathbf{r}_1 & \mathbf{r}_0 & \cdots & \mathbf{r}_{-(n-1)} \\ \vdots & \vdots & \ddots & \vdots \\ \mathbf{r}_n & \mathbf{r}_{n-1} & \cdots & \mathbf{r}_0 \end{bmatrix} = \mathbf{T}^*_{-n} \tag{5.234}$$

is involved. Obviously, these matrices are quite distinct in the multichannel case, and they represent the true difference between the single-channel and the multichannel situations.[9]

Equation (5.232) together with the Paley-Wiener criterion in (5.35)–(5.36) can be used to demonstrate the positivity of $\mathbf{M}_n^*\mathbf{M}_n$, $\mathbf{N}_n^*\mathbf{N}_n$ and $\mathbf{K}_n$. To see this, from (5.232), we have

$$[\mathbf{A}_0^{(n)}, \mathbf{A}_1^{(n)}, \ldots, \mathbf{A}_n^{(n)}] = [(\mathbf{A}_0^{(n)*})^{-1}, \mathbf{0}, \ldots, \mathbf{0}]\mathbf{T}_n^{-1} = (\mathbf{A}_0^{(n)*})^{-1}\mathbf{T}_n^{11},$$

where (use footnote 10, chapter 2)

$$(\mathbf{T}_n^{11})^{-1} = \mathbf{r}_0 - [\mathbf{r}_1, \mathbf{r}_2, \ldots, \mathbf{r}_n]\mathbf{T}_{n-1}^{-1} \begin{bmatrix} \mathbf{r}_1^* \\ \mathbf{r}_2^* \\ \vdots \\ \mathbf{r}_n^* \end{bmatrix}$$

and $\mathbf{T}_n^{11} = (\mathbf{T}_n^{-1})_{1,1}$ represents the $(1,1)$ entry of $\mathbf{T}_n^{-1}$. From the Paley-Wiener criterion $\mathbf{T}_n > 0$, $n = 0 \to \infty$, and hence $\mathbf{T}_n^{11} > 0$, which gives

$$\mathbf{A}_0^{(n)*}\mathbf{A}_0^{(n)} = \mathbf{T}_n^{11} > 0, \quad n = 0 \to \infty. \tag{5.235}$$

Similarly from (5.233),

$$\mathbf{C}_0^{(n)}\mathbf{C}_0^{(n)*} = \mathbf{T}_{-n}^{11} > 0, \quad n = 0 \to \infty, \tag{5.236}$$

where $(\mathbf{T}_{-n}^{11})^{-1}$ represents the $(1,1)$ entry of $\mathbf{T}_{-n}^{-1}$. Comparing the constant terms in (5.201), (5.203), we also have

$$\mathbf{M}_n^*\mathbf{A}_0^{(n)} = \mathbf{A}_0^{(n-1)}, \quad \mathbf{C}_0^{(n)}\mathbf{N}_n = \mathbf{C}_0^{(n-1)}$$

and hence

$$\mathbf{M}_n^*\mathbf{M}_n = \mathbf{I} - \mathbf{s}_n\mathbf{s}_n^* = \mathbf{A}_0^{(n-1)}\left(\mathbf{A}_0^{(n)*}\mathbf{A}_0^{(n)}\right)^{-1}\mathbf{A}_0^{(n-1)*} > 0 \tag{5.237}$$

---

[9]In the single-channel case, $\mathbf{T}_{-n}$ is the complex conjugate of $\mathbf{T}_n$ and hence from (5.233)–(5.234) with $m = 1$, $c_i = a_i$, $i = 0 \to n$, or $C_n(z) = A_n(z)$, $n = 0 \to \infty$. (c_i, $i = 1 \to n$, represent the complex conjugate of the best one-step backward predictor coefficients. See also Problem 2.2.)

$$\mathbf{N}_n^* \mathbf{N}_n = \mathbf{I} - \mathbf{s}_n^* \mathbf{s}_n = \mathbf{C}_0^{(n-1)*} \left(\mathbf{C}_0^{(n)} \mathbf{C}_0^{(n)*} \right)^{-1} \mathbf{C}_0^{(n-1)} > 0 . \qquad (5.238)$$

Since $\mathbf{K}_0 = \mathbf{r}_0 > 0$, to prove the positivity of $\mathbf{K}_n$, it is best to proceed using induction. Thus assume $\mathbf{K}_{n-1} > 0$ and rewrite (5.A.7) as

$$\mathbf{K}_n = \frac{\mathbf{R}_n + \mathbf{R}_n^*}{2} = (\mathbf{R}_n + \mathbf{R}_{n-1}^*) \mathbf{K}_{n-1}^{-1/2} \mathbf{M}_n^* \mathbf{M}_n \mathbf{K}_{n-1}^{-1/2} (\mathbf{R}_n^* + \mathbf{R}_{n-1}) > 0$$
$$(5.239)$$

where the last inequality follows from (5.237) and $\mathbf{K}_{n-1} > 0$. Thus, the Paley-Wiener criterion guarantees the invertibility of $\mathbf{M}_n$, $\mathbf{N}_n$ and $\mathbf{K}_n$.

Maximum Entropy Extension

It is interesting to note that as in the single-channel case $\rho_{n+1}(z) \equiv \mathbf{0}$ in (5.209)–(5.210) corresponds to the maximum entropy extension. To see this, from (5.35) the entropy expression gives

$$\mathcal{H} = \frac{1}{2\pi} \int_{-\pi}^{\pi} \ln \det \mathbf{S}(\theta) d\theta = \ln \det \mathbf{B}(0) \mathbf{B}^*(0) = \ln |\det \mathbf{B}(0)|^2, \qquad (5.240)$$

where $\mathbf{B}(0)$ represents the constant term in the Wiener factor of $\mathbf{S}(\theta)$. Using (5.209)–(5.212) in (5.240), we get

$$\begin{aligned}
\mathcal{H} &= \ln \det \left(\mathbf{A}_n^{-1}(0) \boldsymbol{\Gamma}_l(0) \boldsymbol{\Gamma}_l^*(0) \mathbf{A}_n^{*-1}(0) \right) \\
&= \ln \det \left(\boldsymbol{\Gamma}_r(0) \mathbf{C}_n(0) \mathbf{C}_n^{*-1}(0) \boldsymbol{\Gamma}_r^*(0) \right) \\
&= -\ln |\det \mathbf{A}_n(0)|^2 - \ln |1/\det \boldsymbol{\Gamma}_l(0)|^2 \\
&= -\ln |\det \mathbf{C}_n(0)|^2 - \ln |1/\det \boldsymbol{\Gamma}_r(0)|^2 . \qquad (5.241)
\end{aligned}$$

Since $\mathbf{A}_n(0)$ and $\mathbf{C}_n(0)$ depend only on $\mathbf{r}_0 \to \mathbf{r}_n$, clearly $\mathcal{H}$ is maximized if the second term in (5.241) goes to zero, and thus entropy is maximized if

$$|\det \boldsymbol{\Gamma}_l(0)|^2 = |\det \boldsymbol{\Gamma}_r(0)|^2 = 1 .$$

Since $\boldsymbol{\Gamma}_l(z)$ and $\boldsymbol{\Gamma}_r(z)$ represent bounded matrix functions, letting

$$\boldsymbol{\Gamma}_l(z) = \boldsymbol{\Gamma}_0 + \boldsymbol{\Gamma}_1 z + \cdots + \boldsymbol{\Gamma}_k z^k + \cdots , \quad |z| < 1 , \qquad (5.242)$$

and using (5.P.2), we have

$$\mathbf{I} - \sum_{i=0}^{k} \boldsymbol{\Gamma}_i \boldsymbol{\Gamma}_i^* \geq 0 , \quad k = 0 \to \infty , \qquad (5.243)$$

which for $k = 0$ gives

$$\mathbf{I}_m - \boldsymbol{\Gamma}_0 \boldsymbol{\Gamma}_0^* \geq 0 .$$

Let $\lambda_1, \lambda_2, \ldots, \lambda_m$ denote the eigenvalues of the Hermitian nonnegative definite matrix $\boldsymbol{\Gamma}_0 \boldsymbol{\Gamma}_0^*$. Then every λ_i, $i = 1 \to m$, satisfies $0 \leq \lambda_i \leq 1$ and since $(\boldsymbol{\Gamma}_l(0) = \boldsymbol{\Gamma}_0)$

$$|\det \boldsymbol{\Gamma}_l(0)|^2 = |\det \boldsymbol{\Gamma}_0|^2 = \det \boldsymbol{\Gamma}_0 \boldsymbol{\Gamma}_0^* = \prod_{i=1}^{m} \lambda_i = 1 ,$$

we have $\lambda_i = 1$, $i = 1 \rightarrow m$, and hence $\boldsymbol{\Gamma}_0 \boldsymbol{\Gamma}_0^* = \mathbf{I}$ or $\boldsymbol{\Gamma}_0$ represents a unitary matrix. Returning back to (5.243), this gives

$$\sum_{i=1}^{k} \boldsymbol{\Gamma}_i \boldsymbol{\Gamma}_i^* \equiv 0, \quad k = 1 \rightarrow \infty,$$

implying $\boldsymbol{\Gamma}_i \equiv \mathbf{0}$, $i = 1 \rightarrow \infty$, and hence from (5.242) we have $\boldsymbol{\Gamma}_l(z) = \boldsymbol{\Gamma}_0 = \mathbf{U}$, a unitary matrix. Similarly, $\boldsymbol{\Gamma}_r(z) = \mathbf{V}$, another unitary matrix and from (5.211)–(5.212), we get

$$\boldsymbol{\rho}_{n+1}(e^{j\theta})\boldsymbol{\rho}_{n+1}^*(e^{j\theta}) \equiv \mathbf{0}, \quad -\pi \le \theta \le \pi,$$

or $\boldsymbol{\rho}_{n+1}(z) \equiv \mathbf{0}$. But from (5.197) and (5.103), $\boldsymbol{\rho}_{n+1}(z) \equiv \mathbf{0}$ implies $\mathbf{Z}_{n+1}(z) = \mathbf{R}_n$. Thus the maximum entropy extension corresponds to terminating the $(n+1)^{st}$ line in Fig. 5.3 on its characteristic impedance matrix $\mathbf{R}_n$, and the maximum value of the entropy is given by

$$\mathcal{H}_{ME} = -\ln|\det \mathbf{A}_n(0)|^2 = -\ln|\det \mathbf{C}_n(0)|^2. \tag{5.244}$$

With $\boldsymbol{\rho}_{n+1}(z) \equiv \mathbf{0}$ in (5.208)–(5.210), the maximum entropy spectral extension has the form

$$\mathbf{S}_{ME}(\theta) = \mathbf{A}_n^{-1}(e^{j\theta})\mathbf{A}_n^{-1*}(e^{j\theta}) = \mathbf{C}_n^{-1*}(e^{j\theta})\mathbf{C}_n^{-1}(e^{j\theta}) \tag{5.245}$$

and the first $(n+1)$ autocorrelation matrices $\mathbf{r}_0, \mathbf{r}_1, \ldots, \mathbf{r}_n$ on both sides match exactly. Thus in the limit as $n \rightarrow \infty$,

$$\lim_{n \to \infty} \mathbf{A}_n(z) \stackrel{\triangle}{=} \mathbf{P}(z) \tag{5.246}$$

and

$$\lim_{n \to \infty} \mathbf{C}_n(z) \stackrel{\triangle}{=} \mathbf{Q}(z) \tag{5.247}$$

represent the inverses of the left- and right-Wiener factors of $\mathbf{S}(\theta)$, since

$$\mathbf{S}(\theta) = \mathbf{P}^{-1}(e^{j\theta})\mathbf{P}^{-1*}(e^{j\theta}) = \mathbf{Q}^{-1*}(e^{j\theta})\mathbf{Q}^{-1}(e^{j\theta}).$$

Thus, for example, the left-Wiener factor $\mathbf{B}(z)$ of $\mathbf{S}(\theta)$ can be easily obtained from the coefficients of $\mathbf{P}(z)$ since up to multiplication by a unitary matrix on the right,

$$\mathbf{B}(z) = \mathbf{P}^{-1}(z)$$

or

$$\mathbf{P}(z)\mathbf{B}(z) = \mathbf{I}$$

or

$$\begin{bmatrix} \mathbf{P}_0 & \mathbf{0} & \cdots & \mathbf{0} \\ \mathbf{P}_1 & \mathbf{P}_0 & \cdots & \mathbf{0} \\ \vdots & \vdots & \ddots & \vdots \\ \mathbf{P}_k & \mathbf{P}_{k-1} & \cdots & \mathbf{P}_0 \end{bmatrix} \begin{bmatrix} \mathbf{B}_0 & \mathbf{0} & \cdots & \mathbf{0} \\ \mathbf{B}_1 & \mathbf{B}_0 & \cdots & \mathbf{0} \\ \vdots & \vdots & \ddots & \vdots \\ \mathbf{B}_k & \mathbf{B}_{k-1} & \cdots & \mathbf{B}_0 \end{bmatrix} = \begin{bmatrix} \mathbf{I} & \mathbf{0} & \cdots & \mathbf{0} \\ \mathbf{0} & \mathbf{I} & \cdots & \mathbf{0} \\ \vdots & \vdots & \ddots & \vdots \\ \mathbf{0} & \mathbf{0} & \cdots & \mathbf{I} \end{bmatrix}. \tag{5.248}$$

Hence

$$\mathbf{B}_0 = \mathbf{P}_0^{-1} \tag{5.249}$$

and

$$\mathbf{B}_k = \mathbf{B}_0\left(\mathbf{I} - \sum_{i=0}^{k-1} \mathbf{P}_i\mathbf{B}_{k-i}\right), \quad k \geq 1. \tag{5.250}$$

It may be remarked that if $\mathbf{S}(\theta)$ is given (instead of $\mathbf{r}_0$, $\mathbf{r}_1$, ..., $\mathbf{r}_n$, ...), its Wiener factor can be obtained directly from the Levinson polynomial matrices associated with $\mathbf{S}^{-1}(\theta)$, provided it is integrable and satisfies the Paley-Wiener criterion. The above analysis shows that the limiting form of the left Levinson polynomial matrix $\mathbf{A}_n(z)$ for $\mathbf{S}^{-1}(\theta)$ becomes the right-Wiener factor for $\mathbf{S}(\theta)$ and vice versa. Conversely, $\mathbf{P}(z)$ and $\mathbf{Q}(z)$ in (5.246)–(5.247) represent the left- and right-whitening filters for $\mathbf{S}(\theta)$.

We conclude this section by summarizing the main result in this section regarding the multichannel spectral extension in the following theorem.

Theorem 5.1: Let $\mathbf{T}_n$ represent the Hermitian block Toeplitz matrix

$$\mathbf{T}_n = \begin{bmatrix} \mathbf{r}_0 & \mathbf{r}_1 & \cdots & \mathbf{r}_n \\ \mathbf{r}_1^* & \mathbf{r}_0 & \cdots & \mathbf{r}_{n-1} \\ \vdots & \vdots & \ddots & \vdots \\ \mathbf{r}_n^* & \mathbf{r}_{n-1}^* & \cdots & \mathbf{r}_0 \end{bmatrix} \tag{5.251}$$

generated from the given autocorrelation matrices $\mathbf{r}_0$, $\mathbf{r}_1$, ..., $\mathbf{r}_n$ and Δ_n represent the determinant of $\mathbf{T}_n$. Then every Δ_n, $k = 0 \to n$ is positive.

1. The left and right Levinson polynomial matrices of the first kind $\mathbf{A}_n(z)$ and $\mathbf{C}_n(z)$ given by (5.232)–(5.233) are minimum-phase and satisfy the recursions

$$\mathbf{M}_n^*\mathbf{A}_n(z) = \mathbf{A}_{n-1}(z) - z\mathbf{s}_n\tilde{\mathbf{C}}_{n-1}(z) \tag{5.252}$$

$$\mathbf{C}_n(z)\mathbf{N}_n = \mathbf{C}_{n-1}(z) - z\tilde{\mathbf{A}}_{n-1}(z)\mathbf{s}_n \tag{5.253}$$

that start with $\mathbf{A}_0(z) = \mathbf{C}_0(z) = \mathbf{r}_0^{-1/2}$. The junction reflection coefficient matrices $\mathbf{s}_n$ can be recursively computed from the relation

$$\mathbf{s}_n = \left\{\mathbf{A}_{n-1}(z)\sum_{k=1}^{n}\mathbf{r}_k z^k\right\}_n \mathbf{C}_{n-1}(0), \quad n \geq 1, \tag{5.254}$$

and, moreover, $\mathbf{M}_n$ and $\mathbf{N}_n$ are the unique lower triangular matrices with positive diagonal entries that result from the factorization

$$\mathbf{M}_n^*\mathbf{M}_n = \mathbf{I} - \mathbf{s}_n\mathbf{s}_n^* > 0 \tag{5.255}$$

and

$$\mathbf{N}_n^*\mathbf{N}_n = \mathbf{I} - \mathbf{s}_n^*\mathbf{s}_n > 0. \tag{5.256}$$

2. For every $n \geq 0$, the class of all spectral extensions that interpolates the given autocorrelation matrices $\mathbf{r}_0, \mathbf{r}_1, \ldots, \mathbf{r}_n$ is given by

$$\mathbf{S}(\theta) = \mathbf{B}_l(z)\mathbf{B}_l^*(z)|_{z=e^{j\theta}} = \mathbf{B}_r^*(z)\mathbf{B}_r(z)|_{z=e^{j\theta}} \geq 0$$

$$= \sum_{k=-n}^{n} \mathbf{r}_k e^{jk\theta} + \text{higher-order terms}, \tag{5.257}$$

where $\mathbf{B}_l(z)$ and $\mathbf{B}_r(z)$ represent respectively the left- and right-Wiener factors

$$\mathbf{B}_l(z) = \left(\mathbf{A}_n(z) - z\boldsymbol{\rho}_{n+1}(z)\tilde{\mathbf{C}}_n(z)\right)^{-1} \boldsymbol{\Gamma}_l(z) \tag{5.258}$$

and

$$\mathbf{B}_r(z) = \boldsymbol{\Gamma}_r(z)\left(\mathbf{C}_n(z) - z\tilde{\mathbf{A}}_n(z)\boldsymbol{\rho}_{n+1}(z)\right)^{-1}. \tag{5.259}$$

Here $\boldsymbol{\rho}_{n+1}(z)$ is an arbitrary bounded matrix function that represents the termination after $n+1$ equi-delay multiline extractions and $\boldsymbol{\Gamma}_l(z)$ and $\boldsymbol{\Gamma}_r(z)$ are the left- and right-Wiener factors resulting from the factorizations

$$\mathbf{I} - \boldsymbol{\rho}_{n+1}(e^{j\theta})\boldsymbol{\rho}_{n+1}^*(e^{j\theta}) = \boldsymbol{\Gamma}_l(e^{j\theta})\boldsymbol{\Gamma}_l^*(e^{j\theta}) \tag{5.260}$$

and

$$\mathbf{I} - \boldsymbol{\rho}_{n+1}^*(e^{j\theta})\boldsymbol{\rho}_{n+1}(e^{j\theta}) = \boldsymbol{\Gamma}_r^*(e^{j\theta})\boldsymbol{\Gamma}_r(e^{j\theta}) \tag{5.261}$$

that are guaranteed to exist under the physical-realizability criterion

$$\frac{1}{2\pi} \int_{-\pi}^{\pi} \ln \det \left(\mathbf{I} - \boldsymbol{\rho}_{n+1}(e^{j\theta})\boldsymbol{\rho}_{n+1}^*(e^{j\theta})\right) d\theta > -\infty. \tag{5.262}$$

5.3   Multichannel Rational System Identification

Given any integrable power spectral density matrix

$$\mathbf{S}(\theta) = \sum_{k=-\infty}^{\infty} \mathbf{r}_k e^{jk\theta} \geq 0 \quad.$$

that satisfies the causality requirement, from (5.34)–(5.35), there exist unique functions

$$\mathbf{H}(z) = \sum_{k=0}^{\infty} \mathbf{H}_k z^k$$

and

$$\mathbf{B}(z) = \sum_{k=0}^{\infty} \mathbf{B}_k z^k$$

that are analytic in $|z| < 1$ such that

$$\det \mathbf{H}(z) \neq 0, \quad \det \mathbf{B}(z) \neq 0, \quad |z| < 1,$$

and
$$\mathbf{S}(\theta) = \mathbf{H}(e^{j\theta})\mathbf{H}^*(e^{j\theta}) = \mathbf{B}^*(e^{j\theta})\mathbf{B}(e^{j\theta}).$$

$\mathbf{H}(z)$ and $\mathbf{B}(z)$ are known as the left- and right-Wiener factors of $\mathbf{S}(\theta)$ and these minimum-phase factors are unique up to multiplication from the right and left, respectively, by arbitrary constant unitary matrices. Moreover, from (5.28)–(5.29) together with Schur's theorem (5.68)–(5.69), it follows that associated with every such spectral density matrix $\mathbf{S}(\theta)$, there exists a unique positive matrix function

$$\mathbf{Z}(z) = \mathbf{r}_0 + 2\sum_{k=1}^{\infty} \mathbf{r}_k z^k, \quad |z| < 1, \tag{5.263}$$

such that

$$\mathbf{S}(\theta) = \frac{\mathbf{Z}(e^{j\theta}) + \mathbf{Z}^*(e^{j\theta})}{2} = \mathbf{H}(e^{j\theta})\mathbf{H}^*(e^{j\theta}) = \mathbf{B}^*(e^{j\theta})\mathbf{B}(e^{j\theta}). \tag{5.264}$$

Since $\mathbf{Z}_*(z)|_{z=e^{j\theta}} = \mathbf{Z}^*(e^{j\theta})$, the all z-extension of (5.264) reads

$$\frac{\mathbf{Z}(z) + \mathbf{Z}_*(z)}{2} = \mathbf{H}(z)\mathbf{H}_*(z) = \mathbf{B}_*(z)\mathbf{B}(z). \tag{5.265}$$

The representation in (5.265) shows the one-to-one correspondence between positive matrix functions and minimum-phase matrix transfer functions. In particular, the transfer function $\mathbf{H}(z)$ (or $\mathbf{B}(z)$) is rational iff $\mathbf{Z}(z)$ is a rational positive matrix function.

To prove this, we can make use of either the left-coprime representation or the right-coprime representation of rational matrices. To proceed further, assume $\mathbf{Z}(z)$ is a rational matrix function with left-coprime order p as defined in (5.23). Then

$$\mathbf{Z}(z) = \mathbf{A}^{-1}(z)\mathbf{B}(z), \tag{5.266}$$

where

$$\mathbf{A}(z) = \mathbf{A}_0 + \mathbf{A}_1 z + \cdots + \mathbf{A}_p z^p$$
$$\mathbf{B}(z) = \mathbf{B}_0 + \mathbf{B}_1 z + \cdots + \mathbf{B}_p z^p$$

and

$$\frac{\mathbf{Z}(z) + \mathbf{Z}_*(z)}{2} = \frac{1}{2}\mathbf{A}^{-1}(z)\Big(\mathbf{B}(z)\mathbf{A}_*(z) + \mathbf{A}(z)\mathbf{B}_*(z)\Big)\mathbf{A}_*^{-1}(z).$$

Since $\mathbf{Z}(e^{j\theta}) + \mathbf{Z}^*(e^{j\theta}) \geq 0$, we have $\mathbf{B}(z)\mathbf{A}_*(z) + \mathbf{A}(z)\mathbf{B}_*(z)$ is nonnegative on the unit circle and hence it can be factored into the form $2\boldsymbol{\beta}(z)\boldsymbol{\beta}_*(z)$, where $\boldsymbol{\beta}(z)$ is a minimum-phase matrix polynomial of the form

$$\boldsymbol{\beta}(z) = \boldsymbol{\beta}_0 + \boldsymbol{\beta}_1 z + \cdots + \boldsymbol{\beta}_q z^q,$$

where $q \leq p$. Thus

$$\mathbf{B}(z)\mathbf{A}_*(z) + \mathbf{A}(z)\mathbf{B}_*(z) = 2\boldsymbol{\beta}(z)\boldsymbol{\beta}_*(z) \,.$$

Thus

$$\frac{\mathbf{Z}(z) + \mathbf{Z}_*(z)}{2} = \mathbf{A}^{-1}(z)\boldsymbol{\beta}(z)\boldsymbol{\beta}_*(z)\mathbf{A}_*^{-1}(z) = \mathbf{H}(z)\mathbf{H}_*(z) \,, \qquad (5.267)$$

where

$$\mathbf{H}(z) = \mathbf{A}^{-1}(z)\boldsymbol{\beta}(z)$$

is minimum-phase. Conversely, let $\mathbf{H}(z)$ represent a rational minimum-phase matrix transfer function as in (5.43) with left-coprime representation

$$\mathbf{H}(z) = \mathbf{H}_l(z) = \mathbf{P}^{-1}(z)\mathbf{Q}(z) \,, \qquad (5.268)$$

where

$$\mathbf{P}(z) = \mathbf{I} + \mathbf{P}_1 z + \cdots + \mathbf{P}_p z^p \qquad (5.269)$$

$$\mathbf{Q}(z) = \mathbf{Q}_0 + \mathbf{Q}_1 z + \cdots + \mathbf{Q}_q z^q \,, \quad p \geq q \,. \qquad (5.270)$$

Letting

$$\mathbf{H}(e^{j\theta})\mathbf{H}^*(e^{j\theta}) = \sum_{k=-\infty}^{\infty} \mathbf{r}_k e^{jk\theta} \,,$$

we have

$$\mathbf{Z}(z) = \mathbf{r}_0 + 2\sum_{k=1}^{\infty} \mathbf{r}_k z^k \,, \qquad (5.271)$$

a positive matrix function. From (5.50) these $\mathbf{r}_k$'s satisfy the relation

$$\mathbf{r}_k = -\sum_{i=1}^{p} \mathbf{P}_i \mathbf{r}_{k-i} \,, \quad k \geq q + 1 \,,$$

and since $p \geq q$, substituting this into (5.271), we obtain

$$\mathbf{Z}(z) = \mathbf{r}_0 + 2\sum_{k=1}^{p} \mathbf{r}_k z^k + 2\sum_{k=p+1}^{\infty} \left(-\sum_{i=1}^{p} \mathbf{P}_i \mathbf{r}_{k-i}\right) z^k$$

$$= \mathbf{r}_0 + 2\sum_{k=1}^{p} \mathbf{r}_k z^k - 2\sum_{i=1}^{p} \mathbf{P}_i z^i \sum_{k=p+1}^{\infty} \mathbf{r}_{k-i} z^{k-i} \,. \qquad (5.272)$$

But

$$2\sum_{k=p+1}^{\infty} \mathbf{r}_{k-i} z^{k-i} = 2\sum_{m=p-i+1}^{\infty} \mathbf{r}_m z^m = \mathbf{Z}(z) - \mathbf{r}_0 - 2\sum_{k=1}^{p-i} \mathbf{r}_k z^k \qquad (5.273)$$

and with (5.273) in (5.272), we get

$$\left(\sum_{i=0}^{p} \mathbf{P}_i z^i\right) \mathbf{Z}(z) = \mathbf{r}_0 + \sum_{i=1}^{p} \left(\mathbf{P}_i \mathbf{r}_0 + 2\mathbf{r}_i\right) z^i + 2 \sum_{i=1}^{p} \mathbf{P}_i z^i \sum_{k=1}^{p-i} \mathbf{r}_k z^k$$

$$= \mathbf{r}_0 + \left(\mathbf{P}_1 \mathbf{r}_0 + 2\mathbf{r}_1\right) z + \sum_{i=2}^{p} \left(\mathbf{P}_i \mathbf{r}_0 + 2\mathbf{r}_i + 2\sum_{k=1}^{i-1} \mathbf{P}_k \mathbf{r}_{i-k}\right) z^i$$

$$\stackrel{\triangle}{=} \mathbf{B}(z)$$

or

$$\mathbf{Z}(z) = \mathbf{P}^{-1}(z)\mathbf{B}(z), \tag{5.274}$$

a rational positive matrix function of left-coprime order p.

5.3.1   MARMA SYSTEM CHARACTERIZATION

The Schur algorithm described in section 5.2.1 (see Fig. 5.3) can be used to generate new positive matrix functions $\mathbf{Z}_1(z)$, $\mathbf{Z}_2(z)$, ... that are of no greater complexity (in terms of their order) than the above $\mathbf{Z}(z)$. In fact, proceeding as in (5.93)–(5.99), the new positive matrix function $\mathbf{Z}_1(z)$ so generated is rational, and from (5.97) order reduction happens $(\delta_o(\mathbf{Z}_1(z)) < \delta_o(\mathbf{Z}(z)))$ iff

$$\left(\mathbf{Z}(z) + \mathbf{Z}_*(z)\right)|_{z=0} \mathbf{P}_n^* = \mathbf{0}. \tag{5.275}$$

To investigate the above order reduction condition in the case of rational systems, we can make use of (5.265), (5.268) and (5.275). In that case, $\delta_o(\mathbf{Z}(z)) = p$ and

$$\frac{\mathbf{Z}(z) + \mathbf{Z}_*(z)}{2} = \mathbf{H}(z)\mathbf{H}_*(z) = \mathbf{P}^{-1}(z)\mathbf{Q}(z)\mathbf{Q}_*(z)\mathbf{P}_*^{-1}(z)$$

$$= z^{p-q}\mathbf{P}^{-1}(z)\mathbf{Q}(z)\tilde{\mathbf{Q}}(z)\tilde{\mathbf{P}}^{-1}(z),$$

where

$$\tilde{\mathbf{Q}}(z) \stackrel{\triangle}{=} z^q \mathbf{Q}_*(z) = \mathbf{Q}_q^* + \mathbf{Q}_{q-1}^* z + \cdots + \mathbf{Q}_0^* z^q$$

and

$$\tilde{\mathbf{P}}(z) \stackrel{\triangle}{=} z^p \mathbf{P}_*(z) = \mathbf{P}_p^* + \mathbf{P}_{p-1}^* z + \cdots + \mathbf{P}_1^* z^{p-1} + \mathbf{I} z^p$$

represent the respective reciprocal matrix polynomials. Thus

$$\left(\mathbf{Z}(z) + \mathbf{Z}_*(z)\right)\tilde{\mathbf{P}}(z) = z^{p-q} \cdot 2\mathbf{P}^{-1}(z)\mathbf{Q}(z)\tilde{\mathbf{Q}}(z). \tag{5.276}$$

Letting $z = 0$ in the above expression we get

$$\left(\mathbf{Z}(z) + \mathbf{Z}_*(z)\right)|_{z=0} \mathbf{P}_n^* = \left(\lim_{z\to 0} z^{p-q}\right) 2\mathbf{P}_0^{-1}\mathbf{Q}_0\mathbf{Q}_q^*. \tag{5.277}$$

Since $\mathbf{P}_0^{-1}\mathbf{Q}_0\mathbf{Q}_q^* \neq \mathbf{0}$ (provided $\mathbf{Q}_0\mathbf{Q}_q^* \neq \mathbf{0}$), $z = 0$ is a zero of $(\mathbf{Z}(z) + \mathbf{Z}_*(z))\mathbf{P}_n^*$ of exact order $p - q$, and hence from (5.275)

$$\delta_o(\mathbf{Z}_1(z)) = p - 1 \quad \text{iff} \quad p - q > 0, \tag{5.278}$$

i.e., if $p > q$, the new rational positive matrix function $\mathbf{Z}_1(z)$ is exactly of one order less than that of $\mathbf{Z}(z)$. In this case, we obtain the interesting configuration shown in Fig. 5.2, where $\mathbf{Z}(z)$ is realized as the input impedance of an ideal delay multiline of characteristic impedance $\mathbf{R}_0 = \mathbf{Z}(0)$ closed on the positive matrix function $\mathbf{Z}_1(z)$ whose order is one less than that of $\mathbf{Z}(z)$. Thus $\delta_o(\mathbf{Z}_1(z)) = p - 1$, and this process of order reduction can be repeated if $p - q \geq 2$. In fact from (5.88)–(5.91) and with $\mathbf{Z}_0(z) \equiv \mathbf{Z}(z)$,

$$\begin{aligned}
\mathbf{d}_0(z) &= \mathbf{K}_0^{-1/2}\,(\mathbf{Z}_0(z) - \mathbf{R}_0)\,(\mathbf{Z}_0(z) + \mathbf{R}_0^*)^{-1}\,\mathbf{K}_0^{1/2} \\
&= z\mathbf{d}_1(z) = z\mathbf{K}_0^{1/2}\,(\mathbf{Z}_1(z) + \mathbf{R}_0^*)^{-1}\,(\mathbf{Z}_1(z) - \mathbf{R}_0)\,\mathbf{K}_0^{-1/2} \quad (5.279)
\end{aligned}$$

or

$$(\mathbf{Z}_0(z) - \mathbf{R}_0)\,(\mathbf{Z}_0(z) + \mathbf{R}_0^*)^{-1} = z\mathbf{K}_0\,(\mathbf{Z}_1(z) + \mathbf{R}_0^*)^{-1}\,(\mathbf{Z}_1(z) - \mathbf{R}_0)\,\mathbf{K}_0^{-1}$$
$$\overset{\triangle}{=} \mathbf{D}(z) \tag{5.280}$$

and hence

$$\mathbf{Z}_0(z) = (\mathbf{I} - \mathbf{D}(z))^{-1}\,(\mathbf{R}_0 + \mathbf{D}(z)\mathbf{R}_0^*)\ . \tag{5.281}$$

But using (5.280),

$$\mathbf{I} - \mathbf{D}(z) = \mathbf{K}_0\,(\mathbf{Z}_1(z) + \mathbf{R}_0^*)^{-1}\,\mathbf{M}(z)\mathbf{K}_0^{-1},$$

where

$$\mathbf{M}(z) = \mathbf{R}_0^* + z\mathbf{R}_0 + (1 - z)\mathbf{Z}_1(z)$$

and hence (5.281) simplifies to

$$\mathbf{Z}_0(z) = \mathbf{K}_0\mathbf{M}^{-1}(z)\Big(\mathbf{Z}_1(z)\mathbf{K}_0^{-1}\,(\mathbf{R}_0 + z\mathbf{R}_0^*) + \mathbf{R}_0^*\mathbf{K}_0^{-1}\mathbf{R}_0 - z\mathbf{R}_0\mathbf{K}_0^{-1}\mathbf{R}_0^*\Big)$$

$$\overset{\triangle}{=} \mathbf{K}_0\mathbf{M}^{-1}(z)\mathbf{E}_1(z)\ .$$

This gives

$$\mathbf{Z}_{0*}(z) = \Big((\mathbf{R}_0 + z\mathbf{R}_0^*)\,\mathbf{K}_0^{-1}\mathbf{Z}_{1*}(z) + z\mathbf{R}_0^*\mathbf{K}_0^{-1}\mathbf{R}_0 - \mathbf{R}_0\mathbf{K}_0^{-1}\mathbf{R}_0^*\Big)\mathbf{N}^{-1}(z)\mathbf{K}_0\ ,$$

$$\overset{\triangle}{=} \mathbf{E}_2(z)\mathbf{N}^{-1}(z)\mathbf{K}_0\ ,$$

where

$$\mathbf{N}(z) \overset{\triangle}{=} \mathbf{R}_0^* + z\mathbf{R}_0 - (1 - z)\mathbf{Z}_{1*}(z)\ .$$

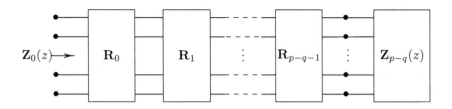

FIGURE 5.4. A multichannel ARMA(p, q) system $\mathbf{H}(z)$ and its irreducible representation. Here $\mathbf{Z}_0(z) + \mathbf{Z}_{0*}(z) = 2\mathbf{H}(z)\mathbf{H}_*(z)$.

As a result,

$$\frac{\mathbf{Z}_0(z) + \mathbf{Z}_{0*}(z)}{2} = \frac{1}{2}\mathbf{K}_0\mathbf{M}^{-1}(z)\Big(\mathbf{E}_1(z)\mathbf{K}_0^{-1}\mathbf{N}(z)$$
$$+\mathbf{M}(z)\mathbf{K}_0^{-1}\mathbf{E}_2(z)\Big)\mathbf{N}^{-1}(z)\mathbf{K}_0,$$

and after a series of straightforward manipulations, we obtain

$$\mathbf{E}_1(z)\mathbf{K}_0^{-1}\mathbf{N}(z) + \mathbf{M}(z)\mathbf{K}_0^{-1}\mathbf{E}_2(z) = 4z\left(\mathbf{Z}_1(z) + \mathbf{Z}_{1*}(z)\right),$$

which gives

$$\mathbf{H}(z)\mathbf{H}_*(z) = \frac{\mathbf{Z}_0(z) + \mathbf{Z}_{0*}(z)}{2}$$
$$= z\left(\mathbf{R}_0 + \mathbf{R}_0^*\right)\mathbf{M}^{-1}(z)\left(\frac{\mathbf{Z}_1(z) + \mathbf{Z}_{1*}(z)}{2}\right)\mathbf{N}^{-1}(z)\left(\mathbf{R}_0 + \mathbf{R}_0^*\right).$$

Since $\mathbf{M}(0)$ and $\mathbf{N}(0)$ are nonsingular, using (5.277), it follows that $z = 0$ is a zero of $(\mathbf{Z}_1(z) + \mathbf{Z}_{1*}(z))$ of order $p - q - 1$, and repeating this entire procedure that led to $\mathbf{Z}_1(z)$, we obtain a new positive matrix function $\mathbf{Z}_2(z)$ with $\delta_o(\mathbf{Z}_2(z)) = \delta_o(\mathbf{Z}_1(z)) - 1$. After $p-q$ such cycles with order reduction, we obtain the configuration shown in Fig. 5.4. Clearly, at every such cycle, the new positive matrix function $\mathbf{Z}_k(z)$ satisfies $\delta_o(\mathbf{Z}_k(z)) = p - k$, $k = 1 \to p - q$, and

$$\frac{\mathbf{Z}_k(z) + \mathbf{Z}_{k*}(z)}{2}\bigg|_{z=0} = \mathbf{0}, \quad k = 0 \to p - q - 1, \tag{5.282}$$

where $\mathbf{Z}_0(z) \equiv \mathbf{Z}(z)$ given in (5.274). From the above analysis, the positive termination $\mathbf{Z}_{p-q}(z)$ in Fig. 5.4 has order q, and as in the single-channel case, because $z = 0$ is *not* a zero of the 'even part'

$$\frac{\mathbf{Z}_{p-q}(z) + \mathbf{Z}_{p-q*}(z)}{2},$$

additional line extractions do not produce any more order reduction and hence

$$\delta_o(\mathbf{Z}_n(z)) = q, \quad \text{for every } n \geq p - q. \quad (5.283)$$

From the basic transformation formulas in (5.101)–(5.109), at the nth stage the corresponding bounded matrix functions $\mathbf{d}_n(z)$ and $\mathbf{d}_{n+1}(z)$ are related through the identities (5.119) or (5.156). However, the complicated forms of the constant matrices $\mathbf{A}$ and $\mathbf{B}$ appearing there (see (5.120)–(5.121)) are not particularly convenient for their easy evaluation, and as shown in section 5.2.4 (see also Appendix 5.B), this can be circumvented by defining a new set of modified bounded matrix functions

$$\boldsymbol{\rho}_0(z) = \mathbf{d}_0(z), \quad \boldsymbol{\rho}_1(z) = \mathbf{d}_1(z)$$

and

$$\boldsymbol{\rho}_n(z) = \mathbf{U}_{n-1}\mathbf{U}_{n-2}\cdots\mathbf{U}_1\mathbf{d}_n(z)\mathbf{V}_1^*\mathbf{V}_2^*\cdots\mathbf{V}_{n-1}^*, \quad n \geq 2, \quad (5.284)$$

where $\mathbf{U}_i, \mathbf{V}_i, i \geq 0$, are a specific sequence of unitary matrices as defined in (5.196). These modified bounded matrix functions satisfy the simpler recursions in (5.198) and (5.199), where $\mathbf{M}_n$ and $\mathbf{N}_n$ are unique lower triangular matrices with positive diagonal entries that result from the factorizations in (5.200). Since the above argument shows that there is no more order reduction in $\mathbf{Z}_n(z)$, or equivalently in $\boldsymbol{\rho}_n(z)$ for $n \geq p - q$, from (5.284), (5.102)–(5.103) we have

$$\delta_o(\boldsymbol{\rho}_{n+1}(z)) = \delta_o(\mathbf{Z}_{n+1}(z)) = \delta_o(\mathbf{Z}_n(z)) = q, \quad \text{for } n \geq p - q,$$

where the last two equalities follow from (5.283). In addition, from (5.102), we also have

$$\delta_o(z\boldsymbol{\rho}_{n+1}(z)) = \delta_o(\mathbf{Z}_n(z)) = q, \quad \text{for } n \geq p - q,$$

and hence

$$\delta_o(\boldsymbol{\rho}_{n+1}(z)) = \delta_0(z\boldsymbol{\rho}_{n+1}(z)) = q, \quad n \geq p - q. \quad (5.285)$$

Equation (5.282) together with (5.285) represent the key feature of a rational L-ARMA(p, q) system $\mathbf{H}_l(z)$ as in (5.268)–(5.270). The associated rational positive matrix function $\mathbf{Z}(z)$ in (5.274) upon line extraction undergoes order reduction for the first $p - q$ stages and thereafter remains *invariant* with respect to its order. Thus from (5.285), bounded matrix functions $\boldsymbol{\rho}_{n+1}(z)$ and $z\boldsymbol{\rho}_{n+1}(z)$ have left-coprime orders equal to q, for any $n \geq p - q$. As a result, every such bounded matrix function $\boldsymbol{\rho}_{n+1}(z)$ has the left-coprime representation

$$\boldsymbol{\rho}_{n+1}(z) = \mathbf{G}^{-1}(z)\mathbf{H}(z), \quad n \geq p - q, \quad (5.286)$$

where

$$\mathbf{G}(z) = \mathbf{I} + \mathbf{G}_1 z + \cdots + \mathbf{G}_q z^q \qquad (5.287)$$

$(det\ \mathbf{G}(z) \neq 0$ in $|z| \leq 1)$ and

$$\mathbf{H}(z) = \mathbf{H}_0 + \mathbf{H}_1 z + \cdots + \mathbf{H}_{q-1} z^{q-1}. \qquad (5.288)$$

Notice that from the second equality in (5.285), the order of the 'numerator' matrix polynomial $\mathbf{H}(z)$ cannot exceed $q - 1$ and hence the 'denominator' matrix polynomial $\mathbf{G}(z)$ must be of order q. This invariant character of the bounded matrix functions $\boldsymbol{\rho}_{n+1}(z)$ with respect to their order beyond a certain stage $(n \geq p - q)$ is characteristic of MARMA systems.

Naturally, the above analysis will hold even if we start with an R-ARMA(p, q) system

$$\mathbf{H}_r(z) = \mathbf{B}_1(z)\mathbf{A}_1^{-1}(z)$$

in (5.268). In that case (5.285) gives

$$\boldsymbol{\rho}_{n+1}(z) = \mathbf{H}_1(z)\mathbf{G}_1^{-1}(z), \quad n \geq p - q, \qquad (5.289)$$

where $\mathbf{G}_1(z)$ and $\mathbf{H}_1(z)$ are two matrix polynomials $(det\ \mathbf{G}_1(z) \neq 0$, in $|z| \leq 1)$ with respective orders q and $q - 1$ as in (5.287)–(5.288).

5.3.2   MARMA PARAMETER ESTIMATION

To make use of the above observations in identifying the unknown system, we can utilize the class of all spectral extension formula derived in the previous section. From Theorem 5.1, the unknown minimum-phase rational system also *must* follow from (5.257)–(5.261) for a specific choice of the rational bounded matrix function $\boldsymbol{\rho}_{n+1}(z)$. Resorting to the left-coprime representation, the left-Wiener factor $\mathbf{B}_l(z)$ in (5.258) identifies the unknown rational system in (5.268) up to multiplication by a constant unitary matrix on its right, and making use of (5.286) there, we obtain

$$\mathbf{B}_l(z) = \left(\mathbf{A}_n(z) - z\mathbf{G}^{-1}(z)\mathbf{H}(z)\tilde{\mathbf{C}}_n(z)\right)^{-1} \boldsymbol{\Gamma}_l(z)$$

$$= \left(\mathbf{G}(z)\mathbf{A}_n(z) - z\mathbf{H}(z)\tilde{\mathbf{C}}_n(z)\right)^{-1} \mathbf{Q}(z), \quad n \geq p - q, \qquad (5.290)$$

where, from (5.260), $\mathbf{Q}(z)$ is a polynomial matrix of order q that satisfies

$$\mathbf{Q}(z)\mathbf{Q}_*(z) = \mathbf{G}(z)\mathbf{G}_*(z) - \mathbf{H}(z)\mathbf{H}_*(z). \qquad (5.291)$$

The formal order of the 'denominator' factor in (5.290) is $n + q$ and since its actual order is only p in (5.268), the coefficients of the higher-order terms z^{p+1}, z^{p+2}, ..., z^{n+q} must be zeros there. This gives rise to $n + q - p$ linear matrix equations in $2q$ unknown matrices $\mathbf{G}_1 \rightarrow \mathbf{G}_q$ $(\mathbf{G}_0 = \mathbf{I})$ and $\mathbf{H}_0 \rightarrow \mathbf{H}_{q-1}$. Clearly, for a solution to exist, we must have $n \geq p+q$, and the

smallest value of n suitable for this purpose is $n = p + q$. In that case, the
$2q$ matrix equations in the $2q$ unknown matrices have the representation

$$\mathbf{XA} = \mathbf{B},\tag{5.292}$$

where the $2q \times 2q$ block matrix

$$\mathbf{A} \triangleq \begin{bmatrix}
\mathbf{A}_p & \mathbf{A}_{p+1} & \cdots & \mathbf{A}_{p+q-1} & \mathbf{A}_{p+q} & \mathbf{0} & \cdots & \mathbf{0} \\
\mathbf{A}_{p-1} & \mathbf{A}_p & \cdots & \mathbf{A}_{p+q-2} & \mathbf{A}_{p+q-1} & \mathbf{A}_{p+q} & \cdots & \mathbf{0} \\
\vdots & \vdots & \cdots & \vdots & \vdots & \vdots & \cdots & \vdots \\
\mathbf{A}_{p-q+1} & \mathbf{A}_{p-q+2} & \cdots & \mathbf{A}_p & \mathbf{A}_{p+1} & \mathbf{A}_{p+2} & \cdots & \mathbf{A}_{p+q} \\
-\mathbf{C}_q^* & -\mathbf{C}_{q-1}^* & \cdots & -\mathbf{C}_1^* & -\mathbf{C}_0^* & \mathbf{0} & \cdots & \mathbf{0} \\
-\mathbf{C}_{q+1}^* & -\mathbf{C}_q^* & \cdots & -\mathbf{C}_2^* & -\mathbf{C}_1^* & -\mathbf{C}_0^* & \cdots & \mathbf{0} \\
\vdots & \vdots & \cdots & \vdots & \vdots & \vdots & \cdots & \vdots \\
-\mathbf{C}_{2q-1}^* & -\mathbf{C}_{2q-2}^* & \cdots & -\mathbf{C}_q^* & -\mathbf{C}_{q-1}^* & -\mathbf{C}_{q-2}^* & \cdots & -\mathbf{C}_0^*
\end{bmatrix}\tag{5.293}$$

$$\mathbf{X} \triangleq \begin{bmatrix} \mathbf{G}_1 & \mathbf{G}_2 & \cdots & \mathbf{G}_q \mid \mathbf{H}_0 & \mathbf{H}_1 & \cdots & \mathbf{H}_{q-1} \end{bmatrix}\tag{5.294}$$

$$\mathbf{B} \triangleq -\begin{bmatrix} \mathbf{A}_{p+1} & \mathbf{A}_{p+2} & \cdots & \mathbf{A}_{p+q} \mid \mathbf{0} & \mathbf{0} & \cdots & \mathbf{0} \end{bmatrix},\tag{5.295}$$

and $\mathbf{A}_k$, $\mathbf{C}_k$, $k = 0 \to p + q$, represent the coefficients of the left and right
Levinson polynomial matrices $\mathbf{A}_{p+q}(z)$ and $\mathbf{C}_{p+q}(z)$ in (5.252) and (5.253),
respectively. Notice that since $\mathbf{B}_l(z)$ in (5.290) exists as a minimum-phase
L-ARMA(p, q) system, (5.292) is guaranteed to possess a unique solution
at the correct stage and, moreover, the unknowns so obtained at that stage
correspond to a bounded matrix function $\boldsymbol{\rho}_{p+q+1}(z)$. Having determined
$\mathbf{G}(z)$ and $\mathbf{H}(z)$ from (5.292), the 'numerator' factor $\mathbf{Q}(z)$ can be determined
using (5.291). From (5.290), the remaining minimum-phase 'denominator'
factor

$$\mathbf{P}(z) = \mathbf{P}_0 + \mathbf{P}_1 z + \mathbf{P}_2 z^2 + \cdots + \mathbf{P}_p z^p\tag{5.296}$$

is given by

$$\mathbf{P}_0 = \mathbf{A}_0\tag{5.297}$$

and

$$\mathbf{P}_i = \begin{cases}
\displaystyle\sum_{k=0}^{i} \mathbf{G}_k \mathbf{A}_{i-k} - \sum_{k=0}^{i-1} \mathbf{H}_k \mathbf{C}_{p+q+k-i+1}^*, & 1 \le i \le q, \\[4mm]
\displaystyle\sum_{k=0}^{q} \mathbf{G}_k \mathbf{A}_{i-k} - \sum_{k=0}^{q-1} \mathbf{H}_k \mathbf{C}_{p+q+k-i+1}^*, & q+1 \le i \le p,
\end{cases}\tag{5.298}$$

with $\mathbf{G}_0 = \mathbf{I}$. The factorization in (5.291) can be carried out in any number of ways, and if $det\,\mathbf{H}_l(z)$ is known in advance to be free of zeros on $|z| = 1$, to exploit the polynomial nature of $\mathbf{Q}(z)$ it is best to consider

$$\mathbf{K}(\theta) \triangleq \left(\mathbf{G}(e^{j\theta})\mathbf{G}^*(e^{j\theta}) - \mathbf{H}(e^{j\theta})\mathbf{H}^*(e^{j\theta})\right)^{-1} = \mathbf{Q}^{-1*}(e^{j\theta})\mathbf{Q}^{-1}(e^{j\theta}).$$
(5.299)

From (5.257)–(5.259) with $\boldsymbol{\rho}_{n+1}(z) \equiv \mathbf{0}$, we have $\mathbf{B}_r(z) = \mathbf{C}_n^{-1}(z)$, and in that case

$$\mathbf{S}(\theta) = \mathbf{C}_n^{-1*}(e^{j\theta})\mathbf{C}_n^{-1}(e^{j\theta}).$$
(5.300)

Comparing (5.299)–(5.300), the unknown $\mathbf{Q}(z)$ can be identified as the order q right Levinson polynomial matrix of the first kind $\mathbf{C}_q(z)$ associated with the spectrum $\mathbf{K}(\theta)$ in (5.299). Thus, upto multiplication by a constant unitary matrix on its right,

$$\mathbf{Q}(z) = \mathbf{C}_q(z)$$
(5.301)

and the unknown minimum-phase rational system $\mathbf{H}_l(z)$ in (5.268) can be represented in the left-coprime form as

$$\mathbf{B}_l(z) = \mathbf{P}^{-1}(z)\mathbf{Q}(z)$$
(5.302)

with $\mathbf{P}(z)$ as in (5.296)–(5.298) and $\mathbf{Q}(z)$ as in (5.291) or (5.301). Finally renormalization may be carried out with respect to $\mathbf{P}_0$ to obtain the normalized minimum-phase factor $\mathbf{H}_l(z)$. From (5.257), the power spectrum associated with the system so obtained interpolates the given autocorrelation matrices $\mathbf{r}_0, \mathbf{r}_1, \ldots, \mathbf{r}_{p+q}$.

The above analysis assumes that p and q are known. These quantities are usually unknown, and as in the single-channel case, the invariant character of the bounded matrix function in (5.286) with respect to its order can be exploited here also to evaluate the model orders p and q.

5.3.3 MODEL ORDER SELECTION

Having determined $\boldsymbol{\rho}_{p+q+1}(z) = \mathbf{G}^{-1}(z)\mathbf{H}(z)$ at stage $n = p + q$, this procedure of line extraction may be repeated at the next stage to obtain the bounded matrix function

$$\boldsymbol{\rho}_{p+q+2}(z) = \mathbf{E}^{-1}(z)\mathbf{F}(z).$$
(5.303)

From (5.286)–(5.288), $\mathbf{E}(z)$ and $\mathbf{F}(z)$ must have the form

$$\mathbf{E}(z) = \mathbf{I} + \mathbf{E}_1 z + \cdots + \mathbf{E}_q z^q$$
(5.304)

and

$$\mathbf{F}(z) = \mathbf{F}_0 + \mathbf{F}_1 z + \cdots \mathbf{F}_{q-1} z^{q-1}.$$
(5.305)

Arguing as in (5.290), the left-coprime Wiener factor in this case is given by

$$\mathbf{B}_l(z) = \Big(\mathbf{E}(z)\mathbf{A}_{p+q+1}(z) - z\mathbf{F}(z)\tilde{\mathbf{C}}_{p+q+1}(z)\Big)^{-1}\mathbf{Q}_1(z),\qquad(5.306)$$

where $\mathbf{Q}_1(z)$ satisfies the recursion

$$\mathbf{Q}_1(z)\mathbf{Q}_{1*}(z) = \mathbf{E}(z)\mathbf{E}_*(z) - \mathbf{F}(z)\mathbf{F}_*(z)\,.$$

Notice that the 'denominator' term in $\mathbf{E}(z)\mathbf{A}_{p+q+1}(z) - z\mathbf{F}(z)\tilde{\mathbf{C}}_{p+q+1}(z)$ in (5.306) has formal order $p + 2q + 1$ and since its actual order is only p, the coefficients of the higher-order terms z^{p+1}, z^{p+2}, ..., z^{p+2q+1} must be zeros there. This gives rise to $2q+1$ linear matrix equations in $2q$ unknown matrices, and the first $2q$ of these equations can be expressed as

$$\mathbf{YC} = \mathbf{D}\,,\qquad(5.307)$$

where the $2q \times 2q$ block matrix

$$\mathbf{C} \triangleq \begin{bmatrix}
\mathbf{A}_p^{(1)} & \mathbf{A}_{p+1}^{(1)} & \cdot\cdot & \mathbf{A}_{p+q-1}^{(1)} & \mathbf{A}_{p+q}^{(1)} & \mathbf{A}_{p+q+1}^{(1)} & \cdot\cdot & \mathbf{0} \\
\mathbf{A}_{p-1}^{(1)} & \mathbf{A}_p^{(1)} & \cdot\cdot & \mathbf{A}_{p+q-2}^{(1)} & \mathbf{A}_{p+q-1}^{(1)} & \mathbf{A}_{p+q}^{(1)} & \cdot\cdot & \mathbf{0} \\
\vdots & \vdots & \cdot\cdot & \vdots & \vdots & \vdots & \cdot\cdot & \vdots \\
\mathbf{A}_{p-q+2}^{(1)} & \mathbf{A}_{p-q+3}^{(1)} & \cdot\cdot & \mathbf{A}_{p+1}^{(1)} & \mathbf{A}_{p+2}^{(1)} & \mathbf{A}_{p+3}^{(1)} & \cdot\cdot & \mathbf{A}_{p+q+1}^{(1)} \\
\mathbf{A}_{p-q+1}^{(1)} & \mathbf{A}_{p-q+2}^{(1)} & \cdot\cdot & \mathbf{A}_p^{(1)} & \mathbf{A}_{p+1}^{(1)} & \mathbf{A}_{p+2}^{(1)} & \cdot\cdot & \mathbf{A}_{p+q}^{(1)} \\
-\mathbf{C}_{q+1}^{(1)*} & -\mathbf{C}_q^{(1)*} & \cdot\cdot & -\mathbf{C}_2^{(1)*} & -\mathbf{C}_1^{(1)*} & -\mathbf{C}_0^{(1)*} & \cdot\cdot & \mathbf{0} \\
-\mathbf{C}_{q+2}^{(1)*} & -\mathbf{C}_{q+1}^{(1)*} & \cdot\cdot & -\mathbf{C}_3^{(1)*} & -\mathbf{C}_2^{(1)*} & -\mathbf{C}_1^{(1)*} & \cdot\cdot & \mathbf{0} \\
\vdots & \vdots & \cdot\cdot & \vdots & \vdots & \vdots & \cdot\cdot & \vdots \\
-\mathbf{C}_{2q-1}^{(1)*} & \mathbf{C}_{2q-2}^{(1)*} & \cdot\cdot & -\mathbf{C}_q^{(1)*} & -\mathbf{C}_{q-1}^{(1)*} & -\mathbf{C}_{q-2}^{(1)*} & \cdot\cdot & -\mathbf{C}_0^{(1)*} \\
-\mathbf{C}_{2q}^{(1)*} & -\mathbf{C}_{2q-1}^{(1)*} & \cdot\cdot & -\mathbf{C}_{q+1}^{(1)*} & -\mathbf{C}_q^{(1)*} & -\mathbf{C}_{q-1}^{(1)*} & \cdot\cdot & -\mathbf{C}_1^{(1)*}
\end{bmatrix}$$

$$(5.308)$$

$$\mathbf{Y} \triangleq \begin{bmatrix} \mathbf{E}_1 & \mathbf{E}_2 & \cdots & \mathbf{E}_q \,\big|\, \mathbf{F}_0 & \mathbf{F}_1 & \cdots & \mathbf{F}_{q-1} \end{bmatrix}\qquad(5.309)$$

$$\mathbf{D} \triangleq -\begin{bmatrix} \mathbf{A}_{p+1}^{(1)} & \mathbf{A}_{p+2}^{(1)} & \cdots & \mathbf{A}_{p+q}^{(1)} \,\big|\, \mathbf{A}_{p+q+1}^{(1)} & \mathbf{0} & \cdots & \mathbf{0} \end{bmatrix},\qquad(5.310)$$

and the last equation (coefficient of z^{p+2q+1}) reads

$$\mathbf{E}_q\mathbf{A}_{p+q+1}^{(1)} - \mathbf{F}_{q-1}\mathbf{C}_0^{(1)*} = \mathbf{0}\,.\qquad(5.311)$$

Here $\mathbf{A}_k^{(1)}$, $\mathbf{C}_k^{(1)}$, $k = 0 \to p + q + 1$, represent the coefficients of the left and right Levinson polynomial matrices $\mathbf{A}_{p+q+1}(z)$ and $\mathbf{C}_{p+q+1}(z)$, respectively. Once again, since (5.306) represents a minimum-phase L-ARMA(p, q) system, $\rho_{p+q+2}(z)$ must exist as a bounded matrix function and hence equation (5.307) is also guaranteed to possess a unique solution at the correct stage. Thus, having obtained $\rho_{p+q+1}(z) = \mathbf{G}^{-1}(z)\mathbf{H}(z)$ and $\rho_{p+q+2}(z) = \mathbf{E}^{-1}(z)\mathbf{F}(z)$, their interdependence through the matrix Schur algorithm in (5.198)–(5.199) can be further exploited to obtain a new model order selection criterion for multichannel rational systems.

With $n = p+q+1$ and substituting $\rho_{p+q+1}(z)$ and $\rho_{p+q+2}(z)$ so obtained into (5.199), we obtain

$$\mathbf{G}^{-1}(z)\mathbf{H}(z) = \left(\mathbf{s}_{p+q+1} + z\mathbf{M}_{p+q+1}^{-1}\mathbf{E}^{-1}(z)\mathbf{F}(z)\mathbf{N}_{p+q+1}\right)$$
$$\times \left(\mathbf{I} + z\mathbf{s}_{p+q+1}^{*}\mathbf{M}_{p+q+1}^{-1}\mathbf{E}^{-1}(z)\mathbf{F}(z)\mathbf{N}_{p+q+1}\right)^{-1}$$

or

$$\left(\mathbf{H}(z) - \mathbf{G}(z)\mathbf{s}_{p+q+1}\right)\mathbf{N}_{p+q+1}^{-1} = z\left(\mathbf{G}(z) - \mathbf{H}(z)\mathbf{s}_{p+q+1}^{*}\right)\mathbf{M}_{p+q+1}^{-1}\mathbf{E}^{-1}(z)\mathbf{F}(z).$$
$$(5.312)$$

Since $\mathbf{E}^{-1}(z)\mathbf{F}(z)$ is a left-coprime representation of $\rho_{p+q+2}(z)$, from (5.8), there exist polynomial matrices $\mathbf{X}_1(z)$ and $\mathbf{Y}_1(z)$ that satisfy

$$\mathbf{E}(z)\mathbf{X}_1(z) + \mathbf{F}(z)\mathbf{Y}_1(z) = \mathbf{I}$$

or

$$\mathbf{X}_1(z) + \mathbf{E}^{-1}(z)\mathbf{F}(z)\mathbf{Y}_1(z) = \mathbf{E}^{-1}(z).\qquad(5.313)$$

Letting

$$\mathbf{X}(z) = z\left(\mathbf{G}(z) - \mathbf{H}(z)\mathbf{s}_{p+q+1}^{*}\right)\mathbf{M}_{p+q+1}^{-1},\qquad(5.314)$$

and multiplying both sides of (5.313) by $\mathbf{X}(z)$ on the left yields

$$\mathbf{X}(z)\mathbf{X}_1(z) + \mathbf{X}(z)\mathbf{E}^{-1}(z)\mathbf{F}(z)\mathbf{Y}_1(z) = \mathbf{X}(z)\mathbf{E}^{-1}(z).\qquad(5.315)$$

Define

$$\mathbf{Y}(z) = \mathbf{X}(z)\mathbf{E}^{-1}(z)\mathbf{F}(z);$$

then (5.315) becomes

$$\mathbf{X}(z)\mathbf{X}_1(z) + \mathbf{Y}(z)\mathbf{Y}_1(z) = \mathbf{X}(z)\mathbf{E}^{-1}(z).\qquad(5.316)$$

From (5.312), since $\mathbf{Y}(z) = \mathbf{X}(z)\mathbf{E}^{-1}(z)\mathbf{F}(z)$ is a polynomial matrix, the left-hand side of (5.316) also represents a polynomial matrix, and hence so too does the right side, i.e.,

$$\mathbf{X}(z)\mathbf{E}^{-1}(z) \overset{\triangle}{=} \mathbf{K}_1(z),\qquad(5.317)$$

where $\mathbf{K}_1(z)$ is a polynomial matrix. Using (5.314), equation (5.317) can be rewritten as

$$z\left(\mathbf{G}(z) - \mathbf{H}(z)\mathbf{s}_{p+q+1}^{*}\right) = \mathbf{K}_1(z)\mathbf{E}(z)\mathbf{M}_{p+q+1},\qquad(5.318)$$

which gives at $z = 0$

$$\mathbf{K}_1(0)\mathbf{E}_0\mathbf{M}_{p+q+1} \equiv \mathbf{0}.$$

Since $\mathbf{E}_0 = \mathbf{I}$ and $\mathbf{M}_{p+q+1}$ is nonsingular, $\mathbf{K}_1(0) \equiv \mathbf{0}$ and hence

$$\mathbf{K}_1(z) = z\mathbf{K}(z), \tag{5.319}$$

where

$$\mathbf{K}(z) = \mathbf{K}_0 + \mathbf{K}_1 z + \cdots + \mathbf{K}_k z^k, \quad k \geq 0. \tag{5.320}$$

Using (5.319), equation (5.318) simplifies to

$$\mathbf{G}(z) - \mathbf{H}(z)\mathbf{s}^*_{p+q+1} = \mathbf{K}(z)\mathbf{E}(z)\mathbf{M}_{p+q+1}. \tag{5.321}$$

Finally, substituting (5.321) into (5.312), we obtain

$$\mathbf{H}(z) - \mathbf{G}(z)\mathbf{s}_{p+q+1} = z\mathbf{K}(z)\mathbf{F}(z)\mathbf{N}_{p+q+1}. \tag{5.322}$$

On comparing the constant terms in (5.321) and (5.322), we get

$$\mathbf{G}_0 - \mathbf{H}_0\mathbf{s}^*_{p+q+1} = \mathbf{K}_0\mathbf{E}_0\mathbf{M}_{p+q+1}$$

$$\mathbf{H}_0 - \mathbf{G}_0\mathbf{s}_{p+q+1} = \mathbf{0},$$

and using $\mathbf{G}_0 = \mathbf{E}_0 = \mathbf{I}$, this gives

$$\mathbf{K}_0\mathbf{M}_{p+q+1} = \mathbf{I} - \mathbf{s}_{p+q+1}\mathbf{s}^*_{p+q+1} = \mathbf{M}^*_{p+q+1}\mathbf{M}_{p+q+1}$$

or

$$\mathbf{K}(0) = \mathbf{K}_0 = \mathbf{M}^*_{p+q+1} \tag{5.323}$$

is nonsingular. However, a priori $\mathbf{K}(z)$ in (5.320)–(5.322) need not be a constant matrix. This follows by noticing that although the order restrictions in (5.321)–(5.322) generate the equations

$$\begin{aligned}
\mathbf{K}_k\mathbf{E}_q &= \mathbf{K}_k\mathbf{F}_{q-1} = \mathbf{0} \\
\mathbf{K}_k\mathbf{E}_{q-1} + \mathbf{K}_{k-1}\mathbf{E}_q &= \mathbf{K}_k\mathbf{F}_{q-2} + \mathbf{K}_{k-1}\mathbf{F}_{q-1} = \mathbf{0} \\
&\vdots \\
\mathbf{K}_k\mathbf{E}_{q-k+1} + \cdots + \mathbf{K}_1\mathbf{E}_q &= \mathbf{K}_k\mathbf{F}_{q-k} + \cdots \mathbf{K}_1\mathbf{F}_{q-1} = \mathbf{0},
\end{aligned} \tag{5.324}$$

these equations can be satisfied without $\mathbf{K}_i$, $i = 1 \rightarrow k$, becoming identically equal to the zero matrix. In that case, *both* $\mathbf{E}_q$ and $\mathbf{F}_{q-1}$ must be necessarily singular and if such is not the case, then from the above equations it follows that $\mathbf{K}_i \equiv \mathbf{0}$, $i = 1 \rightarrow k$, and $\mathbf{K}(z) = \mathbf{K}_0$.

From (5.321)–(5.322), we also obtain

$$\mathbf{H}(z)\left(\mathbf{I} - \mathbf{s}^*_{p+q+1}\mathbf{s}_{p+q+1}\right) = \mathbf{K}(z)\left(\mathbf{E}(z)\mathbf{M}_{p+q+1}\mathbf{s}_{p+q+1} + z\mathbf{F}(z)\mathbf{N}_{p+q+1}\right),$$

$$\mathbf{G}(z)\left(\mathbf{I} - \mathbf{s}_{p+q+1}\mathbf{s}^*_{p+q+1}\right) = \mathbf{K}(z)\left(\mathbf{E}(z)\mathbf{M}_{p+q+1} + z\mathbf{F}(z)\mathbf{N}_{p+q+1}\mathbf{s}^*_{p+q+1}\right)$$

and making use of (5.200) with $n = p + q + 1$, these equations reduce to

$$\mathbf{H}(z) = \mathbf{K}(z)\Big(\mathbf{E}(z)\mathbf{M}_{p+q+1}^{*-1}\mathbf{s}_{p+q+1} + z\mathbf{F}(z)\mathbf{N}_{p+q+1}^{*-1}\Big) \tag{5.325}$$

and

$$\mathbf{G}(z) = \mathbf{K}(z)\Big(\mathbf{E}(z)\mathbf{M}_{p+q+1}^{*-1} + z\mathbf{F}(z)\mathbf{N}_{p+q+1}^{*-1}\mathbf{s}_{p+q+1}^{*}\Big). \tag{5.326}$$

Equations (5.324) together with (5.325)–(5.326) show the interrelationship between the coefficients of the bounded matrix functions $\boldsymbol{\rho}_{p+q+1}(z) = \mathbf{G}^{-1}(z)\mathbf{H}(z)$, $\boldsymbol{\rho}_{p+q+2}(z) = \mathbf{E}^{-1}(z)\mathbf{F}(z)$ and, as in the single-channel case, these identities can be used to design a test criterion for model order selection. Needless to say, the additional equations in (5.324) complicate these conditions to be expressed in a simple form except for the special case when at least $\mathbf{E}_q$ or $\mathbf{F}_{q-1}$ is nonsingular. In that case $\mathbf{K}_i \equiv \mathbf{0}$, $i = 1 \rightarrow k$, and comparing coefficients of like powers of z on both sides of (5.325), we get

$$\epsilon_0 \stackrel{\triangle}{=} \mathbf{E}_q \mathbf{M}_{p+q+1}^{*-1}\mathbf{s}_{p+q+1} + \mathbf{F}_{q-1}\mathbf{N}_{p+q+1}^{*-1} = \mathbf{0} \tag{5.327}$$

for z^q, and using (5.323)

$$\epsilon_k \stackrel{\triangle}{=} \mathbf{H}_k - \mathbf{M}_{p+q+1}^{*}\Big(\mathbf{E}_k\mathbf{M}_{p+q+1}^{*-1}\mathbf{s}_{p+q+1} + \mathbf{F}_{k-1}\mathbf{N}_{p+q+1}^{*-1}\Big) = \mathbf{0}, \tag{5.328}$$

for $k = 1 \rightarrow q - 1$. Similarly, from (5.326) we obtain

$$\epsilon_{q+k-1} \stackrel{\triangle}{=} \mathbf{G}_k - \mathbf{M}_{p+q+1}^{*}\Big(\mathbf{E}_k\mathbf{M}_{p+q+1}^{*-1} + \mathbf{F}_{k-1}\mathbf{N}_{p+q+1}^{*-1}\mathbf{s}_{p+q+1}^{*}\Big) = \mathbf{0}, \tag{5.329}$$

for $k = 1 \rightarrow q$. Following the single-channel case, the scalar functions

$$\epsilon_0(p, q) \stackrel{\triangle}{=} tr\left(\epsilon_0\epsilon_0^{*}\right), \tag{5.330}$$

or, more generally,

$$\epsilon(p, q) \stackrel{\triangle}{=} \sum_{k=0}^{2q-1} tr\left(\epsilon_k\epsilon_k^{*}\right) \tag{5.331}$$

can be used as model order criteria, since from the above discussion, starting with $n = 1$, $m = 0$ and maintaining $n \geq m$, when the indices n and m are updated sequentially, the first value of n and m for which $\epsilon_0(n, m)$ and $\epsilon(n, m)$ goes to zeros occurs at $n = p$ and $m = q$.

It is interesting to note that (5.327) is *always* true irrespective of the exact form of $\mathbf{K}(z)$. To see this, rewrite (5.327) as

$$\epsilon_0 = \mathbf{E}_q \mathbf{M}_{p+q+1}^{*-1}\mathbf{s}_{p+q+1}\mathbf{N}_{p+q+1}^{*} + \mathbf{F}_{q-1} = \mathbf{0} \tag{5.332}$$

and from (5.252)–(5.253) with $n = p + q + 1$, comparing the coefficients of z^{p+q+1} and the constant term respectively, we obtain

$$\mathbf{M}_{p+q+1}^{*}\mathbf{A}_{p+q+1}^{(p+q+1)} = -\mathbf{s}_{p+q+1}\mathbf{C}_0^{(p+q)*}$$

and
$$\mathbf{C}_0^{(p+q+1)}\mathbf{N}_{p+q+1} = \mathbf{C}_0^{(p+q)}.$$

Eliminating $\mathbf{C}_0^{(p+q)}$, we get

$$\mathbf{A}_{p+q+1}^{(p+q+1)} = -\mathbf{M}_{p+q+1}^{*-1}\mathbf{S}_{p+q+1}\mathbf{N}_{p+q+1}^{*}\mathbf{C}_0^{(p+q+1)*}$$

and substituting this into (5.332), we obtain

$$\mathbf{E}_q\mathbf{A}_{p+q+1}^{(p+q+1)} - \mathbf{F}_{q-1}\mathbf{C}_0^{(p+q+1)*} \equiv 0$$

which is the same as (5.311). Thus (5.327) is true at the correct stage $n = p$ and $m = q$, and equation (5.330) can be used *always* as an L-ARMA model order selection criterion. Obviously, a similar procedure can be developed starting with the right-coprime representation of multichannel rational systems (see (5.P.23)–(5.P.24) in Problem 5.9).

The invariant characteristics of a multichannel rational system is shown to be the order of an associated bounded matrix function from a certain stage onwards, and this invariant character has been utilized here to identify the model order and evaluate the system parameters of the matrix transfer function. In this approach, first, the underlying bounded matrix function is computed using the Levinson polynomial matrices of the first kind at a particular stage. In the final step, the bounded matrix function so obtained is used to evaluate simultaneously the 'numerator' and 'denominator' matrix parameters of the unknown system.

5.3.4 SIMULATION RESULTS

The examples in this section deal with the reconstruction of rational multichannel system transfer functions both from their exact output autocorrelation matrices (Figs. 5.5–5.13) as well as from the output data samples (Figs. 5.14–5.18). In the exact case, the system transfer function $\mathbf{H}(z)$ is used to compute the output autocorrelation matrices using the formula

$$\mathbf{r}_k = \frac{1}{2\pi}\int_{-\pi}^{\pi}\mathbf{S}(\theta)e^{-jk\theta}d\theta \tag{5.333}$$

where the power spectral density matrix $\mathbf{S}(\theta)$ is given by either

$$\mathbf{S}(\theta) = \mathbf{H}(e^{j\theta})\mathbf{H}^{*}(e^{j\theta}) \tag{5.334}$$

or

$$\mathbf{S}(\theta) = \mathbf{H}^{*}(e^{j\theta})\mathbf{H}(e^{j\theta}), \tag{5.335}$$

depending on whether the left-coprime or the right-coprime form is desired. From (5.257)–(5.259) and (5.290), since the final identification is carried out either in the left-coprime or the right-coprime representation of rational transfer functions, the form in (5.334) must be employed in (5.333) for left-coprime identification and (5.335) for right-coprime identification. In particular, for real processes, letting $\mathbf{S}(\theta) = \mathbf{R}(\theta) + j\mathbf{X}(\theta)$, where $\mathbf{R}(\theta)$ and $\mathbf{X}(\theta)$ are real matrix functions, (5.333) reduces to (see also Problem 5.1)

$$\mathbf{r}_k = \frac{1}{\pi} \int_0^\pi \left(\mathbf{R}(\theta)\cos k\theta + \mathbf{X}(\theta)\sin k\theta\right)d\theta. \qquad (5.336)$$

Thus, the autocorrelation matrices and hence the Levinson polynomial matrices assoicated with real stationary processes are real.

In the data case, the original system transfer function $\mathbf{H}(z)$ is recursively employed either as in (5.17) or as in (5.22) and then driven by a stationary zero-mean white noise source to generate the required output data vectors

$$\mathbf{x}(n) = [\, x_1(n), \; x_2(n), \; \ldots, \; x_m(n)\,]^T$$

or

$$\mathbf{y}(n) = [\, y_1(n), \; y_2(n), \; \ldots, \; y_m(n)\,],$$

$n = N_0 + 1 \rightarrow N_0 + N$. From (5.26), (5.32), the test statistic

$$\hat{\mathbf{r}}_k = \frac{1}{N} \sum_{n=N_0+1}^{N_0+N-k} \mathbf{x}(n+k)\mathbf{x}^*(n), \quad k \geq 0, \qquad (5.337)$$

or

$$\hat{\mathbf{r}}_k = \frac{1}{N} \sum_{n=N_0+1}^{N_0+N-k} \mathbf{y}^*(n+k)\mathbf{y}(n), \quad k \geq 0, \qquad (5.338)$$

may be used to estimate the sample autocorrelation matrices involved in the identification of the left- and right-coprime factors respectively. The number of data samples N should be large enough to guarantee positive definiteness of the associated block Toeplitz matrix.

Starting with $n = 1$, $m = 0$ and maintaining $n \geq m$, the algorithm proceeds in a sequential manner by examining the hypothesis that the given system is L-ARMA(n, m). (See Problem 5.9 for R-ARMA(p, q) system identification.) Towards this, the left and right Levinson polynomial matrices $\mathbf{A}_{n+m}(z)$, $\mathbf{C}_{n+m}(z)$, $\mathbf{A}_{n+m+1}(z)$ and $\mathbf{C}_{n+m+1}(z)$ are computed recursively using (5.252), (5.253) followed by the matrix functions $\boldsymbol{\rho}_{n+m+1}(z) = \mathbf{G}^{-1}(z)\mathbf{H}(z)$, $\boldsymbol{\rho}_{n+m+2}(z) = \mathbf{E}^{-1}(z)\mathbf{F}(z)$ using (5.292) and (5.307), respectively. If $both$ $\boldsymbol{\rho}_{n+m+1}(z)$ and $\boldsymbol{\rho}_{n+m+2}(z)$ exist as bounded matrix functions, then the model order selection criteria $\epsilon_0(n, m)$ (and if necessary, $\epsilon(n, m)$) can be computed using (5.327)–(5.332). Notice that since

$$\boldsymbol{\rho}_{n+m+1}(0) = \mathbf{H}_0 = \mathbf{s}_{n+m+1}$$

and $\mathbf{M}_{n+m+1}$, $\mathbf{N}_{n+m+1}$ represent the unique lower triangular matrix factors with positive diagonal entries resulting from the factorizations

$$\mathbf{I} - \mathbf{H}_0\mathbf{H}_0^* = \mathbf{M}_{n+m+1}^*\mathbf{M}_{n+m+1} \qquad (5.339)$$

and

$$I - H_0^* H_0 = N_{n+m+1}^* N_{n+m+1} ,\tag{5.340}$$

the error matrix ϵ_0 in (5.327) only involves the coefficients of the bounded matrix functions $\rho_{n+m+1}(z)$ and $\rho_{n+m+2}(z)$ that have been computed from the given information. Thus, (5.327) simplifies to

$$\epsilon_0 = E_m M_{n+m+1}^{*-1} H_0 + F_{m-1} N_{n+m+1}^* \tag{5.341}$$

with M_{n+m+1} and N_{n+m+1} as given by (5.339)–(5.340). The heavy dots on the $\epsilon_0(n, m) = tr(\epsilon_0 \epsilon_0^*)$ curves in all figures indicate the presence of such a stage where $\rho_{n+m+1}(z)$ as well as $\rho_{n+m+2}(z)$ exist as bounded matrix functions (if such is not the case, that particular stage is skipped) and these bounded matrix functions are guaranteed to exist at the correct stage $n = p$, $m = q$. Since the first stage where $\epsilon_0(n, m)$ equals zero occurs only at $n = p$ and $m = q$, sequential updating of n and m is carried out until substantial relative minima is observed to occur for the first time in the value of $\epsilon_0(n, m)$. The corresponding pair (n, m) is then identified with (p, q).

Finally, the parameters of the left-coprime representation

$$B_l(z) = P^{-1}(z) Q(z)$$

in (5.302) of the unknown rational system $H(z)$ can be computed as in (5.292)–(5.302). Similarly, the parameters of the right-coprime representation

$$B_r(z) = Q_1(z) P^{-1}(z)$$

of the unknown rational system $H(z)$ can be computed using (5.P.13)–(5.P.22) and $\epsilon_0(n, m)$ given by (5.P.23)–(5.P.24) can be used as a model order criterion in this case. To facilitate comparison, the reconstructed spectrum (dotted line)

$$K(\theta) = B_l(e^{j\theta}) B_l^*(e^{j\theta}) \quad \text{or} \quad B_r^*(e^{j\theta}) B_r(e^{j\theta})$$

and the actual spectrum (solid line)

$$S(\theta) = H(e^{j\theta}) H^*(e^{j\theta}) \quad \text{or} \quad H^*(e^{j\theta}) H(e^{j\theta})$$

(see also (5.334)–(5.335)) are plotted on the same diagram with all other details as indicated in the figures. Of these, Figs. 5.5–5.9 correspond to reconstruction of 2×2 as well as 3×3 rational system transfer functions using their left-coprime representation in the exact case as described above, and Figs. 5.10–5.13 correspond to the right-coprime formulation in the exact case as described in Problem 5.9. Finally, Figs. 5.14–5.18 refer to the data case where the estimated autocorrelation matrices are used to identify the system parameters in a similar manner. Clearly, the technique presented here can be used to obtain either the left-coprime representation or the right-coprime representation of the given rational matrices. (For an example, see Problem 5.10.)

Once again, the efficacy of the criterion $\epsilon_0(n, m)$ in pinpointing the model order in the multichannel case is remarkable. This superior performance of $\epsilon_0(n, m)$ in model order determination can be attributed to the exploitation of rationality through the order reduction procedure in a systematic manner. The irreducible representation of multichannel rational systems, in terms of certain order-invariant bounded matrix functions, plays a key role in formulating this new model order determination procedure.

In the next section, we examine multichannel nonrational systems and methods to rationalize these systems in an optimal manner.

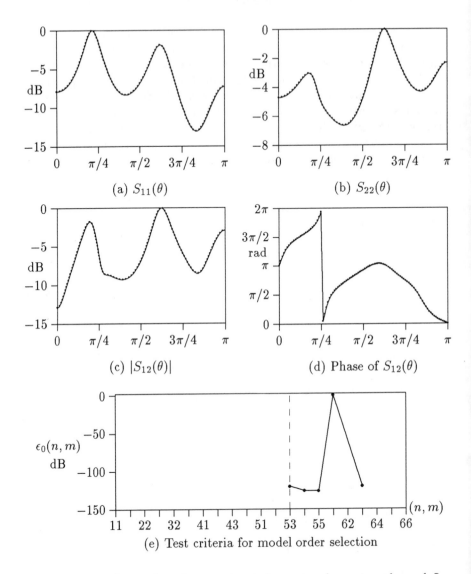

(a) $S_{11}(\theta)$

(b) $S_{22}(\theta)$

(c) $|S_{12}(\theta)|$

(d) Phase of $S_{12}(\theta)$

(e) Test criteria for model order selection

FIGURE 5.5. Original and reconstructed spectra for a two-channel L-ARMA$(5,3)$ model. In (a)–(d), solid lines represent the original model, and dotted lines represent the reconstructed model. (See next page for details.)

FIGURE 5.5 (cont.). The original two-channel L-ARMA$(5, 3)$ system is given by

$$\mathbf{H}(z) \triangleq \mathbf{P}^{-1}(z)\,\mathbf{Q}(z),$$

$$\mathbf{P}(z) = \mathbf{I} + \begin{bmatrix} -0.21844 & 0.36404 \\ -0.17762 & 0.28518 \end{bmatrix} z + \begin{bmatrix} 0.19472 & -0.19278 \\ -0.11445 & 0.06559 \end{bmatrix} z^2$$

$$+ \begin{bmatrix} -0.14883 & 0.29881 \\ 0.03625 & -0.00886 \end{bmatrix} z^3 + \begin{bmatrix} 0.33903 & 0.06655 \\ -0.07965 & -0.06801 \end{bmatrix} z^4$$

$$+ \begin{bmatrix} 0.115498 & 0.12847 \\ -0.07918 & 0.10831 \end{bmatrix} z^5$$

and

$$\mathbf{Q}(z) = \mathbf{I} + \begin{bmatrix} 0.04677 & 0.47289 \\ -0.36062 & 0.33620 \end{bmatrix} z + \begin{bmatrix} -0.03236 & -0.11291 \\ 0.20214 & 0.005223 \end{bmatrix} z^2$$

$$+ \begin{bmatrix} 0.16216 & 0.10151 \\ -0.09692 & 0.13889 \end{bmatrix} z^3,$$

where

$$
\begin{aligned}
det\,\mathbf{P}(z) &= 1 + 0.06674z + 0.26268z^2 - 0.10906z^3 + 0.26110z^4 \\
&\quad + 0.40587z^5 + 0.05266z^6 + 0.056611z^7 - 0.01590z^8 \\
&\quad + 0.044367z^9 + 0.22682z^{10} \\
det\,\mathbf{Q}(z) &= 1 + 0.38297z + 0.15912z^2 + 0.15410z^3 + 0.16611z^4 \\
&\quad - 0.03511z^5 + 0.03236z^6 .
\end{aligned}
$$

The zeros of $det\,\mathbf{P}(z)$ are $1.1905\angle \pm 36.689°$, $1.6207\angle \pm42.463°$, $1.2209\angle \pm112.294°$, $1.4669\angle \pm 114.623°$, $1.3783\angle 180°$, and $2.6789\angle 180°$. The zeros of $det\,\mathbf{Q}(z)$ are $1.9110\angle \pm 59.464°$, $2.1330\angle \pm 70.437°$, $1.3638\angle \pm 146.913°$.

The reconstructed model is given by $\mathbf{P}_p^{-1}(z)\mathbf{Q}_q(z)$, where

$$\mathbf{P}_p(z) = \mathbf{I} + \begin{bmatrix} -0.21844 & 0.36404 \\ -0.17762 & 0.28518 \end{bmatrix} z + \begin{bmatrix} 0.19472 & -0.19278 \\ -0.11445 & 0.06559 \end{bmatrix} z^2$$

$$+ \begin{bmatrix} -0.14883 & 0.2988 \\ 0.03625 & -0.00886 \end{bmatrix} z^3 + \begin{bmatrix} 0.33903 & 0.06644 \\ -0.07965 & -0.0680 \end{bmatrix} z^4$$

$$+ \begin{bmatrix} 0.115498 & 0.12847 \\ -0.07918 & 0.10831 \end{bmatrix} z^5$$

and

$$\mathbf{Q}_q(z) = \begin{bmatrix} 1.0 & 4.8 \times 10^{-15} \\ 4.8 \times 10^{-15} & 1.0 \end{bmatrix} + \begin{bmatrix} 0.046769 & 0.47289 \\ -0.36062 & 0.33620 \end{bmatrix} z$$

$$+ \begin{bmatrix} -0.032361 & -0.11291 \\ 0.202139 & 0.0052226 \end{bmatrix} z^2 + \begin{bmatrix} 0.16216 & 0.01015 \\ -0.096918 & 0.1389 \end{bmatrix} z^3 .$$

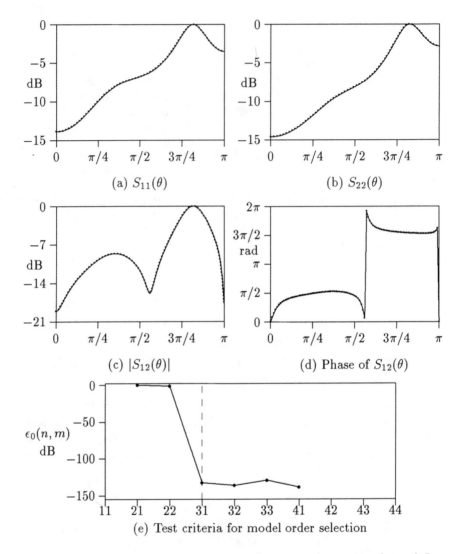

(a) $S_{11}(\theta)$

(b) $S_{22}(\theta)$

(c) $|S_{12}(\theta)|$

(d) Phase of $S_{12}(\theta)$

(e) Test criteria for model order selection

FIGURE 5.6. Original and reconstructed spectra for a two-channel L-ARMA$(3, 1)$ model. In (a)–(d), solid lines represent the original model, and dotted lines represent the reconstructed model. (See next page for details.)

FIGURE 5.6 (cont.). The original two-channel L-ARMA$(3,1)$ system is given by

$$\mathbf{H}(z) \triangleq \mathbf{P}^{-1}(z)\,\mathbf{Q}(z)$$

$$\mathbf{P}(z) \;=\; \mathbf{I} + \begin{bmatrix} 0.04680 & 0.47290 \\ -0.36060 & 0.33620 \end{bmatrix} z + \begin{bmatrix} -0.03240 & -0.11290 \\ 0.20210 & 0.00520 \end{bmatrix} z^2$$

$$+ \begin{bmatrix} 0.16220 & 0.10150 \\ -0.09690 & 0.13890 \end{bmatrix} z^3$$

and

$$\mathbf{Q}(z) \;=\; \mathbf{I} + \begin{bmatrix} -0.85000 & 0.75000 \\ -0.65000 & -0.55000 \end{bmatrix} z \,,$$

where

$$det\,\mathbf{P}(z) \;=\; 1 + 0.38300z + 0.15906z^2 + 0.15417z^3 + 0.16611z^4$$
$$-0.03511z^5 + 0.03237z^6$$

and

$$det\,\mathbf{Q}(z) \;=\; 1 - 1.40000z + 0.95500z^2 \,.$$

The zeros of $det\,\mathbf{P}(z)$ are $1.9109\angle \pm 59.465°$, $2.1329\angle \pm 70.435°$ and $1.3638\angle \pm 146.918°$. The zeros of $det\,\mathbf{Q}(z)$ are $1.0233\angle \pm 44.250°$.

The reconstructed model is given by $\mathbf{P}_p^{-1}(z)\mathbf{Q}_q(z)$, where

$$\mathbf{P}_p(z) \;=\; \mathbf{I} + \begin{bmatrix} 0.046799 & 0.47290 \\ -0.36060 & 0.336199 \end{bmatrix} z + \begin{bmatrix} -0.03240 & -0.11290 \\ 0.20210 & 0.00520 \end{bmatrix} z^2$$

$$+ \begin{bmatrix} 0.16220 & 0.10150 \\ -0.09690 & 0.13890 \end{bmatrix} z^3$$

and

$$\mathbf{Q}_q(z) \;=\; \begin{bmatrix} 1.0 & 9.8 \times 10^{-11} \\ 9.8 \times 10^{-11} & 1.0 \end{bmatrix} + \begin{bmatrix} -0.85001 & 0.74999 \\ -0.64999 & -0.55000 \end{bmatrix} z \,.$$

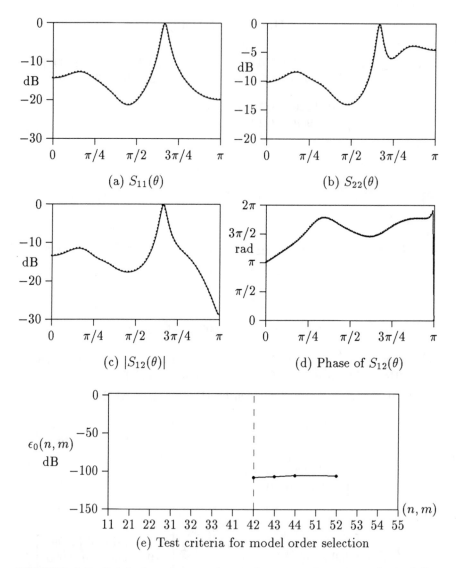

(a) $S_{11}(\theta)$

(b) $S_{22}(\theta)$

(c) $|S_{12}(\theta)|$

(d) Phase of $S_{12}(\theta)$

(e) Test criteria for model order selection

FIGURE 5.7. Original and reconstructed spectra for a two-channel L-ARMA(4, 2) model. In (a)–(d), solid lines represent the original model, and dotted lines represent the reconstructed model. (See next page for details.)

FIGURE 5.7 (cont.). The original two-channel L-ARMA$(4, 2)$ system is given by

$$\mathbf{H}(z) \triangleq \mathbf{P}^{-1}(z)\,\mathbf{Q}(z)$$

$$\mathbf{P}(z) \;=\; \mathbf{I} + \begin{bmatrix} -0.06770 & 0.29700 \\ -0.05950 & 0.22920 \end{bmatrix} z + \begin{bmatrix} 0.34270 & 0.09460 \\ 0.09680 & -0.07180 \end{bmatrix} z^2$$

$$+ \begin{bmatrix} -0.39470 & 0.33810 \\ 0.13160 & 0.06650 \end{bmatrix} z^3 + \begin{bmatrix} 0.31760 & 0.20440 \\ -0.18550 & 0.21070 \end{bmatrix} z^4$$

and

$$\mathbf{Q}(z) \;=\; \mathbf{I} + \begin{bmatrix} -0.29850 & -0.22940 \\ 0.54550 & 0.00403 \end{bmatrix} z + \begin{bmatrix} 0.352201 & -0.19470 \\ -0.25750 & 0.50590 \end{bmatrix} z^2,$$

where

$$det\,\mathbf{P}(z) \;=\; 1 + 0.16150z + 0.27305z^2 - 0.26791z^3 + 0.38060z^4$$
$$+ 0.13174z^5 - 0.02358z^6 - 0.02622z^7 + 0.10483z^8$$

and

$$det\,\mathbf{Q}(z) = 1 - 0.29447z + 0.98203z^2 - 0.10245z^3 + 0.12804z^4.$$

The zeros of $det\,\mathbf{P}(z)$ are $1.3912\angle \pm 32.249°$, $1.4248\angle \pm 59.476°$, $1.0625\angle \pm 119.438°$ and $1.4665\angle \pm 148.690°$. The zeros of $det\,\mathbf{Q}(z)$ are $1.1145\angle \pm 83.332°$ and $2.5075\angle \pm 83.803°$.

The reconstructed model is given by $\mathbf{P}_p^{-1}(z)\mathbf{Q}_q(z)$, where

$$\mathbf{P}_p(z) \;=\; \mathbf{I} + \begin{bmatrix} -0.067699 & 0.29700 \\ -0.059499 & 0.22919 \end{bmatrix} z + \begin{bmatrix} 0.34269 & 0.09460 \\ 0.09680 & -0.07180 \end{bmatrix} z^2$$

$$+ \begin{bmatrix} -0.39470 & 0.33809 \\ 0.13160 & 0.06650 \end{bmatrix} z^3 + \begin{bmatrix} 0.31760 & 0.204399 \\ -0.18550 & 0.21070 \end{bmatrix} z^4$$

and

$$\mathbf{Q}_q(z) \;=\; \begin{bmatrix} 1.0 & -1.2 \times 10^{-11} \\ -1.2 \times 10^{-11} & 1.0 \end{bmatrix} + \begin{bmatrix} -0.29849 & -0.2294 \\ 0.5455 & 0.00403 \end{bmatrix} z$$

$$+ \begin{bmatrix} 0.35219 & -0.19469 \\ -0.25749 & 0.50589 \end{bmatrix} z^2.$$

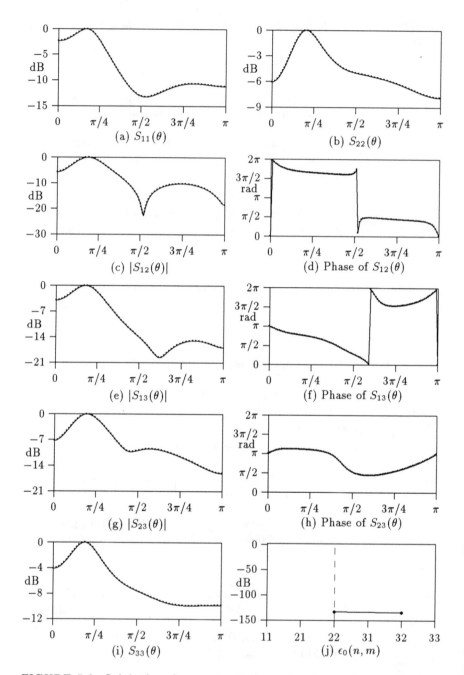

FIGURE 5.8. Original and reconstructed spectra for a three-channel L-ARMA$(2,2)$ model. In (a)–(i), solid lines represent the original model, and dotted lines represent the reconstructed model. (See next page for details.)

FIGURE 5.8 (cont.). The original three-channel L-ARMA$(2,2)$ system is given by

$$\mathbf{H}(z) \triangleq \mathbf{P}^{-1}(z)\,\mathbf{Q}(z)$$

$$\mathbf{P}(z) = \mathbf{I} + \begin{bmatrix} -0.49330 & 0.17050 & -0.0181 \\ -0.21380 & -0.11450 & 0.00220 \\ 0.53460 & 0.12160 & -0.0661 \end{bmatrix} z$$
$$+ \begin{bmatrix} -0.32450 & 0.31050 & -0.1108 \\ -0.29650 & 0.21700 & 0.05820 \\ 0.24840 & -0.05190 & 0.19670 \end{bmatrix} z^2$$

and

$$\mathbf{Q}(z) = \mathbf{I} + \begin{bmatrix} 0.03090 & -0.06910 & -0.0164 \\ -0.09090 & -0.04090 & 0.13640 \\ 0.14360 & -0.00640 & 0.09450 \end{bmatrix} z$$
$$+ \begin{bmatrix} -0.01820 & -0.14550 & 0.1273 \\ 0.31820 & 0.04550 & -0.22730 \\ -0.17270 & 0.11820 & 0.2091 \end{bmatrix} z^2 ,$$

where

$$
\begin{aligned}
det\,\mathbf{P}(z) &= 1 - 0.67390z + 0.23172z^2 - 0.01508z^3 + 0.05364z^4 \\
&\quad + 0.03405z^5 + 0.01203z^6 \\
det\,\mathbf{Q}(z) &= 1 + 0.08450z + 0.23114z^2 - 0.02943z^3 + 0.10121z^4 \\
&\quad + 0.00909z^5 + 0.00910z^6 .
\end{aligned}
$$

The zeros of $det\,\mathbf{P}(z)$ are $1.3935\angle\pm33.705°$, $2.0698\angle\pm90.484°$ and $3.1605\angle$ $\pm143.993°$. The zeros of $det\,\mathbf{Q}(z)$ are $1.8648\angle\pm48.118°$, $3.2094\angle\pm103.239°$ and $1.7520\angle\pm125.186°$.

The reconstructed model is given by $\mathbf{P}_p^{-1}(z)\mathbf{Q}_q(z)$, where

$$\mathbf{P}_p(z) = \mathbf{I} + \begin{bmatrix} -0.49330 & 0.17049 & -0.0181 \\ -0.21380 & -0.11450 & 0.00219 \\ 0.53459 & 0.12159 & -0.06610 \end{bmatrix} z$$
$$+ \begin{bmatrix} -0.32450 & 0.31050 & -0.11080 \\ -0.29650 & 0.21699 & 0.05819 \\ 0.24839 & -0.05190 & 0.19670 \end{bmatrix} z^2$$

and

$$\mathbf{Q}_q(z) = \mathbf{I} + \begin{bmatrix} 0.03090 & -0.06909 & -0.01640 \\ -0.09090 & -0.04090 & 0.13639 \\ 0.14359 & -0.00640 & 0.09449 \end{bmatrix} z$$
$$+ \begin{bmatrix} -0.01820 & -0.14549 & 0.12729 \\ 0.31819 & 0.04549 & -0.22730 \\ -0.17269 & 0.11820 & 0.20910 \end{bmatrix} z^2 .$$

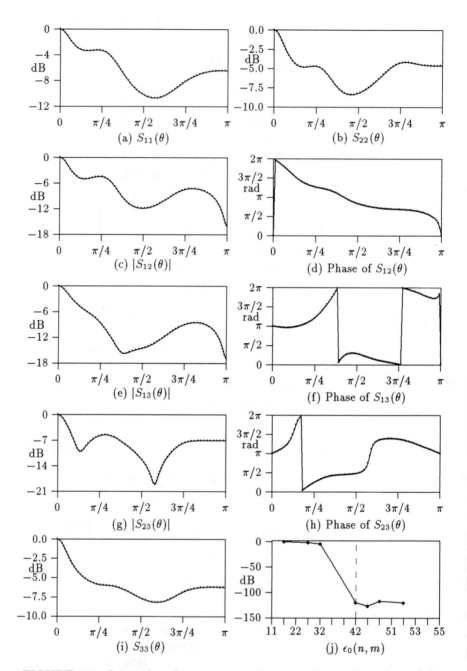

FIGURE 5.9. Original and reconstructed spectra for a three-channel L-ARMA(4, 2) model. In (a)–(i), solid lines represent the original model, and dotted lines represent the reconstructed model. (See next page for details.)

FIGURE 5.9 (cont.). The original three-channel L-ARMA$(4, 2)$ system is given by $\mathbf{H}(z) \overset{\triangle}{=} \mathbf{P}^{-1}(z)\,\mathbf{Q}(z)$, where

$$\mathbf{P}(z) = \mathbf{I} + \begin{bmatrix} -0.225 & -0.091 & -0.038 \\ -0.020 & -0.083 & 0.054 \\ 0.051 & 0.064 & -0.124 \end{bmatrix} z + \begin{bmatrix} -0.030 & 0.147 & 0.124 \\ -0.581 & -0.180 & 0.022 \\ 0.282 & 0.130 & -0.032 \end{bmatrix} z^2$$

$$+ \begin{bmatrix} 0.149 & -0.068 & 0.132 \\ 0.047 & -0.028 & -0.063 \\ -0.137 & -0.067 & 0.069 \end{bmatrix} z^3 + \begin{bmatrix} -0.0002 & -0.063 & 0.050 \\ -0.084 & 0.013 & 0.082 \\ 0.099 & 0.080 & -0.010 \end{bmatrix} z^4$$

and

$$\mathbf{Q}(z) = \mathbf{I} + \begin{bmatrix} 0.032 & 0.207 & -0.432 \\ -0.032 & -0.331 & 0.504 \\ -0.287 & 0.008 & 0.020 \end{bmatrix} z + \begin{bmatrix} 0.278 & 0.073 & -0.170 \\ -0.191 & 0.115 & 0.248 \\ 0.095 & -0.231 & 0.146 \end{bmatrix} z^2,$$

where

$$\begin{aligned} det\,\mathbf{P}(z) &= 1 - 0.4315z - 0.1878z^2 + 0.2142z^3 + 0.0197z^4 - 0.1275z^5 \\ &\quad -0.0414z^6 + 0.015z^7 - 0.0214z^8 - 0.0096z^9 + 0.0007z^{10} \\ &\quad -0.00115z^{11} - 0.0009z^{12} \\ det\,\mathbf{Q}(z) &= 1 - 0.2784z + 0.4007z^2 + 0.0358z^3 + 0.1482z^4 + 0.0047z^5 \\ &\quad +0.0188z^6 \,. \end{aligned}$$

The zeros of $det\,\mathbf{P}(z)$ are $1.2304\angle 0°$, $2.0075\angle \pm 47.547°$, $1.4639\angle \pm 51.564°$, $2.0610\angle \pm 72.389°$, $1.6407\angle \pm 133.237°$, $2.3474\angle \pm 157.953°$ and $1.7489\angle 180°$. The zeros of $det\,\mathbf{Q}(z)$ are $1.4534\angle \pm 53.729°$, $2.5136\angle \pm 84.668°$ and $1.9991\angle \pm 127.547°$.

The reconstructed model is given by $\mathbf{P}_p^{-1}(z)\mathbf{Q}_q(z)$, where

$$\mathbf{P}_p(z) = \mathbf{I} + \begin{bmatrix} -0.225 & -0.091 & -0.038 \\ -0.020 & -0.083 & 0.054 \\ 0.051 & 0.064 & -0.124 \end{bmatrix} z + \begin{bmatrix} -0.030 & 0.147 & 0.124 \\ -0.581 & -0.180 & 0.022 \\ 0.282 & 0.130 & -0.032 \end{bmatrix} z^2$$

$$+ \begin{bmatrix} 0.149 & -0.068 & 0.132 \\ 0.047 & -0.028 & -0.063 \\ -0.137 & -0.067 & 0.069 \end{bmatrix} z^3 + \begin{bmatrix} -0.0002 & -0.063 & 0.05 \\ 0.084 & 0.013 & 0.082 \\ 0.099 & 0.080 & -0.010 \end{bmatrix} z^4$$

and

$$\mathbf{Q}_q(z) = \mathbf{I} + \begin{bmatrix} 0.032 & 0.207 & -0.432 \\ -0.032 & -0.331 & 0.504 \\ -0.287 & 0.008 & 0.020 \end{bmatrix} z + \begin{bmatrix} 0.278 & 0.073 & -0.170 \\ -0.191 & 0.115 & 0.248 \\ 0.095 & -0.231 & 0.146 \end{bmatrix} z^2.$$

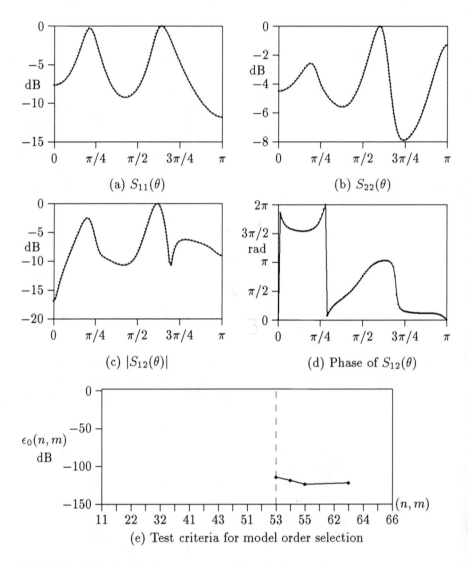

(a) $S_{11}(\theta)$

(b) $S_{22}(\theta)$

(c) $|S_{12}(\theta)|$

(d) Phase of $S_{12}(\theta)$

(e) Test criteria for model order selection

FIGURE 5.10. Original and reconstructed spectra for a two-channel R-ARMA(5, 3) model. In (a)–(d), solid lines represent the original model, and dotted lines represent the reconstructed model. (See next page for details.)

FIGURE 5.10 (cont.). The original two-channel R-ARMA$(5,3)$ system is given by

$$\mathbf{H}(z) \overset{\triangle}{=} \mathbf{Q}_1(z)\mathbf{P}_1^{-1}(z)$$

$$\mathbf{P}_1(z) = \mathbf{I} + \begin{bmatrix} -0.21844 & 0.36404 \\ -0.17762 & 0.28518 \end{bmatrix} z + \begin{bmatrix} 0.19472 & -0.19278 \\ -0.11445 & 0.06559 \end{bmatrix} z^2$$

$$+ \begin{bmatrix} -0.14883 & 0.29881 \\ 0.03625 & -0.00886 \end{bmatrix} z^3 + \begin{bmatrix} 0.33903 & 0.06655 \\ -0.07965 & -0.06801 \end{bmatrix} z^4$$

$$+ \begin{bmatrix} 0.115498 & 0.12847 \\ -0.07918 & 0.10831 \end{bmatrix} z^5$$

and

$$\mathbf{Q}_1(z) = \mathbf{I} + \begin{bmatrix} 0.04677 & 0.47289 \\ -0.36062 & 0.33620 \end{bmatrix} z + \begin{bmatrix} -0.03236 & -0.11291 \\ 0.20214 & 0.005223 \end{bmatrix} z^2$$

$$+ \begin{bmatrix} 0.162155 & 0.101513 \\ -0.096918 & 0.138888 \end{bmatrix} z^3,$$

where

$$\begin{aligned} det\,\mathbf{P}_1(z) &= 1 + 0.06674z + 0.26268z^2 - 0.10906z^3 + 0.26110z^4 \\ &\quad + 0.40587z^5 + 0.05266z^6 + 0.056611z^7 - 0.01590z^8 \\ &\quad + 0.044367z^9 + 0.22682z^{10} \\ det\,\mathbf{Q}_1(z) &= 1 + 0.38297z + 0.15912z^2 + 0.15410z^3 + 0.16611z^4 \\ &\quad - 0.03511z^5 + 0.03236z^6. \end{aligned}$$

The zeros of $det\,\mathbf{P}_1(z)$ are $1.1905\angle \pm 36.689°$, $1.6207\angle \pm 42.463°$, $1.2209\angle \pm 112.294°$, $1.4669\angle \pm 114.623°$, $1.3783\angle 180°$, and $2.6789\angle 180°$. The zeros of $det\,\mathbf{Q}_1(z)$ are $1.9110\angle \pm 59.464°$, $2.1330\angle \pm 70.437°$, $1.3638\angle \pm 146.913°$.

The reconstructed model is given by $\mathbf{Q}_{q_1}(z)\mathbf{P}_{p_1}^{-1}(z)$, where

$$\mathbf{P}_{p_1}(z) = \mathbf{I} + \begin{bmatrix} -0.21844 & 0.36404 \\ -0.17762 & 0.28518 \end{bmatrix} z + \begin{bmatrix} 0.19472 & -0.19278 \\ -0.11445 & 0.06559 \end{bmatrix} z^2$$

$$+ \begin{bmatrix} -0.148830 & 0.29881 \\ 0.036249 & -0.008859 \end{bmatrix} z^3 + \begin{bmatrix} 0.33903 & 0.066439 \\ -0.079649 & -0.06801 \end{bmatrix} z^4$$

$$+ \begin{bmatrix} 0.1154979 & 0.12847 \\ -0.07918 & 0.10831 \end{bmatrix} z^5,$$

and

$$\mathbf{Q}_{q_1}(z) = \mathbf{I} + \begin{bmatrix} 0.04677 & 0.47289 \\ -0.36062 & 0.33620 \end{bmatrix} z + \begin{bmatrix} -0.0323610 & -0.112912 \\ 0.202139 & 0.0052226 \end{bmatrix} z^2$$

$$+ \begin{bmatrix} 0.162155 & 0.010151 \\ -0.0969179 & 0.13889 \end{bmatrix} z^3.$$

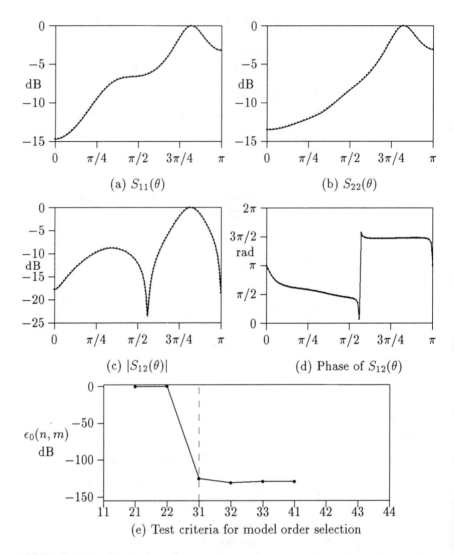

(a) $S_{11}(\theta)$

(b) $S_{22}(\theta)$

(c) $|S_{12}(\theta)|$

(d) Phase of $S_{12}(\theta)$

(e) Test criteria for model order selection

FIGURE 5.11. Original and reconstructed spectra for a two-channel R-ARMA$(3, 1)$ model. In (a)–(d), solid lines represent the original model, and dotted lines represent the reconstructed model. (See next page for details.)

FIGURE 5.11 (cont.). The original two-channel R-ARMA$(3, 1)$ system is given by

$$\mathbf{H}(z) \triangleq \mathbf{Q}_1(z)\mathbf{P}_1^{-1}(z)$$

$$\mathbf{P}_1(z) \;=\; \mathbf{I} + \begin{bmatrix} 0.04680 & 0.47290 \\ -0.36060 & 0.33620 \end{bmatrix} z + \begin{bmatrix} -0.03240 & -0.11290 \\ 0.20210 & 0.00520 \end{bmatrix} z^2$$

$$+ \begin{bmatrix} 0.16220 & 0.10150 \\ -0.09690 & 0.13890 \end{bmatrix} z^3$$

and

$$\mathbf{Q}_1(z) \;=\; \mathbf{I} + \begin{bmatrix} -0.85000 & 0.75000 \\ -0.65000 & -0.55000 \end{bmatrix} z \,,$$

where

$$det\,\mathbf{P}_1(z) \;=\; 1 + 0.38300z + 0.15906z^2 + 0.15417z^3 + 0.16611z^4$$
$$-0.03511z^5 + 0.03237z^6$$

and

$$det\,\mathbf{Q}_1(z) \;=\; 1 - 1.40000z + 0.95500z^2 \,.$$

The zeros of $det\,\mathbf{P}_1(z)$ are $1.911\angle \pm 59.465°$, $2.1329\angle \pm 70.435°$ and $1.364\angle \pm 146.918°$. The zeros of $det\,\mathbf{Q}_1(z)$ are $1.023\angle \pm 44.250°$.

The reconstructed model is given by $\mathbf{Q}_{q_1}(z)\mathbf{P}_{p_1}^{-1}(z)$, where

$$\mathbf{P}_{p_1}(z) \;=\; \mathbf{I} + \begin{bmatrix} 0.046799 & 0.47290 \\ -0.36060 & 0.336199 \end{bmatrix} z + \begin{bmatrix} -0.03240 & -0.11290 \\ 0.20210 & 0.00520 \end{bmatrix} z^2$$

$$+ \begin{bmatrix} 0.16220 & 0.10150 \\ -0.09690 & 0.13890 \end{bmatrix} z^3$$

and

$$\mathbf{Q}_{q_1}(z) \;=\; \begin{bmatrix} 1.0 & -9.8 \times 10^{-11} \\ -9.8 \times 10^{-11} & 1.0 \end{bmatrix} + \begin{bmatrix} -0.85001 & 0.74999 \\ -0.64999 & -0.5500 \end{bmatrix} z \,.$$

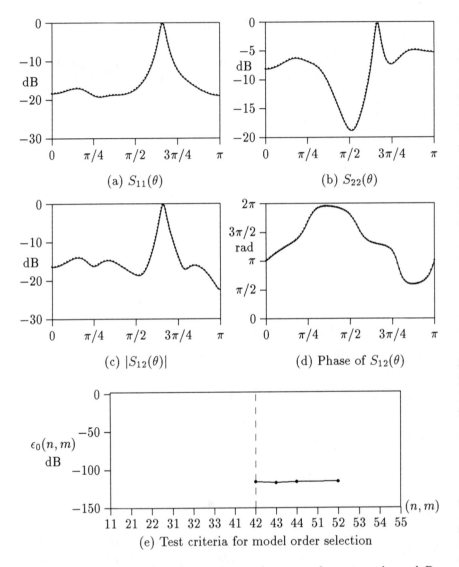

FIGURE 5.12. Original and reconstructed spectra for a two-channel R-ARMA$(4, 2)$ model. In (a)–(d), solid lines represent the original model, and dotted lines represent the reconstructed model. (See next page for details.)

FIGURE 5.12 (cont.). The original two-channel R-ARMA(4, 2) system is given by

$$\mathbf{H}(z) \stackrel{\triangle}{=} \mathbf{Q}_1(z)\mathbf{P}_1^{-1}(z)$$

$$\mathbf{P}_1(z) \;=\; \mathbf{I} + \begin{bmatrix} -0.06770 & 0.29700 \\ -0.05950 & 0.22920 \end{bmatrix} z + \begin{bmatrix} 0.34270 & 0.09460 \\ 0.09680 & -0.07180 \end{bmatrix} z^2$$

$$+ \begin{bmatrix} -0.39470 & 0.33810 \\ 0.13160 & 0.06650 \end{bmatrix} z^3 + \begin{bmatrix} 0.31760 & 0.20440 \\ -0.18550 & 0.21070 \end{bmatrix} z^4$$

and

$$\mathbf{Q}_1(z) \;=\; \mathbf{I} + \begin{bmatrix} -0.29850 & -0.22940 \\ 0.54550 & 0.00403 \end{bmatrix} z + \begin{bmatrix} 0.352201 & -0.19470 \\ -0.25750 & 0.50590 \end{bmatrix} z^2,$$

where

$$det\,\mathbf{P}_1(z) \;=\; 1 + 0.16150z + 0.27305z^2 - 0.26791z^3 + 0.38060z^4$$
$$+0.13174z^5 - 0.02358z^6 - 0.02622z^7 + 0.10483z^8$$

and

$$det\,\mathbf{Q}_1(z) = 1 - 0.29447z + 0.98203z^2 - 0.10245z^3 + 0.12804z^4 \,.$$

The zeros of $det\,\mathbf{P}_1(z)$ are $1.3912\angle \pm 32.249°$, $1.4248\angle \pm 59.476°$, $1.0625\angle \pm119.438°$ and $1.4665\angle \pm 148.690°$. The zeros of $det\,\mathbf{Q}_1(z)$ are $1.1145\angle \pm83.332°$ and $2.5075\angle \pm 83.803°$.

The reconstructed model is given by $\mathbf{Q}_{q_1}(z)\mathbf{P}_{p_1}^{-1}(z)$, where

$$\mathbf{P}_{p_1}(z) \;=\; \mathbf{I} + \begin{bmatrix} -0.067699 & 0.29700 \\ -0.059499 & 0.22919 \end{bmatrix} z + \begin{bmatrix} 0.34269 & 0.09460 \\ 0.09680 & -0.07180 \end{bmatrix} z^2$$

$$+ \begin{bmatrix} -0.39470 & 0.33809 \\ 0.13160 & 0.06650 \end{bmatrix} z^3 + \begin{bmatrix} 0.31760 & 0.204399 \\ -0.18550 & 0.21070 \end{bmatrix} z^4$$

and

$$\mathbf{Q}_{q_1}(z) \;=\; \begin{bmatrix} 1.0 & 4.6 \times 10^{-13} \\ 4.6 \times 10^{-13} & 1.0 \end{bmatrix} + \begin{bmatrix} -0.2985 & -0.2294 \\ 0.5455 & 0.00403 \end{bmatrix} z$$

$$+ \begin{bmatrix} 0.3522 & -0.1947 \\ -0.2575 & 0.5059 \end{bmatrix} z^2.$$

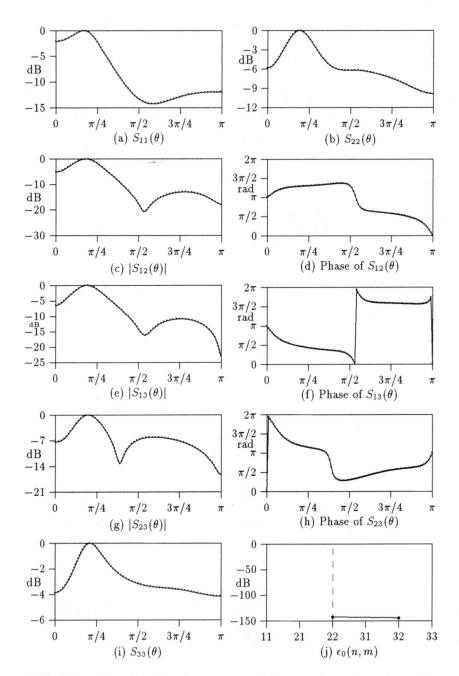

FIGURE 5.13. Original and reconstructed spectra for a three-channel R-ARMA(2, 2) model. In (a)–(i), solid lines represent the original model, and dotted lines represent the reconstructed model. (See next page for details.)

FIGURE 5.13 (cont.). The original three-channel R-ARMA$(2, 2)$ system is given by

$$\mathbf{H}(z) \triangleq \mathbf{Q}_1(z)\mathbf{P}_1^{-1}(z)$$

$$\mathbf{P}_1(z) = \mathbf{I} + \begin{bmatrix} -0.49330 & 0.17050 & -0.0181 \\ -0.21380 & -0.11450 & 0.00220 \\ 0.53460 & 0.12160 & -0.0661 \end{bmatrix} z$$

$$+ \begin{bmatrix} -0.32450 & 0.31050 & -0.1108 \\ -0.29650 & 0.21700 & 0.05820 \\ 0.24840 & -0.05190 & 0.19670 \end{bmatrix} z^2$$

and

$$\mathbf{Q}_1(z) = \mathbf{I} + \begin{bmatrix} 0.03090 & -0.06910 & -0.0164 \\ -0.09090 & -0.04090 & 0.13640 \\ 0.14360 & -0.00640 & 0.09450 \end{bmatrix} z$$

$$+ \begin{bmatrix} -0.01820 & -0.14550 & 0.1273 \\ 0.31820 & 0.04550 & -0.22730 \\ -0.17270 & 0.11820 & 0.2091 \end{bmatrix} z^2,$$

where

$$\begin{aligned} det\,\mathbf{P}_1(z) &= 1 - 0.67390z + 0.23172z^2 - 0.01508z^3 + 0.05364z^4 \\ &\quad + 0.03405z^5 + 0.01203z^6 \\ det\,\mathbf{Q}_1(z) &= 1 + 0.08450z + 0.23114z^2 - 0.02943z^3 + 0.10121z^4 \\ &\quad + 0.00909z^5 + 0.00910z^6 . \end{aligned}$$

The zeros of $det\,\mathbf{P}_1(z)$ are $1.394\angle \pm 33.705°$, $2.070\angle \pm 90.484°$ and $3.161\angle \pm 143.993°$. The zeros of $det\,\mathbf{Q}_1(z)$ are $1.865\angle \pm 48.118°$, $3.209\angle \pm 103.239°$ and $1.752\angle \pm 125.186°$.

The reconstructed model is given by $\mathbf{Q}_{q_1}(z)\mathbf{P}_{p_1}^{-1}(z)$, where

$$\mathbf{P}_{p_1}(z) = \mathbf{I} + \begin{bmatrix} -0.49330 & 0.17049 & -0.0181 \\ -0.21380 & -0.11450 & 0.00219 \\ 0.53459 & 0.12159 & -0.06610 \end{bmatrix} z$$

$$+ \begin{bmatrix} -0.32450 & 0.31050 & -0.11080 \\ -0.29650 & 0.21699 & 0.05819 \\ 0.24839 & -0.05190 & 0.19670 \end{bmatrix} z^2,$$

and

$$\mathbf{Q}_{q_1}(z) = \mathbf{I} + \begin{bmatrix} 0.03090 & -0.06909 & -0.01640 \\ -0.09090 & -0.04090 & 0.13639 \\ 0.14359 & -0.00640 & 0.09449 \end{bmatrix} z$$

$$+ \begin{bmatrix} -0.01820 & -0.14549 & 0.12729 \\ 0.31819 & 0.04549 & -0.22730 \\ -0.17269 & 0.11820 & 0.20910 \end{bmatrix} z^2.$$

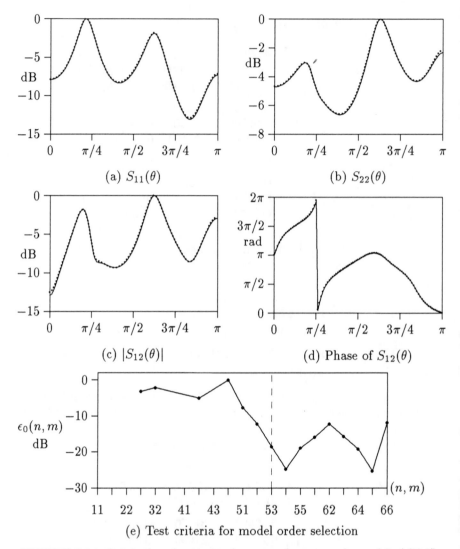

(a) $S_{11}(\theta)$

(b) $S_{22}(\theta)$

(c) $|S_{12}(\theta)|$

(d) Phase of $S_{12}(\theta)$

(e) Test criteria for model order selection

FIGURE 5.14. Original and estimated spectra for a two-channel L-ARMA $(5,3)$ model. In (a)–(d), solid lines represent the original model, and dotted lines represent the estimated model using 3000 samples. (See next page for details.)

FIGURE 5.14 (cont.). The original two-channel L-ARMA$(5,3)$ system is as given in Fig. 5.5. The estimated model using 3000 samples is given by $\hat{\mathbf{P}}_p^{-1}(z)\hat{\mathbf{Q}}_q(z)$, where

$$\hat{\mathbf{P}}_p(z) = \mathbf{I} + \begin{bmatrix} -0.24247 & 0.37239 \\ -0.14842 & 0.23599 \end{bmatrix} z + \begin{bmatrix} 0.19723 & -0.23340 \\ -0.12063 & 0.07362 \end{bmatrix} z^2$$

$$+ \begin{bmatrix} -0.14973 & 0.29189 \\ 0.06101 & -0.01659 \end{bmatrix} z^3 + \begin{bmatrix} 0.35989 & 0.06069 \\ -0.10143 & -0.05663 \end{bmatrix} z^4$$

$$+ \begin{bmatrix} 0.09865 & 0.12809 \\ -0.05586 & 0.11114 \end{bmatrix} z^5$$

and

$$\hat{\mathbf{Q}}_q(z) = \begin{bmatrix} 0.99672 & -0.000705 \\ -0.000705 & 1.00731 \end{bmatrix} + \begin{bmatrix} 0.01642 & 0.48216 \\ -0.32997 & 0.28434 \end{bmatrix} z$$

$$+ \begin{bmatrix} -0.03978 & -0.15788 \\ 0.21218 & -0.01785 \end{bmatrix} z^2 + \begin{bmatrix} 0.17093 & 0.09295 \\ -0.10138 & 0.13297 \end{bmatrix} z^3,$$

where

$$\begin{aligned} det\,\hat{\mathbf{P}}_p(z) &= 1 - 0.00648z + 0.26890z^2 - 0.12734z^3 + 0.27891z^4 \\ &\quad + 0.39039z^5 + 0.01980z^6 + 0.06001z^7 - 0.02401z^8 \\ &\quad + 0.05079z^9 + 0.01812z^{10} \end{aligned}$$

and

$$\begin{aligned} det\,\hat{\mathbf{Q}}_q(z) &= 1.004 + 0.30006z + 0.14153z^2 + 0.13928z^3 + 0.16312z^4 \\ &\quad - 0.03797z^5 + 0.03215z^6. \end{aligned}$$

The zeros of $det\,\hat{\mathbf{P}}_p(z)$ are $1.1907\angle \pm 38.993°$, $1.6015\angle \pm 41.850°$, $1.2189\angle \pm 112.549°$, $1.4678\angle \pm 114.869°$, $1.3451\angle 180°$ and $3.5254\angle 180°$. The zeros of $det\,\hat{\mathbf{Q}}_q(z)$ are $1.8530\angle \pm 57.748°$, $2.1737\angle \pm 69.858°$ and $1.3874\angle \pm 145.757°$.

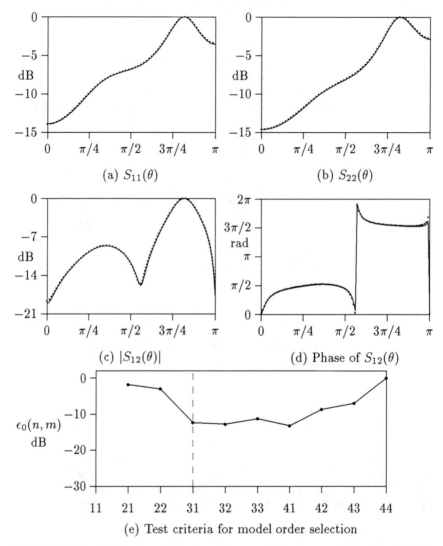

FIGURE 5.15. Original and estimated spectra for a two-channel L-ARMA (3, 1) model. In (a)–(d), solid lines represent the original model, and dotted lines represent the estimated model using 3000 samples. (See next page for details.)

FIGURE 5.15 (cont.). The original two-channel L-ARMA$(3,1)$ system is as given in Fig. 5.6. The estimated model using 3000 samples is given by $\hat{\mathbf{P}}_p^{-1}(z)\hat{\mathbf{Q}}_q(z)$, where

$$\hat{\mathbf{P}}_p(z) = \mathbf{I} + \begin{bmatrix} 0.04674 & 0.46823 \\ -0.36750 & 0.37205 \end{bmatrix} z + \begin{bmatrix} -0.03407 & -0.11633 \\ 0.19999 & 0.02613 \end{bmatrix} z^2$$

$$+ \begin{bmatrix} 0.15983 & 0.09893 \\ -0.08048 & 0.15098 \end{bmatrix} z^3$$

and

$$\hat{\mathbf{Q}}_q(z) = \begin{bmatrix} 1.0526 & 0.00564 \\ 0.00564 & 1.0524 \end{bmatrix} + \begin{bmatrix} -0.85365 & 0.74720 \\ -0.65385 & -0.51223 \end{bmatrix} z,$$

where

$$det\,\hat{\mathbf{P}}_p(z) = 1 + 0.4188z + 0.1815z^2 + 0.1630z^3 + 0.1629z^4$$
$$-0.0301z^5 + 0.0321z^6$$

and

$$det\,\hat{\mathbf{Q}}_q(z) = 1 - 1.43809z + 0.92582z^2.$$

The zeros of $det\,\hat{\mathbf{P}}_p(z)$ are $1.9509\angle\pm59.639°$, $2.0888\angle\pm72.287°$ and $1.3698\angle\pm147.275°$. The zeros of $det\,\hat{\mathbf{Q}}_q(z)$ are $1.0938\angle\pm44.762°$.

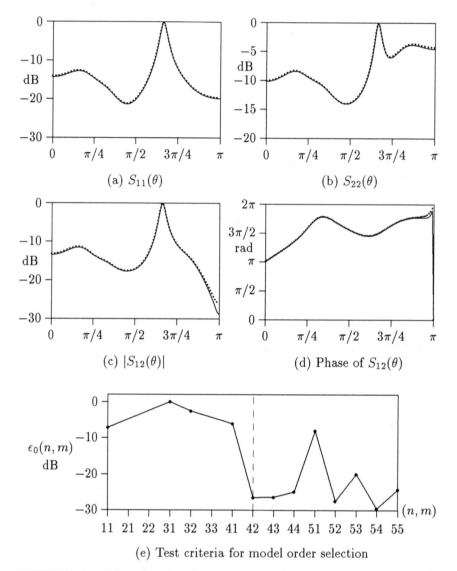

(a) $S_{11}(\theta)$

(b) $S_{22}(\theta)$

(c) $|S_{12}(\theta)|$

(d) Phase of $S_{12}(\theta)$

(e) Test criteria for model order selection

FIGURE 5.16. Original and estimated spectra for a two-channel L-ARMA $(4, 2)$ model. In (a)–(d), solid lines represent the original model, and dotted lines represent the estimated model using 3000 samples. (See next page for details.)

FIGURE 5.16. (cont.). The original two-channel L-ARMA $(4, 2)$ system is as given in Fig. 5.7. The estimated model using 3000 samples is given by $\hat{\mathbf{P}}_p^{-1}(z)\hat{\mathbf{Q}}_q(z)$, where

$$\hat{\mathbf{P}}_p(z) = \mathbf{I} + \begin{bmatrix} -0.06681 & 0.29558 \\ -0.06132 & 0.21668 \end{bmatrix} z + \begin{bmatrix} 0.34607 & 0.10472 \\ 0.10597 & -0.07743 \end{bmatrix} z^2$$

$$+ \begin{bmatrix} -0.39612 & 0.34512 \\ 0.13312 & 0.07803 \end{bmatrix} z^3 + \begin{bmatrix} 0.31631 & 0.20122 \\ -0.18669 & 0.21308 \end{bmatrix} z^4$$

and

$$\hat{\mathbf{Q}}_q(z) = \begin{bmatrix} 1.00277 & -0.00850 \\ -0.00850 & 1.01452 \end{bmatrix} + \begin{bmatrix} -0.30033 & -0.23107 \\ 0.54170 & -0.01716 \end{bmatrix} z$$

$$+ \begin{bmatrix} 0.35124 & -0.18062 \\ -0.25684 & 0.50858 \end{bmatrix} z^2,$$

where

$$det\,\hat{\mathbf{P}}_p(z) = 1 + 0.14987z + 0.27228z^2 - 0.26284z^3 + 0.38226z^4$$
$$+0.12899z^5 - 0.02937z^6 - 0.02208z^7 + 0.10496z^8$$

and

$$det\,\hat{\mathbf{Q}}_q(z) = 1.0173 - 0.31926z + 0.99293z^2 - 0.12027z^3 + 0.13224z^4.$$

The zeros of $det\,\hat{\mathbf{P}}_p(z)$ are $1.3897\angle \pm 32.132°$, $1.4143\angle \pm 59.566°$, $1.0632\angle \pm119.444°$ and $1.4770\angle \pm 148.957°$. The zeros of $det\,\hat{\mathbf{Q}}_q(z)$ are $2.4640\angle \pm82.458°$ and $1.1256\angle \pm 83.300°$.

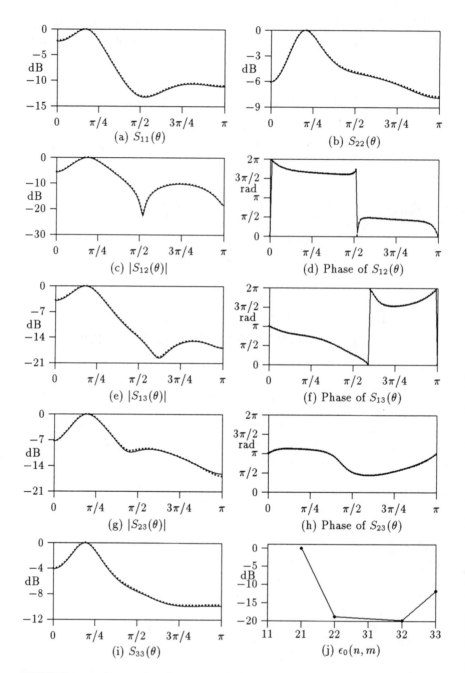

FIGURE 5.17. Original and estimated spectra for a three-channel L-ARMA $(2, 2)$ model. In (a)–(i), solid lines represent the original model, and dotted lines represent the estimated model from 3000 samples. (See next page for details.)

FIGURE 5.17 (cont.). The original three-channel L-ARMA$(2, 2)$ system is as given in Fig. 5.8. The estimated model using 3000 samples is given by $\hat{\mathbf{P}}_p^{-1}(z)\hat{\mathbf{Q}}_q(z)$, where

$$\hat{\mathbf{P}}_p(z) = \mathbf{I} + \begin{bmatrix} -0.49112 & 0.17083 & -0.01952 \\ -0.24299 & -0.14120 & -0.07712 \\ 0.52102 & 0.12139 & -0.08290 \end{bmatrix} z$$

$$+ \begin{bmatrix} -0.32422 & 0.30659 & -0.11095 \\ -0.31875 & 0.22476 & 0.07493 \\ 0.25716 & -0.04078 & 0.22211 \end{bmatrix} z^2$$

and

$$\hat{\mathbf{Q}}_q(z) = \begin{bmatrix} 0.99994 & -0.00108 & 0.00104 \\ -0.00108 & 0.01204 & 0.00409 \\ 0.00104 & 0.00409 & 1.00363 \end{bmatrix}$$

$$+ \begin{bmatrix} 0.02886 & -0.07086 & -0.01626 \\ -0.11812 & -0.07368 & 0.05216 \\ 0.13416 & -0.00818 & 0.07830 \end{bmatrix} z$$

$$+ \begin{bmatrix} -0.01817 & -0.14957 & 0.12565 \\ 0.30787 & 0.07374 & -0.22169 \\ -0.16487 & 0.13823 & 0.23494 \end{bmatrix} z^2,$$

where

$$det\,\hat{\mathbf{P}}_p(z) = 1 - 0.71522z + 0.30546z^2 - 0.03878z^3 + 0.04483z^4 \\ + 0.03691z^5 + 0.01541z^6$$

and

$$det\,\hat{\mathbf{Q}}_q(z) = 1 + 0.03411z + 0.28243z^2 - 0.02886z^3 + 0.11136z^4 \\ + 0.00346z^5 + 0.01136z^6.$$

The zeros of $det\,\hat{\mathbf{P}}_p(z)$ are $1.3953\angle\pm33.904°$, $1.8969\angle\pm86.965°$ and $3.0438\angle \pm 143.805°$. The zeros of $det\,\hat{\mathbf{Q}}_q(z)$ are $1.8568\angle\pm49.801°$, $2.8866\angle\pm96.866°$ and $1.7645\angle \pm 124.744°$.

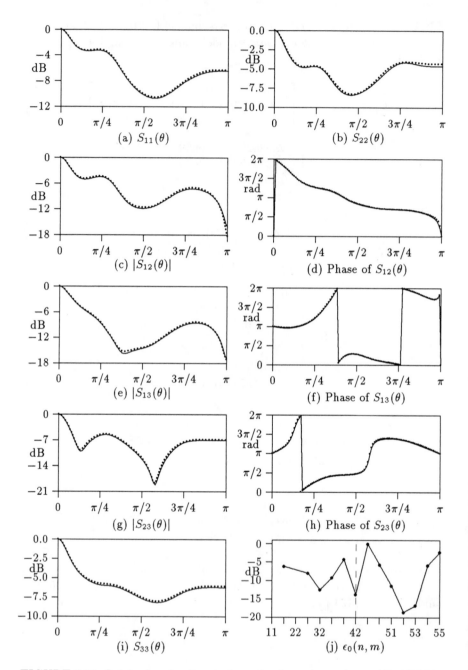

FIGURE 5.18. Original and estimated spectra for a three-channel L-ARMA $(4, 2)$ model. In (a)–(i), solid lines represent the original model, and dotted lines represent the estimated model using 3000 samples. (See next page for details.)

FIGURE 5-18 (cont.). The original three-channel L-ARMA(4, 2) system is as given in Figure 5.9. The estimated model using 3000 samples is given by $\hat{\mathbf{P}}_p^{-1}(z)\hat{\mathbf{Q}}_q(z)$, where

$$
\hat{\mathbf{P}}_p(z) = \mathbf{I} + \begin{bmatrix} -0.18786 & -0.17821 & -0.18396 \\ 0.06292 & -0.47241 & -0.51554 \\ 0.01392 & 0.17570 & 0.02346 \end{bmatrix} z
$$
$$
+ \begin{bmatrix} -0.05763 & 0.14807 & 0.15804 \\ -0.58402 & -0.12184 & 0.30176 \\ 0.28651 & 0.12620 & -0.16934 \end{bmatrix} z^2
$$
$$
+ \begin{bmatrix} 0.16093 & -0.08337 & 0.17421 \\ 0.13868 & -0.07729 & 0.10574 \\ -0.18152 & -0.05769 & 0.02539 \end{bmatrix} z^3
$$
$$
+ \begin{bmatrix} 0.04063 & -0.05139 & 0.04190 \\ 0.04540 & 0.09680 & 0.03947 \\ 0.04540 & 0.03592 & 0.00033 \end{bmatrix} z^4
$$

and

$$
\hat{\mathbf{Q}}_q(z) = \begin{bmatrix} 1.00143 & -0.00323 & 0.00021 \\ -0.00323 & 1.01408 & 0.00401 \\ 0.00021 & 0.00401 & 1.00407 \end{bmatrix}
$$
$$
+ \begin{bmatrix} 0.06457 & 0.11346 & -0.57617 \\ 0.05609 & -0.73552 & -0.07472 \\ -0.32286 & 0.12194 & 0.16948 \end{bmatrix} z
$$
$$
+ \begin{bmatrix} 0.30564 & 0.11486 & -0.21202 \\ 0.02368 & 0.33590 & 0.24896 \\ 0.04199 & -0.28168 & 0.09171 \end{bmatrix} z^2,
$$

where

$$
\begin{aligned}
det\,\hat{\mathbf{P}}_p(z) = {} & 1 - 0.6368z - 0.1712z^2 + 0.1993z^3 + 0.0414z^4 - 0.1035z^5 \\
& -0.4910z^6 + 0.0236z^7 + 0.0009z^8 - 0.0071z^9 + 0.0003z^{10} \\
& +0.0004z^{11} - 0.0003z^{12}
\end{aligned}
$$

and

$$
\begin{aligned}
det\,\hat{\mathbf{Q}}_q(z) = {} & 1.0196 - 0.5012z + 0.3952z^2 - 0.1347z^3 + 0.1667z^4 \\
& -0.0369z^5 + 0.0362z^6 .
\end{aligned}
$$

5.4   Multichannel Rational Approximation of Nonrational Systems

A multichannel nonrational system has nonrational entries in its matrix representation. Thus for example, the system transfer function

$$\mathbf{B}(z) = \begin{bmatrix} e^{-z} & z^2 \\ z & 1/(e^z + 1) \end{bmatrix}$$

is nonrational and in general the 'numerator' and 'denominator' factors in its left-coprime representation

$$\mathbf{B}(z) = \mathbf{P}^{-1}(z)\mathbf{Q}(z) \tag{5.342}$$

and the right-coprime representation

$$\mathbf{B}(z) = \mathbf{Q}_1(z)\mathbf{P}_1^{-1}(z) \tag{5.343}$$

cannot be finite-order matrix polynomials. Thus for nonrational systems

$$\mathbf{P}(z) = \sum_{k=0}^{\infty} \mathbf{P}_k z^k, \quad \mathbf{Q}(z) = \sum_{k=0}^{\infty} \mathbf{Q}_k z^k$$

and the input-output relation

$$\mathbf{X}(z) = \mathbf{B}(z)\mathbf{W}(z)$$

of such systems can be expressed in terms of the infinite recursion (with $\mathbf{P}_0 = \mathbf{I}$)

$$\mathbf{x}(n) = -\sum_{k=1}^{\infty} \mathbf{P}_k \mathbf{x}(n-k) + \sum_{k=0}^{\infty} \mathbf{Q}_k \mathbf{w}(n-k). \tag{5.344}$$

The problem now is, as before, how does one simulate such a system?

Although every term in (5.344) cannot be faithfully represented in a practical setup, the straightforward alternative of truncating the above infinite series is not to be recommended for two reasons: First, even if the original system is stable/minimum-phase, the truncated system need not retain these properties, and secondly if the terms do not decay fast, such a truncation can lead to severe round-off error. The first difficulty can be avoided by approximating the nonrational system by truncating its power series expansion

$$\mathbf{B}(z) = \sum_{k=0}^{\infty} \mathbf{B}_k z^k \tag{5.345}$$

to

$$\mathbf{B}_0 + \mathbf{B}_1 z + \cdots + \mathbf{B}_n z^n.$$

However, it may be possible to realize such a matrix polynomial approximation by making use of a 'simpler' rational approximation such as

$$\mathbf{H}_l(z) = \left(\mathbf{P}_0 + \mathbf{P}_1 z + \cdots + \mathbf{P}_p z^p\right)^{-1}\left(\mathbf{Q}_0 + \mathbf{Q}_1 z + \cdots + \mathbf{Q}_q z^q\right)$$
$$\triangleq \mathbf{P}^{-1}(z)\mathbf{Q}(z) \tag{5.346}$$

or

$$\mathbf{H}_r(z) = \left(\tilde{\mathbf{Q}}_0 + \tilde{\mathbf{Q}}_1 z + \cdots + \tilde{\mathbf{Q}}_q z^q\right)\left(\tilde{\mathbf{P}}_0 + \tilde{\mathbf{P}}_1 z + \cdots + \tilde{\mathbf{P}}_p z^p\right)^{-1}$$
$$\triangleq \mathbf{Q}_1(z)\mathbf{P}_1^{-1}(z) \tag{5.347}$$

as in the case of Padé approximations. Thus the L-ARMA(p, q) system in (5.346) is said to be a Padé approximation to $\mathbf{B}(z) = \sum_{k=0}^{\infty} \mathbf{B}_k z^k$ if at least the first $p + q + 1$ terms in the power series expansion of $\mathbf{B}(z)$ and $\mathbf{P}^{-1}(z)\mathbf{Q}(z)$ match exactly, i.e.,

$$\mathbf{B}(z) - \mathbf{P}^{-1}(z)\mathbf{Q}(z) = O(z^{p+q+1}). \tag{5.348}$$

Similarly, the R-ARMA(p, q) system in (5.347) is said to be a Padé approximation to $\mathbf{B}(z)$ if

$$\mathbf{B}(z) - \mathbf{Q}_1(z)\mathbf{P}_1^{-1}(z) = O(z^{p+q+1}), \tag{5.349}$$

where $\delta_o(\mathbf{P}_1(z)) = p$ and $\delta_o(\mathbf{Q}_1(z)) = q$. If such Padé approximations exist, then they represent the same rational system. This follows from (5.348)–(5.349), since from there

$$\mathbf{P}^{-1}(z)\mathbf{Q}(z) - \mathbf{Q}_1(z)\mathbf{P}_1^{-1}(z) = O(z^{p+q+1})$$

or

$$\mathbf{Q}(z)\mathbf{P}_1(z) - \mathbf{P}(z)\mathbf{Q}_1(z) = O(z^{p+q+1})$$

and since the left side is at most of order z^{p+q}, this gives

$$\mathbf{Q}(z)\mathbf{P}_1(z) - \mathbf{P}(z)\mathbf{Q}_1(z) \equiv 0$$

or

$$\mathbf{H}_l(z) \triangleq \mathbf{P}^{-1}(z)\mathbf{Q}(z) = \mathbf{Q}_1(z)\mathbf{P}_1^{-1}(z) \triangleq \mathbf{H}_r(z),$$

a rational system whose $p+q+1$ terms match with those of the given system $\mathbf{B}(z)$. Since such Padé approximations only involve a finite number of terms $\mathbf{B}_0 \rightarrow \mathbf{B}_{p+q}$ of the actual system, the crucial asymptotic characteristics of the original nonrational system can never be faithfully captured by this approach. Further, such rational approximations need not be stable.

Following the approach developed in the previous chapter for the nonrational case, our solution to this problem tries to impose Padé-like constraints on the power spectrum associated with the given transfer function.

However, for each transfer function $\mathbf{B}(z)$, two distinct spectral density matrices can be defined as

$$\mathbf{S}(\theta) = \mathbf{B}(e^{j\theta})\mathbf{B}^*(e^{j\theta}) = \sum_{k=-\infty}^{\infty} \mathbf{r}_k e^{jk\theta} \tag{5.350}$$

or

$$\mathbf{S}_1(\theta) = \mathbf{B}^*(e^{j\theta})\mathbf{B}(e^{j\theta}) = \sum_{k=-\infty}^{\infty} \tilde{\mathbf{r}}_k e^{jk\theta}, \tag{5.351}$$

and the desired form should be clear from the context as in (5.30), (5.33). For example, if the input-output relation of a nonrational system is of the form $\mathbf{X}(z) = \mathbf{B}(z)\mathbf{W}(z)$, where $\mathbf{W}(z)$ is a stationary white noise process with spectral density matrix $\mathbf{S_W}(\theta) = \mathbf{I}$, then the power spectral density matrix of the nonrational process $\mathbf{x}(n)$ is given by (5.350) and hereonwards, we will denote $\mathbf{S}(\theta)$ to represent either (5.350) or (5.351).

5.4.1   MULTICHANNEL PADÉ-LIKE APPROXIMATION

Let

$$\mathbf{S}(\theta) = \sum_{k=-\infty}^{\infty} \mathbf{r}_k e^{jk\theta} \geq 0 \tag{5.352}$$

represent one of the power spectral density matrices in (5.350)–(5.351) associated with the nonrational matrix transfer function $\mathbf{B}(z)$. A rational L-ARMA(p,q) system $\mathbf{H}_l(z)$ as in (5.346) is said to be a left Padé-like approximation to the above nonrational power spectrum $\mathbf{S}(\theta)$ if

$$\mathbf{H}_l(e^{j\theta})\mathbf{H}_l^*(e^{j\theta}) = \sum_{k=-n}^{n} \mathbf{r}_k e^{jk\theta} + O(e^{j(n+1)\theta}), \tag{5.353}$$

where $n \geq p+q$. Similarly, a rational R-ARMA(p,q) system $\mathbf{H}_r(z)$ as in (5.347) is said to be a right Padé-like approximation to the above $\mathbf{S}(\theta)$ if

$$\mathbf{H}_r^*(e^{j\theta})\mathbf{H}_r(e^{j\theta}) = \sum_{k=-n}^{n} \mathbf{r}_k e^{jk\theta} + O(e^{j(n+1)\theta}), \tag{5.354}$$

where $n \geq p+q$, i.e., the matrix Fourier coefficients of the rational approximation (L-ARMA(p,q) or R-ARMA(p,q)) must match with the matrix Fourier coefficients of the given nonrational spectrum *at least* up to $p+q+1$ terms. In that case, $\mathbf{H}_l(z)$ and/or $\mathbf{H}_r(z)$ represent rational matrices with the property that

$$\mathbf{H}_l(e^{j\theta})\mathbf{H}_l^*(e^{j\theta}) - \sum_{k=-n}^{n} \mathbf{r}_k e^{jk\theta} = \mathbf{H}_r^*(e^{j\theta})\mathbf{H}_r(e^{j\theta}) - \sum_{k=-n}^{n} \mathbf{r}_k e^{jk\theta}$$

$$= O(e^{j(n+1)\theta}), \tag{5.355}$$

where $n \geq p + q$. Notice that p, q are not directly related to the degrees of the entries of $\mathbf{H}_l(z)$ or $\mathbf{H}_r(z)$ and must be obtained from the respective left- and right-coprime representations. Although both $\mathbf{H}_l(z)$ and $\mathbf{H}_r(z)$ 'match' the first $p + q + 1$ autocorrelation matrices of $\mathbf{S}(\theta)$ as in (5.355), their complexities as measured by the respective McMillan degrees can be quite different.

From (5.37)–(5.42), any power spectral density matrix that satisfies the integrability condition (5.34) and the physical-realizability condition (5.35) can be factored as

$$\mathbf{S}(\theta) = \mathbf{B}_l(e^{j\theta})\mathbf{B}_l^*(e^{j\theta}) = \mathbf{B}_r^*(e^{j\theta})\mathbf{B}_r(e^{j\theta}) = \sum_{k=-\infty}^{\infty} \mathbf{r}_k e^{jk\theta}, \qquad (5.356)$$

where $\mathbf{B}_l(z)$ and $\mathbf{B}_r(z)$ represent the left- and right-Wiener factors of $\mathbf{S}(\theta)$ that are minimum-phase together with their inverses. Naturally, if $\mathbf{S}(\theta)$ is nonrational, then $\mathbf{B}_l(z)$, $\mathbf{B}_r(z)$ as well as the corresponding positive matrix function

$$\mathbf{Z}(z) = \mathbf{r}_0 + 2\sum_{k=1}^{\infty} \mathbf{r}_k z^k, \quad |z| < 1, \qquad (5.357)$$

are all nonrational functions. From the class of all spectral extension formula in (5.257)–(5.259) every extension that matches at least the first $p + q + 1$ autocorrelation matrices $\mathbf{r}_0$, $\mathbf{r}_1$, ..., $\mathbf{r}_{p+q}$ is parametrized by an arbitrary bounded matrix function $\boldsymbol{\rho}(z)$ and the associated Wiener factors are given by

$$\mathbf{B}_l(z) = \Big(\mathbf{A}_n(z) - z\boldsymbol{\rho}(z)\tilde{\mathbf{C}}_n(z)\Big)^{-1}\boldsymbol{\Gamma}_l(z) \qquad (5.358)$$

and

$$\mathbf{B}_r(z) = \boldsymbol{\Gamma}_r(z)\Big(\mathbf{C}_n(z) - z\tilde{\mathbf{A}}_n(z)\boldsymbol{\rho}(z)\Big)^{-1}, \qquad (5.359)$$

where $n \geq p + q$. Here $\boldsymbol{\Gamma}_l(z)$ and $\boldsymbol{\Gamma}_r(z)$ represent the minimum-phase factors associated with the factorizations

$$\boldsymbol{\Gamma}_l(e^{j\theta})\boldsymbol{\Gamma}_l^*(e^{j\theta}) = \mathbf{I} - \boldsymbol{\rho}(e^{j\theta})\boldsymbol{\rho}^*(e^{j\theta}) \qquad (5.360)$$

$$\boldsymbol{\Gamma}_r^*(e^{j\theta})\boldsymbol{\Gamma}_r(e^{j\theta}) = \mathbf{I} - \boldsymbol{\rho}^*(e^{j\theta})\boldsymbol{\rho}(e^{j\theta}) \qquad (5.361)$$

that are guaranteed to exist under the causality criterion

$$\frac{1}{2\pi} \int_{-\pi}^{\pi} \ln\det\big(\mathbf{I} - \boldsymbol{\rho}(e^{j\theta})\boldsymbol{\rho}^*(e^{j\theta})\big)d\theta > -\infty. \qquad (5.362)$$

Further, $\mathbf{A}_n(z)$ and $\mathbf{C}_n(z)$ represent the left and right Levinson polynomial matrices of the first kind generated from $\mathbf{r}_0 \rightarrow \mathbf{r}_n$ using (5.252)–(5.253). Clearly if $\mathbf{S}(\theta)$ in (5.356) is nonrational, then it must follow from (5.358)–(5.359) for a specific nonrational $\boldsymbol{\rho}(z)$. From Fig. 5.3, physically, $\boldsymbol{\rho}(z)$ represents the unique renormalized bounded matrix function that remains after extracting $n+1$ multiline sections from the positive matrix function $\mathbf{Z}(z)$ in (5.357).

More interestingly, if an L-ARMA(p,q) rational approximation $\mathbf{H}_l(z)$ as in (5.353) or an R-ARMA(p,q) approximation $\mathbf{H}_r(z)$ as in (5.354) holds here, then that too must follow from (5.358) and (5.359) for a specific *rational bounded* matrix function $\rho(z)$. In that case, since $\delta_o(\boldsymbol{\Gamma}_l(z)) = \delta_o(\boldsymbol{\Gamma}_r(z)) = q$, from (5.360)–(5.361), $\delta_o(\rho(z)) = q$. Let

$$\rho(z) = \mathbf{G}^{-1}(z)\mathbf{H}(z) \tag{5.363}$$

or

$$\rho(z) = \mathbf{H}_1(z)\mathbf{G}_1^{-1}(z) \tag{5.364}$$

depending on whether an L-ARMA(p,q) approximation or an R-ARMA(p,q) approximation is desired. Clearly the orders of $\mathbf{G}(z)$, $\mathbf{H}(z)$, $\mathbf{G}_1(z)$ and $\mathbf{H}_1(z)$ are bounded by q, and (5.358) simplifies to

$$\mathbf{H}_l(z) = \mathbf{P}^{-1}(z)\mathbf{Q}(z), \tag{5.365}$$

where

$$\mathbf{P}(z) = \mathbf{G}(z)\mathbf{A}_n(z) - z\mathbf{H}(z)\tilde{\mathbf{C}}_n(z), \quad n \ge p+q, \tag{5.366}$$

and $\mathbf{Q}(z)$ is given by

$$\mathbf{Q}(z)\mathbf{Q}_*(z) = \mathbf{G}(z)\mathbf{G}_*(z) - \mathbf{H}(z)\mathbf{H}_*(z). \tag{5.367}$$

Similarly (5.359) simplifies to

$$\mathbf{H}_r(z) = \mathbf{Q}_1(z)\mathbf{P}_1^{-1}(z), \tag{5.368}$$

where

$$\mathbf{P}_1(z) = \mathbf{C}_n(z)\mathbf{G}_1(z) - z\tilde{\mathbf{A}}_n(z)\mathbf{H}_1(z), \quad n \ge p+q, \tag{5.369}$$

and $\mathbf{Q}_1(z)$ is given by

$$\mathbf{Q}_{1*}(z)\mathbf{Q}_1(z) = \mathbf{G}_{1*}(z)\mathbf{G}_1(z) - \mathbf{H}_{1*}(z)\mathbf{H}_1(z). \tag{5.370}$$

However, since $\delta_o(\mathbf{P}(z)) = \delta_o(\mathbf{P}_1(z)) = p$ and $\delta_o(\tilde{\mathbf{A}}_n(z)) = \delta_o(\tilde{\mathbf{C}}_n(z)) = n \ge p+q$, from (5.366), (5.369), we must have $\delta_o(\mathbf{H}(z)) = \delta_o(\mathbf{H}_1(z)) \le q-1$. As a result, $\delta_o(\mathbf{G}(z)) = \delta_o(\mathbf{G}_1(z)) = q$ and the bounded matrix function in (5.363)–(5.364) must have the representation

$$\rho(z) = \left(\mathbf{I} + \mathbf{G}_1 z + \cdots + \mathbf{G}_q z^q\right)^{-1} \left(\mathbf{H}_0 + \mathbf{H}_1 z + \cdots + \mathbf{H}_{q-1} z^{q-1}\right) \tag{5.371}$$

or

$$\rho(z) = \left(\tilde{\mathbf{H}}_0 + \tilde{\mathbf{H}}_1 z + \cdots + \tilde{\mathbf{H}}_{q-1} z^{q-1}\right) \left(\mathbf{I} + \tilde{\mathbf{G}}_1 z + \cdots + \tilde{\mathbf{G}}_q z^q\right)^{-1}. \tag{5.372}$$

Note that the forms in (5.371)–(5.372) are identical to the forms in (5.286)–(5.289), but due to quite different considerations.

The formal order of $\mathbf{P}(z)$ and $\mathbf{P}_1(z)$ is still $n + q$, and to respect the multichannel-ARMA(p, q) nature of $\mathbf{H}_l(z)$ and $\mathbf{H}_r(z)$, the coefficients of higher-order terms beyond z^p must be zeros there. This results in $n + q - p$ linear equations in $2q$ unknowns in each case, and since $n \geq p + q$, these equations are at least $2q$ in number. Clearly, the minimum number of these equations is obtained for $n = p + q$. In the case of the left-coprime representation in (5.365), the equations so obtained are exactly the same as in (5.292)–(5.295). Similarly the $2q$ equations resulting from the right-coprime representation in (5.369) are exactly the same as (5.P.17)–(5.P.19). However, unlike the rational case, in the present situation, these equations need not have a solution for every p, q. Even if there exists a solution in some cases, $\mathbf{G}(z)$ or $\mathbf{G}_1(z)$ so obtained need not be minimum-phase and further $\boldsymbol{\rho}(z)$ need not turn out to be a bounded matrix function. If any of the above possibilities occur for some p, q, then there exists no multichannel-ARMA(p, q) approximation $\mathbf{H}_l(z)$ or $\mathbf{H}_r(z)$ as in (5.353) or (5.354) to the given nonrational function. Remarkably, for some p and q, if $\boldsymbol{\rho}(z)$ so obtained turns out to be a bounded matrix function, then $\mathbf{H}_l(z)$ and/or $\mathbf{H}_r(z)$ in (5.365)–(5.368) represent minimum-phase L-ARMA(p, q) and/or R-ARMA(p, q) systems whose first $p + q + 1$ autocorrelation matrices match with those of the given nonrational spectral density matrix $\mathbf{S}(\theta)$. In that case $\mathbf{H}_l(z)$ and/or $\mathbf{H}_r(z)$ are said to be left Padé-like and right Padé-like approximations to $\mathbf{S}(\theta)$, respectively. Clearly the rational approximations $\mathbf{H}_l(z)$ and $\mathbf{H}_r(z)$ are automatically minimum-phase and stable. It may be remarked that, for some p and q, if the left-coprime representation equations (5.292)–(5.295) generate a bounded matrix solution $\boldsymbol{\rho}(z) = \mathbf{G}^{-1}(z)\mathbf{H}(z)$, then necessarily $\mathbf{H}_l(z)$ in (5.365) represents a minimum-phase L-ARMA(p, q) left Padé-like approximation to the given nonrational spectrum $\mathbf{S}(\theta)$. In that situation, $\mathbf{H}_r(z)$ in (5.368)–(5.370) computed with $n = p + q$ and a right-coprime representation of the above $\boldsymbol{\rho}(z)$ also gives rise to an R-ARMA minimum-phase system whose first $p + q + 1$ autocorrelation matrices match with those of $\mathbf{S}(\theta)$. However, the order of $\mathbf{H}_r(z)$ so obtained need not be the same as the above p and q, and hence it may not form a right Padé-like approximation.

To summarize, given the autocorrelation matrices $\mathbf{r}_0$, $\mathbf{r}_1$, ..., $\mathbf{r}_{p+q}$ that form a positive definite sequence, the necessary and sufficient condition to fit an L-ARMA(p, q) model and/or an R-ARMA(p, q) model as in (5.353)–(5.354) is that the system of $2q$ linear equations involving the Levinson polynomial matrices $\mathbf{A}_{p+q}(z)$ and $\mathbf{C}_{p+q}(z)$ in (5.292) and/or (5.P.17) yield bounded solutions for the matrix function $\boldsymbol{\rho}(z)$ in (5.363) and/or (5.364).

The simulation results presented in Figs. 5.19–5.22b in the nonrational case support the conjecture that such optimal $\boldsymbol{\rho}(z)$'s should always exist for every $\mathbf{S}(\theta)$ satisfying (5.34) and (5.35). However, a rigorous proof is still lacking and the issue remains unresolved in the multichannel case also.

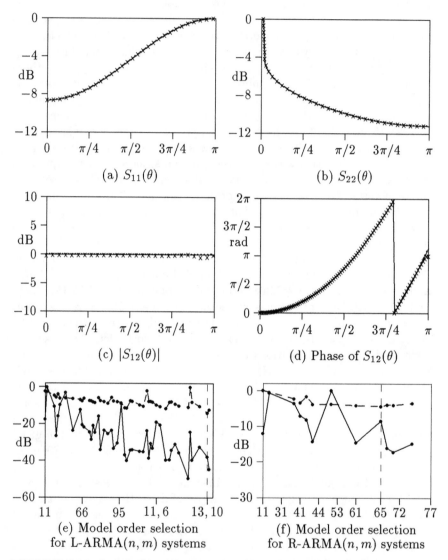

FIGURE 5.19. A two-channel nonrational spectrum and its left and right Padé-like approximations. In (a)–(d), solid lines represent the given spectrum $\mathbf{S}(\theta)$, dotted lines represent the approximated L-ARMA spectrum $\mathbf{H}_l(e^{j\theta})\mathbf{H}_l^*(e^{j\theta})$, and crossed lines represent the approximated R-ARMA spectrum $\mathbf{H}_r^*(e^{j\theta})\mathbf{H}_r(e^{j\theta})$. In (e) and (f), dots on the solid lines indicate the existence of the respective Padé-like approximations, and dots on the dashed lines represent the corresponding spectral percent-error $\eta(n, m)$. (See next page for details.)

FIGURE 5.19 (cont.). The original nonrational spectrum is given by

$$\mathbf{S}(\theta) = \begin{bmatrix} e^{1-\cos\theta} & e^{j\theta^2} \\ e^{-j\theta^2} & 1 - ln\,(sin\,\theta/2) \end{bmatrix}.$$

The approximated L-ARMA(13, 10) model is given by $\mathbf{H}_l(z) = \mathbf{P}^{-1}(z)\mathbf{Q}(z)$ where

$$
\begin{aligned}
\mathbf{P}(z) =\;& \begin{bmatrix} 0.705 & -0.025 \\ 0.000 & 0.808 \end{bmatrix} + \begin{bmatrix} 1.213 & -0.410 \\ 0.043 & -0.737 \end{bmatrix} z + \begin{bmatrix} -0.289 & 0.300 \\ 0.047 & -2.154 \end{bmatrix} z^2 \\[4pt]
&+ \begin{bmatrix} -1.457 & 1.139 \\ -0.067 & 2.178 \end{bmatrix} z^3 + \begin{bmatrix} -0.483 & -0.787 \\ -0.088 & 1.883 \end{bmatrix} z^4 + \begin{bmatrix} 0.414 & -1.094 \\ 0.013 & -2.338 \end{bmatrix} z^5 \\[4pt]
&+ \begin{bmatrix} 0.275 & 0.845 \\ 0.042 & -0.435 \end{bmatrix} z^6 + \begin{bmatrix} 0.020 & 0.373 \\ 0.010 & 1.057 \end{bmatrix} z^7 + \begin{bmatrix} -0.024 & -0.399 \\ -0.003 & -0.166 \end{bmatrix} z^8 \\[4pt]
&+ \begin{bmatrix} -0.0096 & 0.0126 \\ -0.0020 & -0.1510 \end{bmatrix} z^9 + \begin{bmatrix} -0.0019 & 0.063 \\ -0.0005 & 0.0642 \end{bmatrix} z^{10} \\[4pt]
&+ \begin{bmatrix} -0.00025 & -0.0209 \\ -0.7\times10^{-4} & -0.009 \end{bmatrix} z^{11} + \begin{bmatrix} -0.2\times10^{-4} & 0.026 \\ -0.7\times10^{-5} & 0.0005 \end{bmatrix} z^{12} \\[4pt]
&+ \begin{bmatrix} -0.1\times10^{-5} & -0.0001 \\ -0.4\times10^{-6} & -0.7\times10^{-6} \end{bmatrix} z^{13}
\end{aligned}
$$

and

$$
\begin{aligned}
\mathbf{Q}(z) =\;& \begin{bmatrix} 0.999 & 0.000 \\ 0.001 & 0.999 \end{bmatrix} + \begin{bmatrix} 1.147 & -0.293 \\ 0.062 & -0.641 \end{bmatrix} z + \begin{bmatrix} -1.209 & 0.789 \\ 0.007 & -2.776 \end{bmatrix} z^2 \\[4pt]
&+ \begin{bmatrix} -1.521 & 1.915 \\ -0.119 & 1.967 \end{bmatrix} z^3 + \begin{bmatrix} 0.392 & -1.260 \\ -0.006 & 2.719 \end{bmatrix} z^4 + \begin{bmatrix} 0.580 & -2.608 \\ 0.063 & -2.211 \end{bmatrix} z^5 \\[4pt]
&+ \begin{bmatrix} -0.037 & 0.829 \\ -0.013 & -1.028 \end{bmatrix} z^6 + \begin{bmatrix} -0.052 & 1.307 \\ -0.007 & 1.069 \end{bmatrix} z^7 + \begin{bmatrix} 0.010 & -0.323 \\ 0.010 & 0.050 \end{bmatrix} z^8 \\[4pt]
&+ \begin{bmatrix} -0.002 & -0.224 \\ -0.001 & -0.187 \end{bmatrix} z^9 + \begin{bmatrix} -0.002 & 0.061 \\ -0.001 & -0.034 \end{bmatrix} z^{10}.
\end{aligned}
$$

The approximated R-ARMA(6, 5) model is given by $\mathbf{H}_r(z) = \mathbf{Q}_1(z)\mathbf{P}_1^{-1}(z)$ where

$$
\begin{aligned}
\mathbf{P}_1(z) =\;& \begin{bmatrix} 0.627 & 0.000 \\ -0.194 & 0.908 \end{bmatrix} + \begin{bmatrix} -1.240 & 8.935 \\ -0.135 & 0.637 \end{bmatrix} z + \begin{bmatrix} -2.951 & 12.442 \\ 0.110 & -0.516 \end{bmatrix} z^2 \\[4pt]
&+ \begin{bmatrix} -0.561 & 0.955 \\ 0.150 & -0.712 \end{bmatrix} z^3 + \begin{bmatrix} 0.651 & -2.996 \\ 0.043 & -0.201 \end{bmatrix} z^4 + \begin{bmatrix} 0.058 & -0.141 \\ 0.003 & -0.0135 \end{bmatrix} z^5 \\[4pt]
&+ \begin{bmatrix} -0.031 & 0.171 \\ -0.003 & 0.014 \end{bmatrix} z^6
\end{aligned}
$$

and

$$
\begin{aligned}
\mathbf{Q}_1(z) =\;& \begin{bmatrix} 0.999 & 0.0004 \\ 0.000 & 0.999 \end{bmatrix} + \begin{bmatrix} -2.575 & 14.537 \\ -0.614 & 4.044 \end{bmatrix} z + \begin{bmatrix} -3.602 & 12.588 \\ -1.006 & 4.407 \end{bmatrix} z^2 \\[4pt]
&+ \begin{bmatrix} 1.392 & -7.487 \\ -0.160 & 0.265 \end{bmatrix} z^3 + \begin{bmatrix} 1.116 & -4.116 \\ 0.199 & -0.941 \end{bmatrix} z^4 + \begin{bmatrix} -0.436 & 1.917 \\ 0.019 & -0.095 \end{bmatrix} z^5.
\end{aligned}
$$

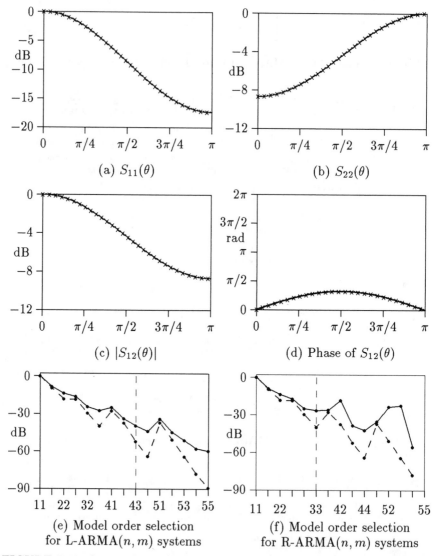

(a) $S_{11}(\theta)$

(b) $S_{22}(\theta)$

(c) $|S_{12}(\theta)|$

(d) Phase of $S_{12}(\theta)$

(e) Model order selection
for L-ARMA(n, m) systems

(f) Model order selection
for R-ARMA(n, m) systems

FIGURE 5.20. A two-channel nonrational spectrum and its left and right Padé-like approximations. In (a)–(d), solid lines represent the given spectrum $\mathbf{S}(\theta)$, dotted lines represent the approximated L-ARMA spectrum $\mathbf{H}_l(e^{j\theta})\mathbf{H}_l^*(e^{j\theta})$, and crossed lines represent the approximated R-ARMA spectrum $\mathbf{H}_r^*(e^{j\theta})\mathbf{H}_r(e^{j\theta})$. In (e) and (f), dots on the solid lines indicate the existence of the respective Padé-like approximations, and dots on the dashed lines represent the corresponding spectral percent-error $\eta(n, m)$. (See next page for details.)

FIGURE 5.20 (cont.). The original nonrational spectrum is given by

$$\mathbf{S}(\theta) = \begin{bmatrix} e^{\sin(3\theta/2)/\sin(\theta/2)} & e^{\cos\theta + j\sin\theta} \\ e^{\cos\theta - j\sin\theta} & e^{1-\cos\theta} \end{bmatrix}.$$

The approximated L-ARMA$(4,3)$ model is given by $\mathbf{H}_l(z) = \mathbf{P}^{-1}(z)\mathbf{Q}(z)$ where

$$\mathbf{P}(z) = \begin{bmatrix} 0.6546 & -0.2519 \\ 0.0000 & 0.6203 \end{bmatrix} + \begin{bmatrix} -0.3666 & -0.1039 \\ -0.0015 & 0.1704 \end{bmatrix} z$$
$$+ \begin{bmatrix} 0.0895 & -0.0192 \\ 0.00075 & 0.0204 \end{bmatrix} z^2 + \begin{bmatrix} -0.0116 & -0.0020 \\ -0.00015 & 0.0013 \end{bmatrix} z^3$$
$$+ \begin{bmatrix} 0.0007 & -0.99 \times 10^{-4} \\ 0.12 \times 10^{-4} & 0.37 \times 10^{-4} \end{bmatrix} z^4$$

and

$$\mathbf{Q}(z) = \begin{bmatrix} 0.9999 & 0.0000 \\ 0.0000 & 0.9999 \end{bmatrix} + \begin{bmatrix} 0.3556 & 0.2187 \\ 0.2052 & -0.2373 \end{bmatrix} z$$
$$+ \begin{bmatrix} 0.0631 & 0.0335 \\ 0.0036 & 0.0206 \end{bmatrix} z^2 + \begin{bmatrix} 0.0044 & 0.0022 \\ 0.0010 & -0.0010 \end{bmatrix} z^3.$$

The approximated R-ARMA$(3,3)$ model is given by $\mathbf{H}_r(z) = \mathbf{Q}_1(z)\mathbf{P}_1^{-1}(z)$ where

$$\mathbf{P}_1(z) = \begin{bmatrix} 0.6187 & 0.0000 \\ 0.0169 & 0.6563 \end{bmatrix} + \begin{bmatrix} -0.3080 & -0.1185 \\ 0.0042 & 0.1635 \end{bmatrix} z$$
$$+ \begin{bmatrix} 0.0598 & 0.0052 \\ 0.0004 & 0.0163 \end{bmatrix} z^2 + \begin{bmatrix} -0.0048 & -0.0011 \\ 0.17 \times 10^{-4} & 0.001 \end{bmatrix} z^3$$

and

$$\mathbf{Q}_1(z) = \begin{bmatrix} 0.9999 & 0.0000 \\ 0.0000 & 0.9999 \end{bmatrix} + \begin{bmatrix} 0.5279 & 0.0472 \\ -0.1789 & -0.3671 \end{bmatrix} z$$
$$+ \begin{bmatrix} 0.1083 & 0.0135 \\ 0.0327 & 0.0314 \end{bmatrix} z^2 + \begin{bmatrix} 0.0093 & 0.0012 \\ -0.0032 & -0.0019 \end{bmatrix} z^3.$$

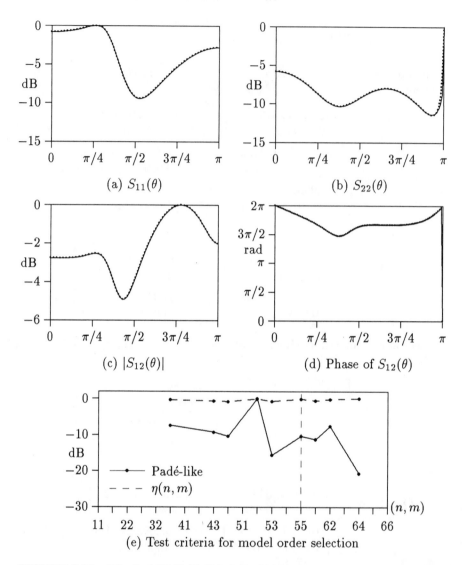

(a) $S_{11}(\theta)$

(b) $S_{22}(\theta)$

(c) $|S_{12}(\theta)|$

(d) Phase of $S_{12}(\theta)$

(e) Test criteria for model order selection

FIGURE 5.21a. The L-ARMA$(5,5)$ left Padé-like approximation $\mathbf{H}_l(z)$ of a nonrational system $\mathbf{H}(z)$. In (a)–(d), solid lines represent the original spectrum $\mathbf{S}(\theta) = \mathbf{H}(e^{j\theta})\mathbf{H}^*(e^{j\theta})$, and dotted lines represent the approximated spectrum $\mathbf{K}(\theta) = \mathbf{H}_l(e^{j\theta})\mathbf{H}_l^*(e^{j\theta})$. In (e), dots on the solid line indicate the existence of left Padé-like approximations, and dots on the dashed line represent the spectral percent-error $\eta(n,m)$. (See next page for $\mathbf{H}(z)$ and $\mathbf{H}_l(z)$.)

FIGURE 5.21a (cont.). The nonrational system transfer function is given by

$$\mathbf{H}(z) = \begin{bmatrix} e^{-z} & \dfrac{1+z}{1-1.4z+2z^2}\, e^z \\ 4+z+z^3 & ln\,(1+z) \end{bmatrix}.$$

The approximated L-ARMA(5, 5) model is given by

$$\mathbf{H}_l(z) = \mathbf{P}^{-1}(z)\mathbf{Q}(z),$$

where

$$\mathbf{P}(z) = \begin{bmatrix} 0.6198 & -0.1154 \\ 0.0000 & 0.2391 \end{bmatrix} + \begin{bmatrix} -0.5489 & -0.2097 \\ 0.1397 & 0.3978 \end{bmatrix} z$$

$$+ \begin{bmatrix} 0.4473 & -0.1132 \\ -0.2724 & 0.1705 \end{bmatrix} z^2 + \begin{bmatrix} 0.0180 & -0.0152 \\ -0.0132 & 0.0076 \end{bmatrix} z^3$$

$$+ \begin{bmatrix} -0.0130 & 0.0021 \\ 0.0071 & -0.0013 \end{bmatrix} z^4 + \begin{bmatrix} -0.0007 & -0.0002 \\ 0.0010 & 0.0002 \end{bmatrix} z^5$$

and

$$\mathbf{Q}(z) = \begin{bmatrix} 0.9975 & 0.0000 \\ 0.0024 & 0.9994 \end{bmatrix} + \begin{bmatrix} -0.1892 & -1.9756 \\ 0.3512 & 1.9365 \end{bmatrix} z$$

$$+ \begin{bmatrix} 0.5910 & 0.5463 \\ -0.0875 & 0.6839 \end{bmatrix} z^2 + \begin{bmatrix} -0.0971 & -1.0358 \\ 0.0117 & 0.7441 \end{bmatrix} z^3$$

$$+ \begin{bmatrix} 0.0887 & 0.0786 \\ -0.0276 & 0.2622 \end{bmatrix} z^4 + \begin{bmatrix} -0.0251 & -0.1850 \\ 0.0295 & 0.1995 \end{bmatrix} z^5.$$

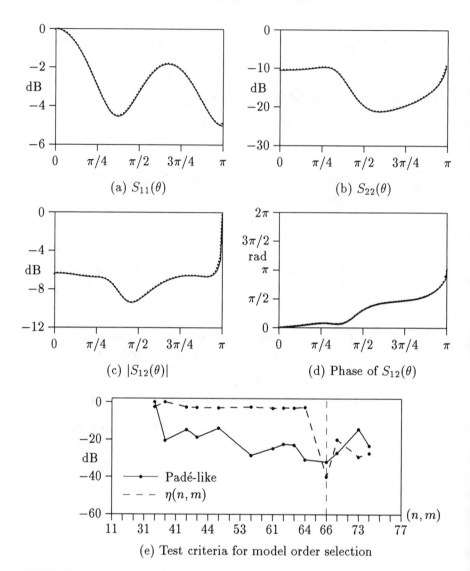

(a) $S_{11}(\theta)$

(b) $S_{22}(\theta)$

(c) $|S_{12}(\theta)|$

(d) Phase of $S_{12}(\theta)$

(e) Test criteria for model order selection

FIGURE 5.21b. The R-ARMA$(6,6)$ right Padé-like approximation $\mathbf{H}_r(z)$ of the nonrational system $\mathbf{H}(z)$ in Fig. 5.21a. In (a)–(d), solid lines represent the original spectrum $\mathbf{S}_1(\theta) = \mathbf{H}^*(e^{j\theta})\mathbf{H}(e^{j\theta})$, and dotted lines represent the approximated spectrum $\mathbf{K}_1(\theta) = \mathbf{H}_r^*(e^{j\theta})\mathbf{H}_r(e^{j\theta})$. In (e), dots on the solid line indicate the existence of right Padé-like approximations, and dots on the dashed line represent the spectral percent-error $\eta(n,m)$. (See next page for $\mathbf{H}_r(z)$.)

FIGURE 5.21b (cont.). The original nonrational system transfer function $\mathbf{H}(z)$ is as given in Fig. 5.21a.

The approximated R-ARMA$(6,6)$ model is given by

$$\mathbf{H}_r(z) = \mathbf{Q}_1(z)\mathbf{P}_1^{-1}(z),$$

where

$$\mathbf{P}_1(z) = \begin{bmatrix} 0.2316 & 0.0000 \\ -0.0785 & 0.6404 \end{bmatrix} + \begin{bmatrix} 0.0599 & -0.2053 \\ -0.0521 & 0.7094 \end{bmatrix} z$$

$$+ \begin{bmatrix} -0.2199 & -0.1981 \\ -0.0537 & 0.2064 \end{bmatrix} z^2 + \begin{bmatrix} -0.0265 & -0.1801 \\ -0.0015 & 0.2064 \end{bmatrix} z^3$$

$$+ \begin{bmatrix} -0.0007 & 0.0020 \\ -0.0379 & 0.2140 \end{bmatrix} z^4 + \begin{bmatrix} 0.0001 & -0.0002 \\ -0.0043 & 0.0041 \end{bmatrix} z^5$$

$$+ \begin{bmatrix} -0.7 \times 10^{-5} & -0.3 \times 10^{-4} \\ 0.0025 & -0.0110 \end{bmatrix} z^6$$

and

$$\mathbf{Q}_1(z) = \begin{bmatrix} 0.9988 & 0.0051 \\ 0.0000 & 0.9945 \end{bmatrix} + \begin{bmatrix} 0.3147 & -0.4228 \\ -0.1443 & 1.8081 \end{bmatrix} z$$

$$+ \begin{bmatrix} -0.8604 & -0.7679 \\ -0.0747 & 1.6409 \end{bmatrix} z^2 + \begin{bmatrix} 0.0446 & -0.0103 \\ -0.0150 & 0.9863 \end{bmatrix} z^3$$

$$+ \begin{bmatrix} -0.0416 & -0.0892 \\ -0.0122 & 0.4339 \end{bmatrix} z^4 + \begin{bmatrix} -0.2215 & -0.1322 \\ -0.0117 & 0.0522 \end{bmatrix} z^5$$

$$+ \begin{bmatrix} -0.0233 & -0.1799 \\ -0.0048 & 0.01301 \end{bmatrix} z^6 .$$

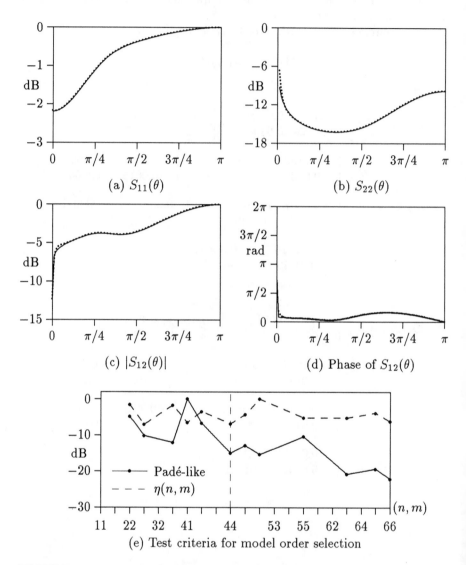

FIGURE 5.22a. The L-ARMA$(4, 4)$ left Padé-like approximation $\mathbf{H}_l(z)$ of a nonrational system $\mathbf{H}(z)$. In (a)–(d), solid lines represent the original spectrum $\mathbf{S}(\theta) = \mathbf{H}(e^{j\theta})\mathbf{H}^*(e^{j\theta})$, and dotted lines represent the approximated spectrum $\mathbf{K}(\theta) = \mathbf{H}_l(e^{j\theta})\mathbf{H}_l^*(e^{j\theta})$. In (e), dots on the solid line indicate the existence of left Padé-like approximations, and dots on the dashed line represent the spectral percent-error $\eta(n, m)$. (See next page for $\mathbf{H}(z)$ and $\mathbf{H}_l(z)$.)

FIGURE 5.22a (cont.). The nonrational system transfer function is given by

$$\mathbf{H}(z) = \begin{bmatrix} 1 + e^{-1/(2-z)} & \dfrac{ze^{-z}}{1 - 2z + 3z^2} \\[2ex] 1 + e^{-z} & ln\,(1 - z) \end{bmatrix}.$$

The approximated L-ARMA(4, 4) model is given by

$$\mathbf{H}_l(z) = \mathbf{P}^{-1}(z)\mathbf{Q}(z),$$

where

$$\mathbf{P}(z) = \begin{bmatrix} 1.0448 & -0.6021 \\ 0.0000 & 0.4318 \end{bmatrix} + \begin{bmatrix} 9.9330 & 0.7298 \\ -7.7929 & -0.5184 \end{bmatrix} z$$

$$+ \begin{bmatrix} -8.3529 & -0.0186 \\ 6.1816 & 0.0116 \end{bmatrix} z^2 + \begin{bmatrix} 0.4060 & -0.0918 \\ -0.2787 & 0.0643 \end{bmatrix} z^3$$

$$+ \begin{bmatrix} 0.5425 & -0.0133 \\ -0.3996 & 0.0093 \end{bmatrix} z^4$$

and

$$\mathbf{Q}(z) = \begin{bmatrix} 0.9972 & 0.0000 \\ 0.0039 & 0.9987 \end{bmatrix} + \begin{bmatrix} 9.7476 & 15.2162 \\ -7.7344 & 11.8767 \end{bmatrix} z$$

$$+ \begin{bmatrix} -9.7829 & 13.1736 \\ 7.2700 & 9.7281 \end{bmatrix} z^2 + \begin{bmatrix} 1.5090 & 1.1451 \\ -1.0970 & -0.8007 \end{bmatrix} z^3$$

$$+ \begin{bmatrix} 0.3435 & 0.8772 \\ -0.2521 & -0.6491 \end{bmatrix} z^4.$$

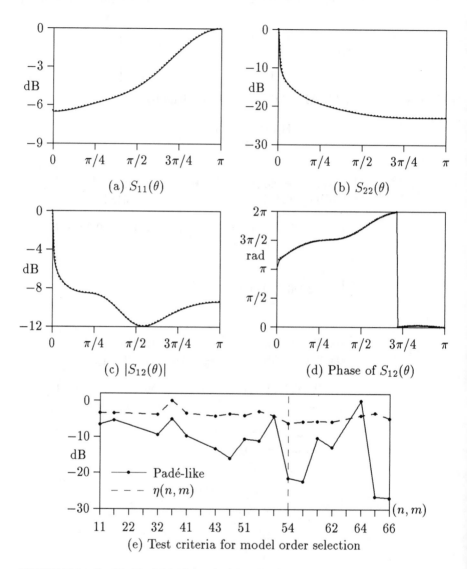

(a) $S_{11}(\theta)$

(b) $S_{22}(\theta)$

(c) $|S_{12}(\theta)|$

(d) Phase of $S_{12}(\theta)$

(e) Test criteria for model order selection

FIGURE 5.22b. The R-ARMA$(5, 4)$ right Padé-like approximation $\mathbf{H}_r(z)$ of the nonrational system $\mathbf{H}(z)$ in Fig. 5.21a. In (a)–(d), solid lines represent the original spectrum $\mathbf{S}_1(\theta) = \mathbf{H}^*(e^{j\theta})\mathbf{H}(e^{j\theta})$, and dotted lines represent the approximated spectrum $\mathbf{K}_1(\theta) = \mathbf{H}_r^*(e^{j\theta})\mathbf{H}_r(e^{j\theta})$. In (e), dots on the solid line indicate the existence of right Padé-like approximations, and dots on the dashed line represent the spectral percent-error $\eta(n, m)$. (See next page for $\mathbf{H}_r(z)$.)

FIGURE 5.22b (cont.). The original nonrational system transfer function
$\mathbf{H}(z)$ is as given in Fig. 5.22a.

The approximated R-ARMA$(5, 4)$ model is given by

$$\mathbf{H}_r(z) = \mathbf{Q}_1(z)\mathbf{P}_1^{-1}(z),$$

where

$$
\mathbf{P}_1(z) = \begin{bmatrix} 0.3879 & 0.0000 \\ 0.0787 & 1.1612 \end{bmatrix} + \begin{bmatrix} -0.5034 & 0.0491 \\ -0.4016 & -2.0142 \end{bmatrix} z
$$

$$
+ \begin{bmatrix} 0.1987 & -0.0508 \\ 0.5778 & 0.7372 \end{bmatrix} z^2 + \begin{bmatrix} -0.0262 & 0.0303 \\ -0.2706 & 0.1932 \end{bmatrix} z^3
$$

$$
+ \begin{bmatrix} -0.0336 & 0.0083 \\ -0.0062 & -0.0561 \end{bmatrix} z^4 + \begin{bmatrix} -0.0040 & -0.0005 \\ 0.0214 & -0.0148 \end{bmatrix} z^5
$$

and

$$
\mathbf{Q}_1(z) = \begin{bmatrix} 0.9999 & 0.0006 \\ 0.0000 & 0.9947 \end{bmatrix} + \begin{bmatrix} -1.7115 & -0.6427 \\ -0.2686 & -1.4032 \end{bmatrix} z
$$

$$
+ \begin{bmatrix} 1.3745 & 0.7000 \\ 0.5530 & 0.3437 \end{bmatrix} z^2 + \begin{bmatrix} -0.7526 & 0.1709 \\ -0.3175 & 0.1043 \end{bmatrix} z^3
$$

$$
+ \begin{bmatrix} 0.1297 & -0.1866 \\ 0.0287 & -0.0022 \end{bmatrix} z^4.
$$

Appendix 5.A

Reflection Coefficient Matrices
and Their Left and Right Factors

In this appendix, with

$$\mathbf{A} = \mathbf{K}_n^{-1/2} \left(\frac{\mathbf{R}_n + \mathbf{R}_{n-1}^*}{2} \right) \mathbf{K}_{n-1}^{-1/2} \triangleq \mathbf{A}_n \qquad (5.A.1)$$

$$\mathbf{B} = \mathbf{K}_n^{-1/2} \left(\frac{\mathbf{R}_n^* + \mathbf{R}_{n-1}}{2} \right) \mathbf{K}_{n-1}^{-1/2} \triangleq \mathbf{B}_n \qquad (5.A.2)$$

as defined in (5.120)–(5.121), and

$$\mathbf{S}_n = \mathbf{K}_{n-1}^{-1/2} (\mathbf{R}_n - \mathbf{R}_{n-1})(\mathbf{R}_n + \mathbf{R}_{n-1}^*)^{-1} \mathbf{K}_{n-1}^{1/2}, \qquad (5.A.3)$$

as in (5.117), we will show that

$$\mathbf{A}_\textsf{I}^{-1}(\mathbf{A}^{-1})^* = \mathbf{I} - \mathbf{S}_n \mathbf{S}_n^* \qquad (5.A.4)$$

and

$$\mathbf{B}^{-1}(\mathbf{B}^{-1})^* = \mathbf{I} - \mathbf{S}_n^* \mathbf{S}_n . \qquad (5.A.5)$$

Proof: From (5.A.1),

$$\mathbf{A}^{-1}(\mathbf{A}^{-1})^* = 4\mathbf{K}_{n-1}^{1/2}(\mathbf{R}_n + \mathbf{R}_{n-1}^*)^{-1} \mathbf{K}_n (\mathbf{R}_n^* + \mathbf{R}_{n-1})^{-1} \mathbf{K}_{n-1}^{1/2}$$

$$= 2\mathbf{K}_{n-1}^{1/2}(\mathbf{R}_n + \mathbf{R}_{n-1}^*)^{-1}(\mathbf{R}_n + \mathbf{R}_n^*)(\mathbf{R}_n^* + \mathbf{R}_{n-1})^{-1} \mathbf{K}_{n-1}^{1/2}. \qquad (5.A.6)$$

On the other hand, from (5.A.3)

$$
\begin{aligned}
\mathbf{I} &- \mathbf{S}_n \mathbf{S}_n^* \\
&= \mathbf{I} - \mathbf{K}_{n-1}^{-1/2}(\mathbf{R}_n - \mathbf{R}_{n-1})(\mathbf{R}_n + \mathbf{R}_{n-1}^*)^{-1} \mathbf{K}_{n-1}^{1/2} \\
&\quad \times \mathbf{K}_{n-1}^{1/2}(\mathbf{R}_n^* + \mathbf{R}_{n-1})^{-1}(\mathbf{R}_n^* - \mathbf{R}_{n-1}^*) \mathbf{K}_{n-1}^{-1/2} \\
&= \mathbf{K}_{n-1}^{1/2} \left\{ \mathbf{K}_{n-1}^{-1} - \mathbf{K}_{n-1}^{-1}(\mathbf{R}_n - \mathbf{R}_{n-1})(\mathbf{R}_n + \mathbf{R}_{n-1}^*)^{-1} \right. \\
&\quad \left. \times \mathbf{K}_{n-1}(\mathbf{R}_n^* + \mathbf{R}_{n-1})^{-1}(\mathbf{R}_n^* - \mathbf{R}_{n-1}^*) \mathbf{K}_{n-1}^{-1} \right\} \mathbf{K}_{n-1}^{1/2} \\
&= \mathbf{K}_{n-1}^{1/2} \left\{ 2(\mathbf{R}_{n-1} + \mathbf{R}_{n-1}^*)^{-1} - 2(\mathbf{R}_{n-1} + \mathbf{R}_{n-1}^*)^{-1}(\mathbf{R}_n - \mathbf{R}_{n-1}) \right. \\
&\quad \times (\mathbf{R}_n + \mathbf{R}_{n-1}^*)^{-1}(\mathbf{R}_{n-1} + \mathbf{R}_{n-1}^*)(\mathbf{R}_n^* + \mathbf{R}_{n-1})^{-1} \\
&\quad \left. \times (\mathbf{R}_n^* - \mathbf{R}_{n-1}^*)(\mathbf{R}_{n-1} + \mathbf{R}_{n-1}^*)^{-1} \right\} \mathbf{K}_{n-1}^{1/2} \\
&\triangleq 2\mathbf{K}_{n-1}^{1/2} \mathbf{W}_1 \mathbf{K}_{n-1}^{1/2},
\end{aligned}
$$

where

$\mathbf{W}_1$

$$
\begin{aligned}
&= (\mathbf{R}_{n-1} + \mathbf{R}_{n-1}^*)^{-1} - (\mathbf{R}_{n-1} + \mathbf{R}_{n-1}^*)^{-1}(\mathbf{R}_n - \mathbf{R}_{n-1})(\mathbf{R}_n + \mathbf{R}_{n-1}^*)^{-1} \\
&\quad \times (\mathbf{R}_{n-1} + \mathbf{R}_{n-1}^*)(\mathbf{R}_n^* + \mathbf{R}_{n-1})^{-1}(\mathbf{R}_n^* - \mathbf{R}_{n-1}^*)(\mathbf{R}_{n-1} + \mathbf{R}_{n-1}^*)^{-1} \\
&= (\mathbf{R}_{n-1} + \mathbf{R}_{n-1}^*)^{-1}\{(\mathbf{R}_{n-1} + \mathbf{R}_{n-1}^*) - (\mathbf{R}_n - \mathbf{R}_{n-1})(\mathbf{R}_n + \mathbf{R}_{n-1}^*)^{-1} \\
&\quad \times (\mathbf{R}_{n-1} + \mathbf{R}_{n-1}^*)(\mathbf{R}_n^* + \mathbf{R}_{n-1})^{-1}(\mathbf{R}_n^* - \mathbf{R}_{n-1}^*)\}(\mathbf{R}_{n-1} + \mathbf{R}_{n-1}^*)^{-1} \\
&= (\mathbf{R}_{n-1} + \mathbf{R}_{n-1}^*)^{-1}\big[(\mathbf{R}_{n-1} + \mathbf{R}_{n-1}^*) \\
&\quad - \{(\mathbf{R}_n + \mathbf{R}_{n-1}^*) - (\mathbf{R}_{n-1} + \mathbf{R}_{n-1}^*)\} \\
&\quad \times (\mathbf{R}_n + \mathbf{R}_{n-1}^*)^{-1}(\mathbf{R}_{n-1} + \mathbf{R}_{n-1}^*)(\mathbf{R}_n^* + \mathbf{R}_{n-1})^{-1} \\
&\quad \times (\mathbf{R}_n^* - \mathbf{R}_{n-1}^*)\big](\mathbf{R}_{n-1} + \mathbf{R}_{n-1}^*)^{-1} \\
&= (\mathbf{R}_{n-1} + \mathbf{R}_{n-1}^*)^{-1}\{(\mathbf{R}_{n-1} + \mathbf{R}_{n-1}^*) - (\mathbf{R}_{n-1} + \mathbf{R}_{n-1}^*)(\mathbf{R}_n^* + \mathbf{R}_{n-1})^{-1} \\
&\quad \times (\mathbf{R}_n^* - \mathbf{R}_{n-1}^*) + (\mathbf{R}_{n-1} + \mathbf{R}_{n-1}^*)(\mathbf{R}_n + \mathbf{R}_{n-1}^*)^{-1}(\mathbf{R}_{n-1} + \mathbf{R}_{n-1}^*) \\
&\quad \times (\mathbf{R}_n^* + \mathbf{R}_{n-1})^{-1}(\mathbf{R}_n^* - \mathbf{R}_{n-1}^*)\}(\mathbf{R}_{n-1} + \mathbf{R}_{n-1}^*)^{-1} \\
&= \{\mathbf{I} - (\mathbf{R}_n^* + \mathbf{R}_{n-1})^{-1}(\mathbf{R}_n^* - \mathbf{R}_{n-1}^*) + (\mathbf{R}_n + \mathbf{R}_{n-1}^*)^{-1} \\
&\quad \times (\mathbf{R}_{n-1} + \mathbf{R}_{n-1}^*)(\mathbf{R}_n^* + \mathbf{R}_{n-1})^{-1}(\mathbf{R}_n^* - \mathbf{R}_{n-1}^*)\}(\mathbf{R}_{n-1} + \mathbf{R}_{n-1}^*)^{-1} \\
&= (\mathbf{R}_n + \mathbf{R}_{n-1}^*)^{-1}\{(\mathbf{R}_n + \mathbf{R}_{n-1}^*) - (\mathbf{R}_n + \mathbf{R}_{n-1}^*)(\mathbf{R}_n^* + \mathbf{R}_{n-1})^{-1} \\
&\quad \times (\mathbf{R}_n^* - \mathbf{R}_{n-1}^*) + (\mathbf{R}_{n-1} + \mathbf{R}_{n-1}^*)(\mathbf{R}_n^* + \mathbf{R}_{n-1})^{-1}(\mathbf{R}_n^* - \mathbf{R}_{n-1}^*)\} \\
&\quad \times (\mathbf{R}_{n-1} + \mathbf{R}_{n-1}^*)^{-1} \\
&= (\mathbf{R}_n + \mathbf{R}_{n-1}^*)^{-1}\{(\mathbf{R}_n + \mathbf{R}_{n-1}^*) - (\mathbf{R}_n + \mathbf{R}_{n-1}^* - \mathbf{R}_{n-1} - \mathbf{R}_{n-1}^*) \\
&\quad \times (\mathbf{R}_n^* + \mathbf{R}_{n-1})^{-1}(\mathbf{R}_n^* - \mathbf{R}_{n-1}^*)\}(\mathbf{R}_{n-1} + \mathbf{R}_{n-1}^*)^{-1} \\
&= (\mathbf{R}_n + \mathbf{R}_{n-1}^*)^{-1}\big[(\mathbf{R}_n + \mathbf{R}_{n-1}^*) - (\mathbf{R}_n - \mathbf{R}_{n-1})(\mathbf{R}_n^* + \mathbf{R}_{n-1})^{-1} \\
&\quad \times \{(\mathbf{R}_n^* + \mathbf{R}_{n-1}) - (\mathbf{R}_{n-1} + \mathbf{R}_{n-1}^*)\}\big](\mathbf{R}_{n-1} + \mathbf{R}_{n-1}^*)^{-1} \\
&= (\mathbf{R}_n + \mathbf{R}_{n-1}^*)^{-1}\{(\mathbf{R}_n + \mathbf{R}_{n-1}^*) - (\mathbf{R}_n - \mathbf{R}_{n-1}) \\
&\quad + (\mathbf{R}_n - \mathbf{R}_{n-1})(\mathbf{R}_n^* + \mathbf{R}_{n-1})^{-1}(\mathbf{R}_{n-1} + \mathbf{R}_{n-1}^*)\}(\mathbf{R}_{n-1} + \mathbf{R}_{n-1}^*)^{-1} \\
&= (\mathbf{R}_n + \mathbf{R}_{n-1}^*)^{-1}\{\mathbf{I} + (\mathbf{R}_n - \mathbf{R}_{n-1})(\mathbf{R}_n^* + \mathbf{R}_{n-1})^{-1}\} \\
&= (\mathbf{R}_n + \mathbf{R}_{n-1}^*)^{-1}\{\mathbf{R}_n^* + \mathbf{R}_{n-1} + \mathbf{R}_n - \mathbf{R}_{n-1}\}(\mathbf{R}_n^* + \mathbf{R}_{n-1})^{-1} \\
&= (\mathbf{R}_n + \mathbf{R}_{n-1}^*)^{-1}(\mathbf{R}_n + \mathbf{R}_n^*)(\mathbf{R}_n^* + \mathbf{R}_{n-1})^{-1}
\end{aligned}
$$

and hence

$\mathbf{I} - \mathbf{S}_n \mathbf{S}_n^*$

$$
= 2\mathbf{K}_{n-1}^{1/2}(\mathbf{R}_n + \mathbf{R}_{n-1}^*)^{-1}(\mathbf{R}_n + \mathbf{R}_n^*)(\mathbf{R}_n^* + \mathbf{R}_{n-1})^{-1}\mathbf{K}_{n-1}^{1/2}, \qquad (5.A.7)
$$

which is identical to (5.A.6) and thus proves (5.A.4). Similarly, from (5.A.5),

$$\mathbf{B}^{-1}(\mathbf{B}^{-1})^* = 4\mathbf{K}_{n-1}^{1/2}(\mathbf{R}_n^* + \mathbf{R}_{n-1})^{-1}\mathbf{K}_n(\mathbf{R}_n + \mathbf{R}_{n-1}^*)^{-1}\mathbf{K}_{n-1}^{1/2}$$

$$= 2\mathbf{K}_{n-1}^{1/2}(\mathbf{R}_n^* + \mathbf{R}_{n-1})^{-1}(\mathbf{R}_n + \mathbf{R}_n^*)(\mathbf{R}_n + \mathbf{R}_{n-1}^*)^{-1}\mathbf{K}_{n-1}^{1/2} . \qquad (5.A.8)$$

But from (5.A.3),

$$\mathbf{I} - \mathbf{S}_n^*\mathbf{S}_n$$

$$\begin{aligned}
= \ & \mathbf{I} - \mathbf{K}_{n-1}^{1/2}(\mathbf{R}_n^* + \mathbf{R}_{n-1})^{-1}(\mathbf{R}_n^* - \mathbf{R}_{n-1}^*)\mathbf{K}_{n-1}^{-1/2} \\
& \times\mathbf{K}_{n-1}^{-1/2}(\mathbf{R}_n - \mathbf{R}_{n-1})(\mathbf{R}_n + \mathbf{R}_{n-1}^*)^{-1}\mathbf{K}_{n-1}^{1/2} \\
= \ & \mathbf{K}_{n-1}^{1/2}\left\{\mathbf{K}_{n-1}^{-1} - 2(\mathbf{R}_n^* + \mathbf{R}_{n-1})^{-1}(\mathbf{R}_n^* - \mathbf{R}_{n-1}^*)(\mathbf{R}_{n-1} + \mathbf{R}_{n-1}^*)^{-1} \right. \\
& \left. \times (\mathbf{R}_n - \mathbf{R}_{n-1})(\mathbf{R}_n + \mathbf{R}_{n-1}^*)^{-1}\right\}\mathbf{K}_{n-1}^{1/2} \\
= \ & 2\mathbf{K}_{n-1}^{1/2}\left\{(\mathbf{R}_{n-1} + \mathbf{R}_{n-1}^*)^{-1} - (\mathbf{R}_n^* + \mathbf{R}_{n-1})^{-1}(\mathbf{R}_n^* - \mathbf{R}_{n-1}^*) \right. \\
& \left. \times (\mathbf{R}_{n-1} + \mathbf{R}_{n-1}^*)^{-1}(\mathbf{R}_n - \mathbf{R}_{n-1})(\mathbf{R}_n + \mathbf{R}_{n-1}^*)^{-1}\right\}\mathbf{K}_{n-1}^{1/2} \\
= \ & 2\mathbf{K}_{n-1}^{1/2}(\mathbf{R}_n^* + \mathbf{R}_{n-1})^{-1}\left\{(\mathbf{R}_n^* + \mathbf{R}_{n-1})(\mathbf{R}_{n-1} + \mathbf{R}_{n-1}^*)^{-1}(\mathbf{R}_n + \mathbf{R}_{n-1}^*) \right. \\
& \left. -(\mathbf{R}_n^* - \mathbf{R}_{n-1}^*)(\mathbf{R}_{n-1} + \mathbf{R}_{n-1}^*)^{-1}(\mathbf{R}_n - \mathbf{R}_{n-1})\right\}(\mathbf{R}_n + \mathbf{R}_{n-1}^*)^{-1}\mathbf{K}_{n-1}^{1/2} \\
\overset{\triangle}{=} \ & 2\mathbf{K}_{n-1}^{1/2}(\mathbf{R}_n^* + \mathbf{R}_{n-1})^{-1}\mathbf{W}_2(\mathbf{R}_n + \mathbf{R}_{n-1}^*)^{-1}\mathbf{K}_{n-1}^{1/2} ,
\end{aligned}$$

where

$$\begin{aligned}
\mathbf{W}_2 & \\
= \ & \left\{(\mathbf{R}_n^* - \mathbf{R}_{n-1}^*) + (\mathbf{R}_{n-1} + \mathbf{R}_{n-1}^*)\right\}(\mathbf{R}_{n-1} - \mathbf{R}_{n-1}^*)^{-1}(\mathbf{R}_n + \mathbf{R}_{n-1}^*) \\
& -(\mathbf{R}_n^* - \mathbf{R}_{n-1}^*)(\mathbf{R}_{n-1} + \mathbf{R}_{n-1}^*)^{-1}(\mathbf{R}_n - \mathbf{R}_{n-1}) \\
= \ & (\mathbf{R}_n^* - \mathbf{R}_{n-1}^*)(\mathbf{R}_{n-1} + \mathbf{R}_{n-1}^*)^{-1}(\mathbf{R}_n + \mathbf{R}_{n-1}^*) + (\mathbf{R}_n + \mathbf{R}_{n-1}^*) \\
& -(\mathbf{R}_n^* - \mathbf{R}_{n-1}^*)(\mathbf{R}_{n-1} + \mathbf{R}_{n-1}^*)^{-1}(\mathbf{R}_n - \mathbf{R}_{n-1}) \\
= \ & (\mathbf{R}_n^* - \mathbf{R}_{n-1}^*)(\mathbf{R}_{n-1} + \mathbf{R}_{n-1}^*)^{-1}(\mathbf{R}_n - \mathbf{R}_{n-1}) + \mathbf{R}_n^* - \mathbf{R}_{n-1}^* \\
& +\mathbf{R}_n + \mathbf{R}_{n-1}^* - (\mathbf{R}_n^* - \mathbf{R}_{n-1}^*)(\mathbf{R}_{n-1} + \mathbf{R}_{n-1}^*)^{-1}(\mathbf{R}_n - \mathbf{R}_{n-1}) \\
= \ & \mathbf{R}_n + \mathbf{R}_n^*
\end{aligned}$$

and hence

$$\mathbf{I} - \mathbf{S}_n^*\mathbf{S}_n$$

$$= 2\mathbf{K}_{n-1}^{1/2}(\mathbf{R}_n^* + \mathbf{R}_{n-1})^{-1}(\mathbf{R}_n + \mathbf{R}_n^*)(\mathbf{R}_n + \mathbf{R}_{n-1}^*)^{-1}\mathbf{K}_{n-1}^{1/2} , \qquad (5.A.9)$$

which coincides with (5.A.8) and this verifies (5.A.5).

Appendix 5.B

Renormalization of $d_n(z)$

The bounded matrix functions $d_n(z)$, $n \geq 0$, satisfy the left and right recursions in (5.119) and (5.156) where A and B are as given in (5.A.1)–(5.A.2). Since A and B are crucial to the evaluation of the Levinson recursions (5.192)–(5.195), naturally they must be determined in advance. However, their forms in (5.A.1)–(5.A.2) are not particularly convenient for easy evaluation, since these expressions involve the characteristic impedance matrices R_n and R_{n-1}. The junction reflection coefficient matrices S_n also relate A and B through (5.A.4)–(5.A.5), and these factorizations are unique only up to arbitrary unitary matrices. Interestingly, the freedom present in (5.A.4)–(5.A.5) in terms of arbitrary unitary matrices can be utilized to renormalize $d_n(z)$ so that the corresponding A and B matrices turn out to be lower triangular with positive diagonal entries. Such factors, of course, can be quickly obtained by performing Gauss factorization on $I - S_n S_n^*$ as well as $I - S_n^* S_n$, and this makes it possible to compute the equivalent A and B matrices directly from the junction reflection coefficient matrices.

To see this, since A and B are functions of n, letting $A = A_n$ and $B = B_n$, (5.119) gives

$$d_n(z) = (A_n + zd_{n+1}(z)B_n S_n^*)^{-1} (A_n S_n + zd_{n+1}(z)B_n) \qquad (5.B.1)$$

which for $n = 1$ reads

$$d_1(z) = (A_1 + zd_2(z)B_1 S_1^*)^{-1} (A_1 S_1 + zd_2(z)B_1) . \qquad (5.B.2)$$

From (5.A.4)–(5.A.5) we also have

$$A_1^{-1}(A_1^{-1})^* = I - S_1 S_1^* \overset{\triangle}{=} M_1^* M_1 \qquad (5.B.3)$$

and

$$B_1^{-1}(B_1^{-1})^* = I - S_1^* S_1 \overset{\triangle}{=} N_1^* N_1 , \qquad (5.B.4)$$

where M_1 and N_1 represent the unique lower triangular factorization with positive diagonal entries in (5.B.3)–(5.B.4). Then

$$A_1^{-1} = M_1^* U_1 , \quad B_1^{-1} = N_1^* V_1 , \qquad (5.B.5)$$

where U_1 and V_1 are two specific unitary matrices. Thus $A_1 = U_1^*(M_1^*)^{-1}$ and $B_1 = V_1^*(N_1^*)^{-1}$ and substituting these into (5.B.2) we get

$$d_1(z) = \left(U_1^* M_1^{*-1} + zd_2(z)V_1^* N_1^{*-1} S_1^*\right)^{-1}$$

$$\times \left(U_1^* M_1^{*-1} S_1 + zd_2(z)V_1^* N_1^{*-1}\right)$$

$$= \left(I + zM_1^* U_1 d_2(z)V_1^* N_1^{*-1} S_1^*\right)^{-1}$$

$$\times \left(\mathbf{S}_1 + z\mathbf{M}_1^*\mathbf{U}_1\mathbf{d}_2(z)\mathbf{V}_1^*\mathbf{N}_1^{*-1}\right) . \tag{5.B.6}$$

Define

$$\rho_2(z) \overset{\triangle}{=} \mathbf{U}_1\mathbf{d}_2(z)\mathbf{V}_1^* \tag{5.B.7}$$

and since $\mathbf{U}_0 = \mathbf{V}_0 = \mathbf{I}$, let

$$\rho_1(z) = \mathbf{U}_0\mathbf{d}_1(z)\mathbf{V}_0^* = \mathbf{d}_1(z) . \tag{5.B.8}$$

From (5.85), $\rho_2(z)$ is also bounded and (5.B.6) simplifies to

$$\rho_1(z) = \left(\mathbf{I} + z\mathbf{M}_1^*\rho_2(z)\mathbf{N}_1^{*-1}\mathbf{s}_1^*\right)^{-1} \left(\mathbf{s}_1 + z\mathbf{M}_1^*\rho_2(z)\mathbf{N}_1^{*-1}\right) , \tag{5.B.9}$$

where

$$\mathbf{s}_1 \overset{\triangle}{=} \rho_1(0) = \mathbf{d}_1(0) = \mathbf{S}_1 . \tag{5.B.10}$$

To make further progress, define

$$\alpha \overset{\triangle}{=} \mathbf{U}_1\mathbf{A}_2\mathbf{U}_1^* \tag{5.B.11}$$

$$\beta \overset{\triangle}{=} \mathbf{V}_1\mathbf{B}_2\mathbf{V}_1^* \tag{5.B.12}$$

and

$$\mathbf{s}_2 \overset{\triangle}{=} \rho_2(0) = \mathbf{U}_1\mathbf{d}_2(0)\mathbf{V}_1^* = \mathbf{U}_1\mathbf{S}_2\mathbf{V}_1^* . \tag{5.B.13}$$

Then

$$\alpha_2^{-1}(\alpha_2^{-1})^* = \mathbf{U}_1\mathbf{A}_2^{-1}(\mathbf{A}_2^{-1})^*\mathbf{U}_1^* = \mathbf{U}_1(\mathbf{I} - \mathbf{S}_2\mathbf{S}_2^*)\mathbf{U}_1^*$$

$$= \mathbf{I} - \mathbf{s}_2\mathbf{s}_2^* \overset{\triangle}{=} \mathbf{M}_2^*\mathbf{M}_2 \tag{5.B.14}$$

$$\beta_2^{-1}(\beta_2^{-1})^* = \mathbf{V}_1\mathbf{B}_2^{-1}(\mathbf{B}_2^{-1})^*\mathbf{V}_1^* = \mathbf{V}_1(\mathbf{I} - \mathbf{S}_2^*\mathbf{S}_2)\mathbf{V}_1^*$$

$$= \mathbf{I} - \mathbf{s}_2^*\mathbf{s}_2 \overset{\triangle}{=} \mathbf{N}_2^*\mathbf{N}_2 , \tag{5.B.15}$$

where $\mathbf{M}_2$ and $\mathbf{N}_2$ are unique lower triangular factors with positive diagonal entries in (5.B.14)–(5.B.15). Thus there exist two unitary matrices $\mathbf{U}_2$ and $\mathbf{V}_2$ such that

$$\alpha_2^{-1} = \mathbf{M}_2^*\mathbf{U}_2 , \quad \beta_2^{-1} = \mathbf{N}_2^*\mathbf{V}_2 \tag{5.B.16}$$

or, in terms of these matrices, from (5.B.11)–(5.B.12) we obtain

$$\mathbf{A}_2 = \mathbf{U}_1^*\mathbf{U}_2^*\mathbf{M}_2^{*-1}\mathbf{U}_1 \tag{5.B.17}$$

$$\mathbf{B}_2 = \mathbf{V}_1^*\mathbf{V}_2^*\mathbf{N}_2^{*-1}\mathbf{V}_1 . \tag{5.B.18}$$

Substituting these into (5.B.1) with $n = 2$, we get

$$\mathbf{d}_2(z) = \left(\mathbf{U}_1^*\mathbf{U}_2^*\mathbf{M}_2^{*-1}\mathbf{U}_1 + z\mathbf{d}_3(z)\mathbf{V}_1^*\mathbf{V}_2^*\mathbf{N}_2^{*-1}\mathbf{V}_1\mathbf{S}_2^*\right)^{-1}$$

$$\times \left(\mathbf{U}_1^*\mathbf{U}_2^*\mathbf{M}_2^{*-1}\mathbf{U}_1\mathbf{S}_2 + z\mathbf{d}_3(z)\mathbf{V}_1^*\mathbf{V}_2^*\mathbf{N}_2^{*-1}\mathbf{V}_1\right)$$

$$= \left(\mathbf{M}_2^{*-1}\mathbf{U}_1 + z\mathbf{U}_2\mathbf{U}_1\mathbf{d}_3(z)\mathbf{V}_1^*\mathbf{V}_2^*\mathbf{N}_2^{*-1}\mathbf{V}_1\mathbf{S}_2^*\right)^{-1}$$

$$\times \left(\mathbf{M}_2^{*-1}\mathbf{U}_1\mathbf{S}_2 + z\mathbf{U}_2\mathbf{U}_1\mathbf{d}_3(z)\mathbf{V}_1^*\mathbf{V}_2^*\mathbf{N}_2^{*-1}\mathbf{V}_1\right). \qquad (5.B.19)$$

Define as before

$$\boldsymbol{\rho}_3(z) = \mathbf{U}_2\mathbf{U}_1\mathbf{d}_3(z)\mathbf{V}_1^*\mathbf{V}_2^*. \qquad (5.B.20)$$

Then (5.B.19) simplifies to

$$\mathbf{d}_2(z) = \mathbf{U}_1^* \left(\mathbf{M}_2^{*-1} + z\boldsymbol{\rho}_3(z)\mathbf{N}_2^{*-1}\mathbf{V}_1\mathbf{S}_2^*\mathbf{U}_1^*\right)^{-1}$$

$$\times \left(\mathbf{M}_2^{*-1}\mathbf{U}_1\mathbf{S}_2\mathbf{V}_1^* + z\boldsymbol{\rho}_3(z)\mathbf{N}_2^{*-1}\right)\mathbf{V}_1,$$

and using (5.B.7) and (5.B.13), the above expression reduces to

$$\boldsymbol{\rho}_2(z) = \left(\mathbf{I} + z\mathbf{M}_2^*\boldsymbol{\rho}_3(z)\mathbf{N}_2^{*-1}\mathbf{s}_2^*\right)^{-1} \left(\mathbf{s}_2 + z\mathbf{M}_2^*\boldsymbol{\rho}_3(z)\mathbf{N}_2^{*-1}\right). \qquad (5.B.21)$$

Once again as in (5.B.9), after this peculiar renormalization, the iteration of the bounded matrix function $\boldsymbol{\rho}_2(z)$ also makes use of only the lower triangular factors $\mathbf{M}_2$ and $\mathbf{N}_2$ with positive diagonal entries. This procedure is perfectly general and to complete this discussion, following (5.B.7)–(5.B.20), define

$$\boldsymbol{\rho}_n(z) = \mathbf{U}_{n-1}\cdots\mathbf{U}_2\mathbf{U}_1\mathbf{d}_n(z)\mathbf{V}_1^*\mathbf{V}_2^*\cdots\mathbf{V}_{n-1}^*, \quad n \geq 2 \qquad (5.B.22)$$

$$\mathbf{U} \stackrel{\triangle}{=} \mathbf{U}_{n-1}\cdots\mathbf{U}_2\mathbf{U}_1 \qquad (5.B.23)$$

and

$$\mathbf{V} \stackrel{\triangle}{=} \mathbf{V}_{n-1}\cdots\mathbf{V}_2\mathbf{V}_1. \qquad (5.B.24)$$

Then

$$\mathbf{s}_n \stackrel{\triangle}{=} \boldsymbol{\rho}_n(0) = \mathbf{U}\mathbf{S}_n\mathbf{V}^* \qquad (5.B.25)$$

and let

$$\boldsymbol{\alpha}_n \stackrel{\triangle}{=} \mathbf{U}\mathbf{A}_n\mathbf{U}^*, \quad \boldsymbol{\beta}_n \stackrel{\triangle}{=} \mathbf{V}\mathbf{B}_n\mathbf{V}^*$$

so that

$$\boldsymbol{\alpha}_n^{-1}(\boldsymbol{\alpha}_n^{-1})^* = \mathbf{U}\mathbf{A}_n^{-1}(\mathbf{A}_n^{-1})^*\mathbf{U}^* = \mathbf{U}(\mathbf{I} - \mathbf{S}_n\mathbf{S}_n^*)\mathbf{U}^*$$

$$= \mathbf{I} - \mathbf{s}_n\mathbf{s}_n^* \stackrel{\triangle}{=} \mathbf{M}_n^*\mathbf{M}_n \qquad (5.B.26)$$

$$\boldsymbol{\beta}_n^{-1}(\boldsymbol{\beta}_n^{-1})^* = \mathbf{V}\mathbf{B}_n^{-1}(\mathbf{B}_n^{-1})^*\mathbf{V}^* = \mathbf{V}(\mathbf{I} - \mathbf{S}_n^*\mathbf{S}_n)\mathbf{V}^*$$

$$= \mathbf{I} - \mathbf{s}_n^*\mathbf{s}_n \stackrel{\triangle}{=} \mathbf{N}_n^*\mathbf{N}_n, \qquad (5.B.27)$$

where $\mathbf{M}_n$ and $\mathbf{N}_n$ are, as before, the unique lower triangular matrix factors with positive diagonal entries in (5.B.26) and (5.B.27). Notice that $\mathbf{M}_n$ and $\mathbf{N}_n$ are not the lower triangular factors resulting from the factorizations in

(5.A.4)–(5.A.5), but instead the ones involving the renormalized reflection coefficient matrix $\mathbf{s}_n$ in (5.B.25). Clearly

$$\alpha_n^{-1} = \mathbf{UA}_n^{-1}\mathbf{U}^* = \mathbf{M}_n^*\mathbf{U}_n \quad \text{and} \quad \beta_n^{-1} = \mathbf{VB}_n^{-1}\mathbf{V}^* = \mathbf{N}_n^*\mathbf{V}_n, \quad (5.B.28)$$

where $\mathbf{U}_n$ and $\mathbf{V}_n$ are two specific unitary matrices and hence

$$\mathbf{U}_n\mathbf{UA}_n = (\mathbf{M}_n^*)^{-1}\mathbf{U} \qquad (5.B.29)$$

$$\mathbf{V}_n\mathbf{VB}_n = (\mathbf{N}_n^*)^{-1}\mathbf{V}. \qquad (5.B.30)$$

Making use of (5.B.29)–(5.B.30) in (5.B.1), we get

$$\mathbf{d}_n(z) = \left(\mathbf{U}^*\mathbf{U}_n^*\mathbf{M}_n^{*-1}\mathbf{U} + z\mathbf{d}_{n+1}(z)\mathbf{V}^*\mathbf{V}_n^*\mathbf{N}_n^{*-1}\mathbf{VS}_n^*\right)^{-1}$$

$$\times \left(\mathbf{U}^*\mathbf{U}_n^*\mathbf{M}_n^{*-1}\mathbf{US}_n + z\mathbf{d}_{n+1}(z)\mathbf{V}^*\mathbf{V}_n^*\mathbf{N}_n^{*-1}\mathbf{V}\right)$$

$$= \mathbf{U}^* \left(\mathbf{M}_n^{*-1} + z\mathbf{U}_n\mathbf{Ud}_{n+1}(z)\mathbf{V}^*\mathbf{V}_n^*\mathbf{N}_n^{*-1}\mathbf{s}_n^*\right)^{-1}$$

$$\times \left(\mathbf{M}_n^{*-1}\mathbf{s}_n + z\mathbf{U}_n\mathbf{Ud}_{n+1}(z)\mathbf{V}^*\mathbf{V}_n^*\mathbf{N}_n^{*-1}\right)\mathbf{V} \qquad (5.B.31)$$

where we have made use of (5.B.25). But from (5.B.22),

$$\boldsymbol{\rho}_{n+1}(z) = \mathbf{U}_n\mathbf{U}_{n-1}\cdots\mathbf{U}_1\mathbf{d}_{n+1}(z)\mathbf{V}_1^*\cdots\mathbf{V}_{n-1}^*\mathbf{V}_n^* = \mathbf{U}_n\mathbf{Ud}_{n+1}(z)\mathbf{V}^*\mathbf{V}_n^*$$

and hence (5.B.31) simplifies to

$$\mathbf{Ud}_n(z)\mathbf{V}^* = \boldsymbol{\rho}_n(z)$$

$$= \left(\mathbf{I} + z\mathbf{M}_n^*\boldsymbol{\rho}_{n+1}(z)\mathbf{N}_n^{*-1}\mathbf{s}_n^*\right)^{-1} \left(\mathbf{s}_n + z\mathbf{M}_n^*\boldsymbol{\rho}_{n+1}(z)\mathbf{N}_n^{*-1}\right). \qquad (5.B.32)$$

To summarize, given a positive matrix function $\mathbf{Z}(z)$, and the corresponding bounded matrix function $\mathbf{d}(z)$ as in (5.76) or (5.84), the bounded matrix functions $\mathbf{d}_1(z)$, $\mathbf{d}_2(z)$, ..., $\mathbf{d}_n(z)$ generated as in (5.90)–(5.105) satisfy the recursion (5.B.1) with $\mathbf{A}_n$ and $\mathbf{B}_n$ as given in (5.A.1)–(5.A.2). The unique set of unitary matrix pairs $(\mathbf{U}_1, \mathbf{V}_1)$, $(\mathbf{U}_2, \mathbf{V}_2)$, ..., $(\mathbf{U}_n, \mathbf{V}_n)$ generated as in (5.B.5), (5.B.16) and (5.B.28) can be used to define a new set of bounded matrix functions $\boldsymbol{\rho}_n(z)$ as in (5.B.22). These bounded matrix functions satisfy the recursion

$$\boldsymbol{\rho}_n(z) = \left(\mathbf{I} + z\mathbf{M}_n^*\boldsymbol{\rho}_{n+1}(z)\mathbf{N}_n^{*-1}\mathbf{s}_n^*\right)^{-1} \left(\mathbf{s}_n + z\mathbf{M}_n^*\boldsymbol{\rho}_{n+1}(z)\mathbf{N}_n^{*-1}\right), \qquad (5.B.33)$$

where $\mathbf{M}_n$ and $\mathbf{N}_n$ are the unique lower triangular matrix factors with positive diagonal entries that satisfy the relation

$$\mathbf{M}_n^*\mathbf{M}_n = \mathbf{I} - \mathbf{s}_n\mathbf{s}_n^* \qquad (5.B.34)$$

and

$$\mathbf{N}_n^*\mathbf{N}_n = \mathbf{I} - \mathbf{s}_n^*\mathbf{s}_n, \qquad (5.B.35)$$

where

$$\mathbf{s}_n \overset{\triangle}{=} \boldsymbol{\rho}_n(0). \tag{5.B.36}$$

In a similar manner the right-inverse recursion in (5.156) simplifies into

$$\boldsymbol{\rho}_n(z) = \left(\mathbf{s}_n + z\mathbf{M}_n^{-1}\boldsymbol{\rho}_{n+1}(z)\mathbf{N}_n\right)\left(\mathbf{I} + z\mathbf{s}_n^*\mathbf{M}_n^{-1}\boldsymbol{\rho}_{n+1}(z)\mathbf{N}_n\right)^{-1}. \tag{5.B.37}$$

Notice that the renormalized bounded matrix function (5.B.22) and the recursions (5.B.33)–(5.B.37) are direct generalizations of the single-channel version in (2.199)–(2.203).

Problems

1. Let $\mathbf{S}(\theta)$ represent the power spectral density matrix of a real vector process.
   (a) Show that $\mathbf{S}(\theta) = \mathbf{S}^T(-\theta)$.
   (b) Write $\mathbf{S}(\theta) = \mathbf{R}(\theta) + j\mathbf{X}(\theta)$. Notice that $\mathbf{R}(\theta)$ and $\mathbf{X}(\theta)$ are real matrices. Show that

$$\mathbf{R}(\theta) = \mathbf{R}^T(\theta) = \mathbf{R}(-\theta)$$

$$\mathbf{X}(\theta) = \mathbf{X}^T(-\theta) = -\mathbf{X}(-\theta)$$

   and

$$\mathbf{r}_k = \frac{1}{2\pi}\int_{-\pi}^{\pi}\mathbf{S}(\theta)e^{-jk\theta}d\theta = \frac{1}{\pi}\int_{0}^{\pi}\left(\mathbf{R}(\theta)\cos k\theta + \mathbf{X}(\theta)\sin k\theta\right)d\theta.$$

   Thus, for real processes, $\mathbf{R}(\theta)$ and $\mathbf{X}(\theta)$ are even and odd functions of θ, respectively. Further, the autocorrelation matrices of such processes are real.

2. Let $\mathbf{Z}(z)$ represent a 2×2 symmetric positive matrix function. Show that the off-diagonal entry $Z_{12}(z)$ can be expressed as the difference of two positive functions.

3. Let $\mathbf{Z}(z)$ represent a positive matrix function and $\mathbf{R}$ any constant matrix with a positive Hermitian part, i.e.,

$$\mathbf{K} = \frac{\mathbf{R} + \mathbf{R}^*}{2} > 0.$$

   Show that

$$\begin{aligned} \mathbf{d}(z) &\overset{\triangle}{=} \mathbf{K}^{-1/2}\left(\mathbf{Z}(z) - \mathbf{R}\right)\left(\mathbf{Z}(z) + \mathbf{R}^*\right)^{-1}\mathbf{K}^{1/2} \\ &= \mathbf{K}^{1/2}\left(\mathbf{Z}(z) + \mathbf{R}^*\right)^{-1}\left(\mathbf{Z}(z) - \mathbf{R}\right)\mathbf{K}^{-1/2} \end{aligned} \tag{5.P.1}$$

   and it represents a bounded matrix function, where $\mathbf{K}^{1/2}$ denotes *any* Hermitian square root of $\mathbf{K}$.

4. Show that the power series $\mathbf{d}(z) = \sum_{k=0}^{\infty} \mathbf{d}_k z^k$, $|z| < 1$, defines a bounded matrix function iff

$$\mathbf{I} - \mathbf{D}_k \mathbf{D}_k^* \geq 0, \quad k = 0 \to \infty, \tag{5.P.2}$$

where

$$\mathbf{D}_k = \begin{bmatrix} \mathbf{d}_0 & \mathbf{0} & \cdots & \mathbf{0} \\ \mathbf{d}_1 & \mathbf{d}_0 & \cdots & \mathbf{0} \\ \vdots & \vdots & \ddots & \vdots \\ \mathbf{d}_k & \mathbf{d}_{k-1} & \cdots & \mathbf{d}_0 \end{bmatrix} \tag{5.P.3}$$

represents the lower triangular block Toeplitz matrix generated from the matrix coefficients $\mathbf{d}_0, \mathbf{d}_1, \ldots, \mathbf{d}_k$.

5. Show that the renormalized junction reflection coefficient matrix $\mathbf{s}_n$ in (5.231) can be alternatively expressed as

$$\mathbf{s}_n = \mathbf{A}_{n-1}(0) \left\{ \left(\sum_{k=1}^{n} \mathbf{r}_k z^k \right) \mathbf{C}_{n-1}(z) \right\}_n, \tag{5.P.4}$$

where $\{\ \}_n$ denotes the coefficient of z^n in $\{\ \}$.

6. Given any positive matrix function $\mathbf{Z}(z)$, let $\mathbf{Z}_k(z)$, $k \geq 1$, represent the sequence of positive matrix functions generated by the Schur multiline extraction procedure. Show that the characteristic impedance matrix $\mathbf{R}_n \triangleq \mathbf{Z}_n(0)$ at the nth stage can be expressed as

$$\mathbf{R}_n = (\mathbf{I} - \mathbf{J}_n)^{-1} \left(\mathbf{R}_{n-1} + \mathbf{J}_n \mathbf{R}_{n-1}^* \right), \tag{5.P.5}$$

where

$$\mathbf{J}_n \triangleq \mathbf{K}_{n-1}^{1/2} \mathbf{S}_n \mathbf{K}_{n-1}^{-1/2}, \quad \mathbf{K}_{n-1} \triangleq \frac{\mathbf{R}_{n-1} + \mathbf{R}_{n-1}^*}{2} \tag{5.P.6}$$

and $\mathbf{S}_n$ represents the junction reflection coefficient matrix at the nth stage. Here $\mathbf{K}_{n-1}^{1/2}$ is *any* Hermitian square root of $\mathbf{K}_{n-1}$.

7. (a) Let Δ_n represent the determinant of the block Toeplitz matrix $\mathbf{T}_n$ in (5.251) and assume that the Paley-Wiener criterion is satisfied. Thus $\Delta_n = |\mathbf{T}_n| > 0$ and with symbols as defined in the text, show that

$$\Delta_n = |\mathbf{T}_{-n}|.$$

(b) Show that

$$\frac{\Delta_n}{\Delta_{n-1}} = \frac{1}{|\mathbf{T}_n^{11}|} = \frac{1}{|\mathbf{A}_0^{(n)}|^2} = \frac{|\mathbf{M}_n^* \mathbf{M}_n|}{|\mathbf{A}_0^{(n-1)}|^2} = |\mathbf{M}_n^* \mathbf{M}_n| \frac{\Delta_{n-1}}{\Delta_{n-2}}$$

$$= |\mathbf{R}_0| \prod_{k=1}^{n} |\mathbf{M}_k^* \mathbf{M}_k| = |\mathbf{R}_0| \prod_{k=1}^{n} |\mathbf{I} - \mathbf{s}_k \mathbf{s}_k^*|.$$

Thus Δ_n/Δ_{n-1} forms a monotone decreasing sequence with a positive limit.

(c) Show that

$$|\mathbf{B}_0|^2 = exp\left(\frac{1}{2\pi}\int_{-\pi}^{\pi} ln\, det\, \mathbf{S}(\theta)d\theta\right) = |\mathbf{R}_0|\prod_{k=1}^{\infty}|\mathbf{I} - \mathbf{s}_k\mathbf{s}_k^*|, \quad (5.P.7)$$

where $\mathbf{B}_0$ represents the constant term in the Wiener factor of $\mathbf{S}(\theta)$. Thus the Paley-Wiener criterion implies $\mathbf{I} - \mathbf{s}_n\mathbf{s}_n^* > 0$, $n = 0 \to \infty$, and $|\mathbf{B}_0| > 0$.

(d) Let $|\mathbf{I} - \mathbf{s}_k\mathbf{s}_k^*| = 1 - \lambda_k^2$. Then show that Paley-Wiener criterion is equivalent to

$$\lambda_k^2 < 1 \quad \text{and} \quad \sum_{k=1}^{\infty}\lambda_k^2 < \infty.$$

This is a direct generalization of (3.80).

8. **Block Toeplitz Inverses**: Let $\mathbf{T}_n$ represent the positive Hermitian block Toeplitz matrix generated from $\mathbf{r}_0, \mathbf{r}_1, \ldots, \mathbf{r}_n$ as in (5.251) and, further, let

$$\mathbf{A}_n(z) \overset{\triangle}{=} \mathbf{A}_0 + \mathbf{A}_1 z + \cdots + \mathbf{A}_n z^n$$
$$\mathbf{C}_n(z) \overset{\triangle}{=} \mathbf{C}_0 + \mathbf{C}_1 z + \cdots + \mathbf{C}_n z^n$$

represent the left and right Levinson polynomial matrices of the first kind as given in (5.252)–(5.253).

(a) Show that

$$\mathbf{T}_n^{-1} = \mathbf{A}_0\mathbf{A}_0^* - \mathbf{C}_n\mathbf{C}_n^* = \mathbf{C}_0\mathbf{C}_0^* - \mathbf{A}_n\mathbf{A}_n^*, \quad (5.P.8)$$

where

$$\mathbf{A}_0 \overset{\triangle}{=} \begin{bmatrix} \mathbf{A}_0^* & \mathbf{0} & \cdots & \mathbf{0} \\ \mathbf{A}_1^* & \mathbf{A}_0^* & \cdots & \mathbf{0} \\ \vdots & \vdots & \ddots & \vdots \\ \mathbf{A}_n^* & \mathbf{A}_{n-1}^* & \cdots & \mathbf{A}_0^* \end{bmatrix} \quad (5.P.9)$$

$$\mathbf{A}_n \overset{\triangle}{=} \begin{bmatrix} \mathbf{0} & \mathbf{0} & \cdots & \mathbf{0} & \mathbf{0} \\ \mathbf{A}_n & \mathbf{0} & \cdots & \mathbf{0} & \mathbf{0} \\ \vdots & \vdots & \cdots & \vdots & \vdots \\ \mathbf{A}_1 & \mathbf{A}_2 & \cdots & \mathbf{A}_n & \mathbf{0} \end{bmatrix} \quad (5.P.10)$$

$$\mathbf{C}_0 \overset{\triangle}{=} \begin{bmatrix} \mathbf{C}_0^* & \mathbf{0} & \cdots & \mathbf{0} \\ \mathbf{C}_1^* & \mathbf{C}_0^* & \cdots & \mathbf{0} \\ \vdots & \vdots & \ddots & \vdots \\ \mathbf{C}_n^* & \mathbf{C}_{n-1}^* & \cdots & \mathbf{C}_0^* \end{bmatrix} \quad (5.P.11)$$

and

$$\mathbf{C}_n \triangleq \begin{bmatrix} \mathbf{0} & \mathbf{0} & \cdots & \mathbf{0} & \mathbf{0} \\ \mathbf{C}_n & \mathbf{0} & \cdots & \mathbf{0} & \mathbf{0} \\ \vdots & \vdots & \cdots & \vdots & \vdots \\ \mathbf{C}_1 & \mathbf{C}_2 & \cdots & \mathbf{C}_n & \mathbf{0} \end{bmatrix}. \tag{5.P.12}$$

(b) Show that the inverse of $\mathbf{T}_{-n}$ in (5.234) is also of the form (5.P.8), with the lower (and upper) triangular block Toeplitz matrices in (5.P.8) replaced by their respective upper (and lower) triangular counterparts. (See also (3.114) and (3.P.9).)

9. **Right-coprime Identification**: Let

$$\mathbf{H}_r(z) = \mathbf{Q}_1(z)\mathbf{P}_1^{-1}(z) \tag{5.P.13}$$

denote the right-coprime representation of a multichannel rational system with

$$\mathbf{P}_1(z) = \mathbf{I} + \tilde{\mathbf{P}}_1 z + \cdots \tilde{\mathbf{P}}_p z^p \tag{5.P.14}$$

and

$$\mathbf{Q}_1(z) = \tilde{\mathbf{Q}}_0 + \tilde{\mathbf{Q}}_1 z + \cdots + \tilde{\mathbf{Q}}_q z^q. \tag{5.P.15}$$

(a) Using (5.259), (5.261) and (5.289) show that

$$\tilde{\mathbf{P}}_i = \begin{cases} \left(\displaystyle\sum_{k=0}^{i} \mathbf{C}_{i-k}\mathbf{G}_k - \sum_{k=0}^{i-1} \mathbf{A}^*_{p+q+k-i+1}\mathbf{H}_k \right) \mathbf{C}_0^{-1}, & 1 \leq i \leq q \\[4mm] \left(\displaystyle\sum_{k=0}^{q} \mathbf{C}_{i-k}\mathbf{G}_k - \sum_{k=0}^{q-1} \mathbf{A}^*_{p+q+k-i+1}\mathbf{H}_k \right) \mathbf{C}_0^{-1}, & q+1 \leq i \leq p, \end{cases} \tag{5.P.16}$$

where $\mathbf{G}_k$, $k = 1 \rightarrow q$, and $\mathbf{H}_k$, $k = 0 \rightarrow q-1$, satisfy the linear equations

$$\mathbf{D}\mathbf{X}_1 = \mathbf{B}_1. \tag{5.P.17}$$

Here $\mathbf{D}$ is a $2q \times 2q$ block matrix given by

$$\begin{bmatrix} \mathbf{C}_p & \mathbf{C}_{p-1} & \cdots & \mathbf{C}_{p-q+1} & -\mathbf{A}^*_q & -\mathbf{A}^*_{q-1} & \cdots & -\mathbf{A}^*_{2q-1} \\ \mathbf{C}_{p+1} & \mathbf{C}_p & \cdots & \mathbf{C}_{p-q+2} & -\mathbf{A}^*_{q-1} & -\mathbf{A}^*_q & \cdots & -\mathbf{A}^*_{2q-2} \\ \vdots & \vdots & \cdots & \vdots & \vdots & \vdots & \cdots & \vdots \\ \mathbf{C}_{p+q-1} & \mathbf{C}_{p+q-2} & \cdots & \mathbf{C}_p & -\mathbf{A}^*_1 & -\mathbf{A}^*_2 & \cdots & -\mathbf{A}^*_q \\ \mathbf{C}_{p+q} & \mathbf{C}_{p+q-1} & \cdots & \mathbf{C}_{p+1} & -\mathbf{A}^*_0 & -\mathbf{A}^*_1 & \cdots & -\mathbf{A}^*_{q-1} \\ \mathbf{0} & \mathbf{C}_{p+q} & \cdots & \mathbf{C}_{p+2} & \mathbf{0} & -\mathbf{A}^*_0 & \cdots & -\mathbf{A}^*_{q-2} \\ \vdots & \vdots & \cdots & \vdots & \vdots & \vdots & \cdots & \vdots \\ \mathbf{0} & \mathbf{0} & \cdots & \mathbf{C}_{p+q} & \mathbf{0} & \mathbf{0} & \cdots & -\mathbf{A}^*_0 \end{bmatrix} \tag{5.P.18}$$

and

$$
\mathbf{X}_1 \triangleq
\begin{bmatrix}
\mathbf{G}_1 \\
\mathbf{G}_2 \\
\vdots \\
\mathbf{G}_q \\
\hline
\mathbf{H}_0 \\
\mathbf{H}_1 \\
\vdots \\
\mathbf{H}_{q-1}
\end{bmatrix}
, \quad
\mathbf{B}_1 \triangleq
\begin{bmatrix}
\mathbf{C}_{p+1} \\
\mathbf{C}_{p+2} \\
\vdots \\
\mathbf{C}_{p+q} \\
0 \\
0 \\
0 \\
\vdots \\
0
\end{bmatrix}
. \tag{5.P.19}
$$

$\mathbf{A}_k$, $\mathbf{C}_k$, $k = 0 \to p+q$, represent the coefficients of the left and right Levinson polynomial matrices of the first kind and order $p + q$ generated from the autocorrelation matrices $\mathbf{r}_0$, $\mathbf{r}_1$, ..., $\mathbf{r}_{p+q}$ associated with the given system. Further,

$$
\mathbf{Q}_1(z) = \mathbf{Q}(z)\mathbf{C}_0^{-1} \tag{5.P.20}
$$

and $\mathbf{Q}(z)$ satisfies the equation

$$
\mathbf{Q}_*(z)\mathbf{Q}(z) = \mathbf{G}_{1*}(z)\mathbf{G}_1(z) - \mathbf{H}_{1*}(z)\mathbf{H}_1(z), \tag{5.P.21}
$$

where $\mathbf{G}_1(z) = \sum_{k=0}^{q} \mathbf{G}_k z^k$ and $\mathbf{H}_1(z) = \sum_{k=0}^{q-1} \mathbf{H}_k z^k$ with $\mathbf{G}_0 = \mathbf{I}$. Here $\mathbf{H}_1(z)\mathbf{G}_1^{-1}(z)$ represents $\rho_{p+q+1}(z)$. Notice that up to a unitary factor, $\mathbf{Q}(z)$ in (5.P.20)–(5.P.21) can be identified as the left Levinson polynomial matrix $\mathbf{A}_q(z)$ of the first kind and order q associated with the power spectral density matrix (see also (5.299)–(5.301))

$$
\mathbf{K}_1(\theta) \triangleq \left(\mathbf{G}_1^*(e^{j\theta})\mathbf{G}_1(e^{j\theta}) - \mathbf{H}_1^*(e^{j\theta})\mathbf{H}_1(e^{j\theta}) \right)^{-1}. \tag{5.P.22}
$$

(b) Let $\rho_{p+q+2}(z) = \mathbf{F}_1(z)\mathbf{E}_1^{-1}(z)$ represent the bounded matrix function at stage $n = p + q + 1$. Show that

$$
\epsilon_0 = \mathbf{M}_{p+q+1}\mathbf{s}_{p+q+1}\mathbf{N}_{p+q+1}^{-1}\mathbf{E}_q + \mathbf{F}_{q-1} \equiv 0 \tag{5.P.23}
$$

with $\mathbf{s}_{p+q+1} = \mathbf{H}_0$ as given in (5.P.19) and $\mathbf{M}_{p+q+1}$, $\mathbf{N}_{p+q+1}$ as given by (5.339)–(5.340). Thus

$$
\epsilon_0(n, m) = tr\left(\epsilon_0 \epsilon_0^*\right) \tag{5.P.24}
$$

can be used as a test criterion to identify the right-coprime orders of multichannel systems.

10. Show that left- and right-coprime representations of the minimum-phase rational matrix

$$
\mathbf{H}(z) =
\begin{bmatrix}
\dfrac{3 + z}{3 - 0.5z - z^2} & \dfrac{z}{1.5 - z} \\[3mm]
\dfrac{1.4 - z}{3 - z + z^3} & \dfrac{3 + z}{4 + 2z + z^2}
\end{bmatrix}
$$

are given by

$$\mathbf{H}_l(z) = \mathbf{P}^{-1}(z)\mathbf{Q}(z),$$

where

$$\mathbf{P}(z) = \mathbf{I} + \begin{bmatrix} -0.9966 & 0.0018 \\ -0.0067 & 0.1665 \end{bmatrix} z$$

$$+ \begin{bmatrix} -0.1941 & 0.0093 \\ -0.2497 & 0.0775 \end{bmatrix} z^2 + \begin{bmatrix} 0.2801 & -0.00033 \\ 0.0438 & 0.2493 \end{bmatrix} z^3$$

$$+ \begin{bmatrix} -0.00051 & -0.0020 \\ 0.0836 & 0.1661 \end{bmatrix} z^4 + \begin{bmatrix} 0.0012 & -0.0031 \\ -0.0012 & 0.0838 \end{bmatrix} z^5$$

$$\mathbf{Q}(z) = \begin{bmatrix} 0.9993 & 0.4667 \\ 0.4667 & 0.7801 \end{bmatrix} + \begin{bmatrix} -0.4957 & 0.2697 \\ -0.2715 & -0.0501 \end{bmatrix} z$$

$$+ \begin{bmatrix} -0.2715 & -0.2867 \\ -0.3058 & -0.2118 \end{bmatrix} z^2 + \begin{bmatrix} 0.0021 & -0.2073 \\ -0.1664 & -0.0171 \end{bmatrix} z^3,$$

and

$$\mathbf{H}_r(z) = \mathbf{Q}_1(z)\mathbf{P}_1^{-1}(z),$$

where

$$\mathbf{P}_1(z) = \mathbf{I} + \begin{bmatrix} 0.1667 & 0.00 \\ -0.7027 & -0.1667 \end{bmatrix} z + \begin{bmatrix} -0.1667 & 0.00 \\ -0.3514 & -0.0833 \end{bmatrix} z^2$$

$$+ \begin{bmatrix} 0.3333 & 0.00 \\ -0.1757 & -0.1667 \end{bmatrix} z^3 + \begin{bmatrix} 0.1667 & 0.00 \\ 0.00 & 0.00 \end{bmatrix} z^4$$

$$\mathbf{Q}_1(z) = \begin{bmatrix} 1.2178 & 0.3500 \\ 0.3500 & 0.5625 \end{bmatrix} + \begin{bmatrix} 0.3741 & 0.5500 \\ -0.4703 & -0.1875 \end{bmatrix} z$$

$$+ \begin{bmatrix} -0.2949 & 0.2556 \\ -0.2568 & -0.1250 \end{bmatrix} z^2 + \begin{bmatrix} 0.0090 & 0.1667 \\ 0.00 & -0.1513 \end{bmatrix} z^3.$$

(*Hint*: Refer to discussions on section 5.3.4 and Fig. 5.23 below.)

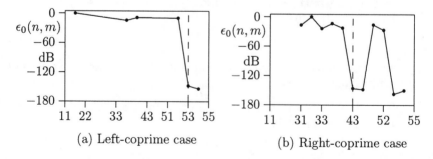

(a) Left-coprime case   (b) Right-coprime case

FIGURE 5.23. Model order selection criterion: The actual $\epsilon_0(n, m)$ curves for Problem 5.10.

References

[1] N. Wiener, "Generalized harmonic analysis," *Acta Math.*, vol. 55, pp. 117–258, 1930.

[2] M. Marcus and H. Minc, *Introduction to Linear Algebra*, New York: Dover Publications, 1965.

[3] F. R. Gantmacher, *The Theory of Matrices*, vol. 1, New York: Chelsea, 1977.

[4] U. Grenander and G. Szegö, *Toeplitz Forms and Their Applications*, New York: Chelsea, 1984.

[5] I. Gikhman and V. Skorokhod, *Introduction to the Theory of Random Processes*, Philadelphia: W. B. Saunders, 1965.

[6] N. Wiener and P. Masani, "The prediction theory of multivariate processes, part I: the regularity condition," *Acta Math.*, vol. 98, 1957.

[7] D. C. Youla and N. N. Kazanjian, "Bauer-type factorization of positive matrices and the theory of matrix polynomials orthogonal on the unit circle," *IEEE Trans. Circuit Systs.*, vol. CAS-25, pp. 57–69, February 1978.

[8] N. N. Kazanjian, "Bauer-type factorization of positive matrices and the theory of matrix polynomials orthogonal on the unit circle," Ph.D. Dissertation, Polytechnic Institute of New York, Brooklyn, New York, 1977.

[9] W. Rudin, *Real and Complex Analysis*, New York: McGraw-Hill, 1974.

[10] R. P. Boas, *Entire Functions*, New York: Academic Press, 1954.

[11] N. I. Ahiezer and M. Krein, *Some Questions in the Theory of Moments*, American Math. Soc., Math. Mono., vol. 2, American Mathematical Society, Providence, R. I. 1962.

[12] D. C. Youla, "The FEE: A new tunable high-resolution spectral estimator," Part I, Technical note, no. 3, Department of Electrical Engineering, Polytechnic Institute of New York, Brooklyn, New York, 1980: also RADC Rep. RADC-TR-81-397, AD A114996, February 1982.

[13] C. L. Dodgson, "Condensation of determinant," *Proc. Royal Soc. London*, vol. 15, pp. 150–155, 1866.

[14] D. P. Robbins and H. Ru, "Determinants and alternating sign matrices," *Advances in Mathematics*, vol. 62, pp. 169–184, 1986.

[15] N. R. Goodman, "Statistical analysis based on a certain multivariate complex Gaussian distribution (an introduction)," *Ann. Math. Stat.*, vol. 34, pp. 152–177, 1963.

[16] A. S. Householder, *The Theory of Matrices in Numerical Analysis*, New York: Blaisdell Pub. Co., 1964.

[17] G. Herglotz, "Über Potenzreihen mit positivem reellem Teil im Einheitskreis," *Leipziger Berichte*, vol. 63, 1911.

[18] J. Schur, "Über Potensreihen, die im Innern des Einheitskreises Beschrankt Sind," *Journal für Reine und Angewandte Mathematik*, vol. 147, pp. 205–232, 1917; also vol. 148, pp. 122–145, 1918.

[19] V. Belevitch, *Classical Network Theory*, San Francisco: Holden-Day, 1968.

[20] D. C. Youla, *Lecture Notes on Network Theory*, Department of Electrical Engineering, Polytechnic University, 1985.

[21] J. D. Rhodes, *Theory of Electrical Filters*, New York: John Wiley and Sons, 1976.

[22] D. C. Youla, J. D. Rhodes, and P. C. Marston, "Driving point synthesis of resistor-terminated cascades composed of lumped lossless passive 2 ports commensurate TEM lines," *IEEE Trans. Circuit Theory*, vol. CT-19, no. 6, pp. 648–664, November 1972.

[23] B. Brune, "Synthesis of a finite two-terminal network whose driving-point impedance is a prescribed function of frequency," *J. Math. Phys.*, vol. 10, no. 3, pp. 191–236, 1931.

[24] E. Guillemin, *Synthesis of Passive Networks*, New York: John Wiley and Sons, 1965.

[25] D. C. Youla, "Interpolatory multichannel spectral estimation, Part I, general theory and the FEE," Technical report, Department of Electrical Engineering, Polytechnic Institute of New York, Brooklyn, New York, 1980; also RADC Rep. RADC-TR-81-397, AD A114996, February 1982.

[26] P. Dienes, *The Taylor Series*, New York: Dover Publications, 1957.

[27] G. M. Goluzin, *Geometric Theory of Functions*, American Math. Soc., Math. Mono., vol. 26, American Mathematical Society, Providence, R. I. 1969.

[28] D. C. Youla, "Correspondence with H. Carlin concerning essential singularities of propagation functions," Department of Electrical Engineering, Polytechnic University, New York, 1972.

[29] R. M. Foster, "A reactance theorem," *Bell System Tech. J.*, vol. 3, pp. 259–267, April 1924.

[30] P. I. Richards, "A special class of Functions with positive real part in a half plane," *Duke Mathematical Journal*, vol. 14, pp. 777–786, September 1947.

[31] S. U. Pillai, D. C. Youla, and T. I. Shim, "A new technique for ARMA system identification and rational approximation," in ONR Annual Report, Department of Electrical Engineering, Polytechnic University, October 1990.

[32] M. Marden, *The Geometry of the Zeros of a Polynomial in a Complex Variable*, Amer. Math. Soc., Math. Survey, no. 3, American Mathematical Society, Providence, R. I. 1949.

[33] E. I. Jury and B. H. Bharucha, "Notes on the stability criterion for linear discrete systems," *IRE Trans. Automat. Contr.*, vol. AC-6, no. 1, pp. 88–90, February 1961.

[34] E. J. Routh, "A treatise on the stability of a given state of motion," London, pp. 74–81, 1877.

[35] L. Ya Geronimus, *Polynomials Orthogonal on a Circle and Their Applications*, American Math. Soc., Transaction, no. 104, American Mathematical Society, Providence, R. I. 1954.

[36] J. A. Shohat and J. D. Tamarkin, *The Problem of Moments*, American Math. Soc., Math. Surveys, no. 104, American Mathematical Society, Providence, R. I. 1970, pp. 6–7, 30.

[37] E. C. Titchmarsh, *The Theory of Functions*, 2nd ed., Oxford University Press, Oxford, 1986.

[38] J. P. Burg, "Maximum entropy analysis," presented at the 37th Annual Meeting, *Soc. Explor. Geophysics.*, Oklahoma City, Oklahoma, 1967.

[39] J. P. Burg, *Maximum Entropy Spectral Analysis*, Ph.D. Dissertation, Stanford University, Stanford, CA, May 1975.

[40] K. Knopp, *Theory and Application of Infinite Series*, New York: Dover Publications, 1990.

[41] L. Ya Geronimus, *Orthogonal Polynomials*, New York: Consultants Bureau Enterprises, Inc., 1961.

[42] I. C. Gohberg and A. A. Semuncul, "On the inversion of finite Toeplitz matrices and their continual analoga," *Mat. Issled.*, vol. 7, no. 2, pp. 201–223, 1972 (Russian).

[43] I. S. Iohvidov, *Hankel and Toeplitz Matrices and Forms: Algebraic Theory*, Translated by G. P. A. Thijsse, Boston: Birkhauser, 1982.

[44] I. H. Sublette, *On the Synthesis of Driving Point Impedances*, Ph.D. Dissertation, University of Pennsylvania, PA, 1957.

[45] D. C. Youla, "Notes on Ikeno's theorem," Department of Electrical Engineering, Polytechnic University, New York, 1989.

[46] H. S. Wall, *Analytic Theory of Continued Fractions*, New York: Chelsea, 1973.

[47] M. R. Spiegel, *Mathematical Handbook of Formulas and Tables*, Schaum's Outline Series in Mathematics, pp. 32, New York: McGraw-Hill, 1968.

[48] S. U. Pillai, T. I. Shim, and H. Benteftifa, "A new spectrum extension method that maximizes the multistep minimum prediction error — Generalization of the maximum entropy concept," *IEEE Trans. Signal Processing*, vol. SP-40, no. 1, pp. 142–158, January 1992.

[49] G. A. Baker, Jr., *Essentials of Pade Approximants*, New York: Academic Press, 1975.

[50] D. S. Mazel, J. S. Geronimo, and M. H. Hayes, III, "On the geometric sequence of reflection coefficients," *IEEE Trans. Acoust. Speech, and Signal Processing*, vol. ASSP-38, no. 10, pp. 1810–1812, October 1990.

[51] W. Gersch, "Estimation of the autoregressive parameters of a mixed autoregressive moving-average time series," *IEEE Trans. Autom. Control*, vol. AC-15, pp. 583–588, October 1970.

[52] J. Cadzow, "ARMA modeling of time series," *IEEE Trans. Patt. Anal. and Mach. Intel.*, vol. PAMI-4, pp. 124–128, March 1982.

[53] J. Cadzow, "Spectral estimation: an overdetermined rational model equation approach," *Proc. IEEE*, vol. 70, no. 9, pp. 907–939, September 1982.

[54] S. M. Kay and S. L. Marple, Jr., "Spectrum analysis – a modern perspective," *Proc. IEEE*, vol. 69, no. 11, pp. 1380–1419, November 1981.

[55] S. M. Kay, *Modern Spectral Estimation: Theory and Application*, Englewood Cliffs, N.J.: Prentice-Hall, 1988.

[56] A. Papoulis, *Probability, Random Variables and Stochastic Processes*, 2nd ed., New York: McGraw-Hill, 1984, pp.421–429.

[57] J. C. Chow, "On estimating the orders of an autoregressive moving-average process with uncertain observations," *IEEE Trans. Autom. Control*, vol. AC-17, pp. 707–709, October 1972.

[58] P. Stoica, "On a procedure for testing the orders of time series," *IEEE Trans. Autom. Control*, vol. AC-26, no. 2, pp. 572–573, April 1981.

[59] Y. T. Chan and J. C. Wood, "A new order determination technique for ARMA processes," *IEEE Trans. Acoust., Speech, Signal Processing*, vol. ASSP-32, no. 3, pp. 517–521, June 1984.

[60] J. J. Fuchs, "ARMA order estimation via matrix perturbation theory," *IEEE Trans. Autom. Control*, vol. AC-32, no. 4, pp. 358–361, April 1987.

[61] D. C. Youla and P. Tissi, "*n*-port synthesis via reactance extraction – part I," *IEEE International Convention Record*, part 7, New York: IEEE Press, 1966.

[62] M. L. Van Blaricum and R. Mittra, "Problems and solutions associated with Prony's method for processing transient data," *IEEE Trans. on Ant. Prop.*, vol. AP-26, pp. 174–182, January 1978.

[63] B. D. O. Anderson, "An algebraic solution to the spectral factorization problem," *IEEE Trans. Autom. Control*, vol. AC-12, no. 4, pp. 410–414, August 1967.

[64] B. D. O. Anderson, K. L. Hitz, and N. D. Diem, "Recursive algorithm for spectral factorization," *IEEE Trans. Circuits Systs.*, vol. CAS-21, no. 6, pp. 742–750, November 1974.

[65] G. Wilson, "Factorization of the covariance generating function of a pure moving average process," *SIAM J. Numer. Anal.*, vol. 6, no. 1, pp. 1–7, March 1969.

[66] E. Parzen, "Some recent advances in time series modeling," *IEEE Trans. Autom. Control*, vol. AC-19, pp. 723–730, December 1974.

[67] D. Graupe, D. J. Krause, and J. B. Moore, "Identification of autoregressive moving-average parameters of time series," *IEEE Trans. Autom. Control*, vol. AC-20, pp. 104–107, February 1975.

[68] C. R. Rao, *Linear Statistical Inference and Its Applications*, 2nd ed., New York: Wiley, 1973.

[69] H. Akaike, "Statistical predictor identification," *Ann. Inst. Statist. Math.*, vol. 22, pp. 203–217, 1970.

[70] H. Akaike, "Maximum likelihood identification of gaussian autoregressive moving average models," *Biometrika*, vol. 60, pp. 255-265, 1973.

[71] H. Akaike, "Markovian representation of stochastic processes and its application to the analysis of autoregressive moving average processes," *Ann. Inst. Statist. Math.*, vol. 26, pp. 363–387, 1974A.

[72] H. Akaike, "A new look at the statistical model identification," *IEEE Trans. Autom. Control*, vol. AC-19, pp. 716–723, December 1974B.

[73] P. Newbold, "The exact likelihood identification of Gaussian autoregressive moving average models," *Biometrika*, vol. 60, pp. 423–426, 1974.

[74] C. F. Ansley, "An algorithm for the exact likelihood of a mixed autoregressive-moving average process," *Biometrika*, vol. 66, no. 1, pp. 59–65, 1979.

[75] G. E. P. Box and G. M. Jenkins, *Time Series Analysis: Forecasting and Control*, San Francisco: Holden-Day, 1976.

[76] K. J. Astrom and T. Soderstrom, "Uniqueness of the maximum likelihood estimates of the parameters of an ARMA model," *IEEE Trans. Autom. Control*, vol. AC-19, pp. 769–773, December 1974.

[77] K. J. Astrom and T. Soderstrom, "Uniqueness of the maximum likelihood estimiates of ARMA model parameters – an elementary proof," *IEEE Trans. Autom. Control*, vol. AC-27, no. 3, pp. 736–738, June 1982.

[78] T. W. Anderson, "Estimation for autoregressive moving average models in the time and frequency domains," *Ann. Statist.*, vol. 5, pp. 842–865, 1977.

[79] J. Makhoul, "Linear prediction: a tutorial review," *Proc. IEEE*, vol. 63, pp. 561–580, April 1975.

[80] S. Li and B. W. Dickinson, "An efficient method to compute consistent estimates of the AR parameters of an ARMA model," *IEEE Trans. Autom. Control*, vol. AC-31, pp. 275–278, March 1986.

[81] S. Li and B. W. Dickinson, "Application of the lattice filter to robust estimation of AR and ARMA models." *IEEE Trans. Acoust., Speech, Signal Processing*, vol. ASSP-36, pp. 502–512, April 1988.

[82] S. Li, Y. Zhu, and B. W. Dickinson, "A comparison of two linear methods of estimating the parameters of ARMA models," *IEEE Trans. Autom. Control*, vol. AC-34, no. 8, pp. 915–917, August 1989.

[83] S. Kullback, *Information Theory and Statistics*, New York: Wiley, 1959.

[84] T. Matsuoka and T. J. Ulrych, "Information theory measures with application to model identification," *IEEE Trans. Acoust., Speech, Signal Processing*, vol. ASSP-34, no. 3, pp. 511–517, June 1986.

[85] G. Schwartz, "Estimating the dimension of a model," *Ann. Statist.*, vol. 6, pp. 461–464, 1978.

[86] J. Rissanen, "Modelling by shortest data description," *Automatica*, vol. 14, pp. 465-471, 1978.

[87] G. M. Jenkins and D. G. Watts, *Spectral Analysis and Its Applications*, San Francisco: Holden-Day, 1968.

[88] G. E. P. Box and D. A. Pierce, "Distribution of residual autocorrelations in autoregressive-integrated moving average time series models," *J. Amer. Statist. Assoc.*, vol. 64, 1970.

[89] S. U. Pillai, T. I. Shim, and D. C. Youla, "A new technique for ARMA-system identification and rational approximation," *IEEE Trans. Signal Processing*, vol. SP-41, no. 3, pp. 1281–1304, March 1993.

[90] J. Cadzow, "High performance spectral estimation–a new ARMA method," *IEEE Trans. Acoust., Speech, Signal Processing*, vol. ASSP-28, pp. 524–529, October 1980.

[91] P. Tuan, "The estimation of parameters for autoregressive moving-average models from sample autocovariances," *Biometrika*, vol. 66, pp. 555–560, 1979.

[92] W. Gersh, "Estimation of the autoregressive parameters of a mixed autoregressive moving-average time series," *IEEE Trans. Autom. Control*, vol. AC-15, pp. 583–588, October 1970.

[93] R. L. Kashyap, "Inconsistency of the AIC rule for estimating the order of autoregressive models," *IEEE Trans. Autom. Control*, vol. AC-25, pp. 996–998, October 1980.

[94] A. I. Markushevich, *Theory of functions of a Complex Variable*, New York: Chelsea, 1985

[95] E. B. Saff and R. S. Varga, "On the zeros and poles of Padé approximants to e^z, II," in *Padé and Rational Approximation*, E. B. Saff and R. S. Varga, eds., pp. 195–214, New York: Academic Press, 1977.

[96] D. C. Youla, "On the regular all-pass nature of diagonal Padé approximations to e^{-p}," Department of Electrical Engineering, Polytechnic University, New York, 1991.

[97] A. N. Khovanskii, *The Application of Continued Fractions and Their Generalization to Problems in Approximation Theory*, Groningen, the Netherlands: P. Noordhoff, 1963.

[98] L. Euler, "Commentatio in fractionem continuam qua illustris La Grange potestates binomiales expressit," *Mémoires Acad. Impér. Sci. Petersb.*, vol. 6, pp. 3–11, 1813–1814.

[99] J. Wallis, *Arithmetica Infinitorum*, 1655.

[100] S. U. Pillai, T. I. Shim, and D. C. Youla, "Nonrational systems and their rational approximation," NSF Annual Report, Electrical Engineering Department, Polytechnic University, Brooklyn, New York, May 1992.

[101] T. I. Shim, *New Techniques for ARMA-System Identification and Rational Approximation*, Ph.D. Dissertation, Polytechnic University, Brooklyn, New York, June 1992.

[102] B. McMillan, "Introduction to formal realizability theory," Bell Telephone System, Monograph, 1994, May 1952.

[103] D. C. Youla and P. Tissi, "An explicit formula for the degree of a rational matrix," *Electrophysics Memo*, Polytechnic Institute of Brooklyn, New York, June 1965.

[104] C. T. Chen, *Introduction to Linear Systems Theory*, New York: Holt, Rinehart and Winston, 1970.

[105] D. C. Youla, J. J. Bongiorno, Jr., and H. A. Jabr, "Modern Wiener-Hopf design of optimal controllers – Part I: The single-input-output case," *IEEE Trans. Autom. Control*, vol. AC-21, no. 1, pp. 3–13, February 1976.

[106] B. D. O. Anderson, "An algebraic solution to the spectral factorization problem," *IEEE Trans. Autom. Control*, vol. AC-12, no. 4, pp. 410–414, Aug. 1967.

[107] D. C. Youla, "Cascade synthesis of passive n-ports," Technical Documentary Rep., no. RADC-TDR-64-332, August 1964.

[108] D. C. Youla, "An introduction to coupled-line network theory – Part I," Memo 54, Microwave Research Group, Polytechnic Institute of Brooklyn, New York, October, 1961.

Index